MARINE
BIOFOULING
Colonization Processes
and Defenses

MARINE BIOFOULING
Colonization Processes and Defenses

Alexander I. Railkin

Translators
Tatiana A. Ganf, Ph.D.
Oleg G. Manylov

CRC Press
Taylor & Francis Group
Boca Raton London New York

CRC Press is an imprint of the
Taylor & Francis Group, an **informa** business

CRC Press
Taylor & Francis Group
6000 Broken Sound Parkway NW, Suite 300
Boca Raton, FL 33487-2742

© 2004 by Taylor & Francis Group, LLC
CRC Press is an imprint of Taylor & Francis Group, an Informa business

First issued in paperback 2019

No claim to original U.S. Government works

ISBN-13: 978-0-367-45441-8 (pbk)
ISBN-13: 978-0-8493-1419-3 (hbk)

Visit the Taylor & Francis Web site at
http://www.taylorandfrancis.com

and the CRC Press Web site at
http://www.crcpress.com

Library of Congress Card Number 2003055802

Library of Congress Cataloging-in-Publication Data

Railkin, Alexander I.
  Marine biofouling : colonization processes and defenses / by Alexander I. Railkin ; translators, Tatiana A. Ganf and Oleg G. Manylov.
    p. cm.
  Includes bibliographical references (p.).
  ISBN 0-8493-1419-4 (alk. paper)
    1. Marine fouling organisms. 2. Fouling. I. Title.

  QH91.8.M3R35 2003
  578.6'5'09162--dc22

                                                     2003055802

Preface
to the American Edition

In the sea medium, the accumulation of organisms can be observed at the water–solid body interface. Biomasses developing on hard surfaces often exceed those on soft-ground bottom communities by tens and hundreds of times. Such a concentration of organisms points to their ecological and economic significance.

Communities inhabiting hard substrates make a significant contribution to the productivity and stability of coastal ecosystems. They play an important role in self-purification of reservoirs, because they include organisms filtering great volumes of water when feeding and sedimenting suspended particles. Settling on external and internal surfaces of man-made structures, foulers hamper their exploitation, causing vast losses. In a number of cases, they are sources of bioinvasion by harmful organisms, as was the case recently when zebra mussels colonized the Great Lakes in the United States.

Concentration of organisms occurs due to colonization processes that are generally similar on surfaces of underwater rocks, hard ground, coral reefs, macroalgae, invertebrate and vertebrate animals, ship hulls, and other objects. Communities inhabiting hard substrates are similar in structure. Their basis is created by attached forms. Based on the above common characteristic, hard-substrate communities are united into one ecological group in the book and are considered together.

This book, published in Russia in 1998, was designed to explain the causes of vast biomasses concentrating on submerged hard substrates. The second task was an attempt at a quantitative description of the colonization processes resulting in such concentration. The third task, associated with the first two, was analysis of the common causes of colonization of man-made structures and discussion of approaches to protection from biofouling, including ecologically safe methods.

Solution of the above problems demanded a detailed consideration of the main processes leading to colonization of various natural and artificial hard substrates: transport of dispersal forms (microorganisms, larvae, spores, etc.) by the current, and subsequent settlement, attachment, development, and growth. This analysis made it possible to explain the causes of concentration of micro- and macroorganisms on the water–hard body interface. In addition, the concept of processes necessary and sufficient for colonization of any hard surfaces was formulated, and mathematical models of the main colonization processes were constructed. On the basis of comparative consideration of industrial antifouling measures and natural defense against epibiosis the principles of ecologically safe protection of man-made structures from biofouling and mathematical models of biofouling control were suggested.

The wide range of problems presented in the book are rarely considered within the limits of one monograph and are not covered sufficiently in university courses.

These are, in particular, locomotor reactions, taxes and drift of larvae, their sensory organs, mechanisms of settlement and attachment of microorganisms, animal larvae, and macroalgal spores, the impact of currents on colonization processes and spatial distribution of organisms on hard substrates, mechanisms of great biomass concentration on hard substrates, protection of macroalgae and animals from epibionts, industrial protection from biofouling, and problems of ecologically safe biofouling control. The book presents a great number of Russian-language works which are not widely known to non-Russian readers.

Taking the above into consideration, the author hopes that this monograph will be useful not only for biologists and engineers, state officials and experts who are interested in and concerned with the problems of marine biology, aquaculture, protection from biofouling, and maintenance of environment, but also for students and postgraduates specializing in the problems of marine ecology, zoology, botany, and microbiology.

Compared to the Russian edition, this monograph is thoroughly revised and supplemented. Considerable help in preparation of the U.S. edition was afforded by A.S. Elfimov, Ph.D. (Russia), G.G. Volsky, Ph.D. (Russia), S. Maack (Germany), N.V. Usov (Russia), Prof. S.A. Karpov (Russia), and especially S.V. Dobretsov, Ph.D. (Russia), to whom the author expresses his sincere gratitude. Owing to the high qualification and talent of the artist L. Reznik (U.S.) and the computer graphics specialists A.O. Domoratsky (Russia) and E.I. Egorova (Russia), the book is well illustrated.

Alexander I. Railkin
Saint-Petersburg

Author

Alexander I. Railkin, Dr. Sci., is Director of the Marine Laboratory (Marine Filial) of the Biological Research Institute of the Saint Petersburg State University (SPbSU) in Russia. He graduated from this university in 1971. He was a post-graduate student (1971–1974), junior research worker (1974–1980), senior research worker (1980–1990), leading research worker (1990–1998), and, since 1998, has been Director of the Marine Laboratory (Marine Filial) at SPbSU. He published 1 book and over 100 papers. He has five Russian patents.

His current research interests are colonization processes, larval behavior, role of hydrodynamic factors in formation and development of benthic communities, and ecologically safe protection from biofouling. Simultaneously, Dr. Railkin is an assistant professor at the Faculty of Biology and Soils of SPbSU. He gives master's level lectures on marine biofouling, experimental zoology, and ecology of protists.

Dr. Railkin is a member of the Russian Protozoological Society and the Saint Petersburg Society of Naturalists. He is a member of two doctorate dissertation boards and the Research Board on Biodeterioration of the Russian Academy of Sciences.

Contents

Chapter 1
Communities on Submerged Hard Bodies... 1
  1.1 Organisms and Communities Inhabiting the Surfaces of Hard Bodies.......... 1
  1.2 The Phenomenon of Concentration of Organisms on the Surfaces
     of Hard Bodies.. 9
  1.3 Biofouling as a Source of Technical Obstacles.. 14

Chapter 2
Biofouling as a Process... 25
  2.1 Colonization 25
  2.2 Primary Succession... 28
  2.3 Recovery Successions. Self-Assembly of Communities............................ 35

Chapter 3
Temporary Planktonic Existence.. 41
  3.1 Release of Propagules into Plankton.. 41
  3.2 Buoyancy and Locomotion of Propagules 43
  3.3 Taxes and Vertical Distribution of Larvae... 48
  3.4 Offshore and Oceanic Drift 52

Chapter 4
Settlement of Larvae.. 57
  4.1 The Reasons for Passing to Periphytonic Existence................................ 57
  4.2 Taxes and Distribution of Larvae during Settlement 59
  4.3 Sensory Systems Participating in Substrate Selection 63
  4.4 Selectivity during Settlement... 69

Chapter 5
Induction and Stimulation of Settlement by a Hard Surface 75
  5.1 Types of Induction and Stimulation of Settlement 75
  5.2 Distant Chemical Induction 77
  5.3 Contact Heterospecific Chemical Induction.. 79
  5.4 Conspecific Chemical Induction and Aggregations 81
  5.5 Stimulation of Settlement, Attachment, and Metamorphosis
     by Microfouling 85
  5.6 The Influence of Physical Surface Factors on Settlement 93
  5.7 Combined Influence of Surface Factors on Settlement. The Hierarchy
     of Factors.. 96
  5.8 Settlement on the Surfaces of Technical Objects.................................. 100

Chapter 6

Attachment, Development, and Growth...103
  6.1 Attachment of Microorganisms 103
  6.2 Mechanisms of Attachment of Larvae and Spores of Macroorganisms.....112
  6.3 Natural Inductors of Settlement, Attachment, and Metamorphosis............125
  6.4 Universal Mechanisms of Attachment..129
  6.5 Growth and Colonization of the Hard Surface 133

Chapter 7

Fundamentals of the Quantitative Theory of Colonization 143
  7.1 Mathematical Models of Accumulation 143
  7.2 Mathematical Models of Feeding and Growth...152
  7.3 Gradient Distribution of Foulers over Surfaces in a Flow..........................156

Chapter 8

General Regularities of Biofouling 169
  8.1 Causes, Mechanisms, and Limits of Biofouling Concentration
      on Hard Surfaces...169
  8.2 Evolution of Hard-Substrate Communities 175

Chapter 9

Protection of Man-Made Structures against Biofouling.......................................179
  9.1 Physical Protection...179
  9.2 Commercial Chemobiocidal Protection...182
  9.3 Ecological Consequences of Toxicant Application 189

Chapter 10

Ecologically Safe Protection from Biofouling..195
  10.1 Defense against Epibionts..195
  10.2 Natural and Industrial Anticolonization Protection....................................204
  10.3 Repellent Protection...207
  10.4 Antiadhesive Protection 212
  10.5 Biocidal Protection...215
  10.6 Prospects of Developing Ecologically Safe
       Anticolonization Protection 221

Chapter 11

The General Model of Protection against Biofouling 227

Chapter 12

Conclusion 231

References 235

Chemicals Index 281

Taxonomic Index 285

Subject Index 291

1 Communities on Submerged Hard Bodies

1.1 ORGANISMS AND COMMUNITIES INHABITING THE SURFACES OF HARD BODIES

In seas and oceans, especially along the coasts, there are many hard bodies, both at the bottom and within the water column. One group is made up of non-living natural substrates: underwater rocks, reefs, hard ground, clastic rocks, stones, tree trunks, etc. In another group, a more active one both chemically and physically, there are living organisms: macroalgae and animals, whose surfaces are inhabited by numerous epibionts. The third group includes material constructed of metal, plastic, concrete, and wood: ships, pipelines, cables, piles, etc. They may be chemically inert or, on the contrary, aggressive, if they are protected from biofouling by toxic substances.

The underwater world of hard surfaces is rather diversified, both in its species composition and in the abundance of organisms. It includes various types of microorganisms, invertebrates, and macroalgae. It is rather heterogeneous because it is represented by communities developing on various hard substrates under different ecological conditions.

V.N.N. Marfenin (1993a) writes:

> Among bottom biocenoses, the systems of hard grounds are the most variable ones. They are populated both by seston feeders, utilizing suspended particles, zoo- and phytoplankton, and by algae (within the photic zone). Among them, numerous commensals, predators, and saprophages find shelter and food. Animals from other biotopes frequently come to spawn there. And all of this exists owing to the hard ground, which creates a reliable surface for colonization, and the water movement over the substrate, which brings food to the animals (p. 131).

Coral reefs are well known hard-substrate communities (Odum, 1983; Naumov et al., 1985; Sorokin, 1993; Valiela, 1995). The calcareous foundation of the reef may go down many hundreds of meters, sometimes more than a kilometer. It consists of skeletons of dead organisms, mainly corals, sedentary reef-forming polychaetes, and coralline algae. The total area of the live coral reefs in the Indian, Pacific, and Atlantic oceans is about 600,000 km^2 (Sorokin, 1993). In principle, practically any region of the Tropical zone is suitable for coral life. Therefore, some experts believe that the corals could occupy an area 15 to 20 times greater (Naumov et al., 1985). Coral reefs are among the most productive areas in the world (Valiela, 1984, 1995). On the hard substrates of the reef, the biomass of zoobenthos may exceed the biomass of nearby soft grounds by one to three orders of magnitude (Sorokin, 1993). A vast number of animal and plant species inhabit the reef. The population of a single reef usually includes over a hundred species of polychaetes, crustaceans, mollusks, and echinoderms.

The plant and animal population of the benthos, plankton, and nekton may serve as a hard substrate for communities of epibionts, which are extremely widespread (Wahl, 1989, 1997; Wahl and Mark, 1999). It is difficult to find species of attached animals and plants or slow moving animals which do not carry other organisms on their surface. The specific features of communities developing on animals and macroalgae are mainly determined by the way of life and other properties of the basibiont organisms, serving as support for epibionts. Many seaweeds are little fouled or not fouled at all. Of attached animals, only sponges are little fouled, and also some corals and ascidians. All those organisms release bioactive substances that inhibit colonization and development of epibionts on them (see Chapter 10). Fast-swimming animals, such as fishes and dolphins, are also little fouled, which may be partly accounted for by the toxins contained in their mucous covers (see Pawlik, 1992).

Of practical importance are communities developing on the surfaces of industrial objects: ships, port and hydrotechnical structures, pipes, fishing nets, and other movable and stationary structures. They are rather heterogeneous. Some of them (nets, piles, moorings, etc.) have chemically inert surfaces and are subject to intensive colonization by marine organisms. Others, such as ships, are protected from fouling by toxic substances. As toxins in the paint are exhausted, the ship hull gradually gets fouled. The communities of macroorganisms developing on such surfaces have low diversity, owing to the dominance of the few macroalgal and invertebrate species most resistant to the toxic paints and life on the surface of a moving ship.

Different hard substrates, both natural and artificial, in accordance with their integral properties, can be divided into neutral, attractive, repellent, toxic, and bio-cidal. The peculiarities of colonization of different types of surfaces by the dispersal forms are considered in Chapters 4 to 10.

Communities developing on hard substrates on or near the bottom and in the water column, in spite of certain differences in their structure and species composition, are similar in general, because they develop in the same ecological environment, on the interface between hard surfaces and water, usually under conditions of increased water exchange as compared to communities on soft ground. The following life forms are characteristic of communities inhabiting hard substrates: sessile organisms, borers, and vagile forms (Railkin, 1998a).

In hard-substrate communities, sessile forms usually dominate in abundance and biomass, and act as edificators, i.e., determine the community structure and its microenvironment. These include macroforms such as sponges, hydroids, corals, sessile polychaetes, barnacles, mussels, bryozoans, sea cucumbers, ascidians (Figure 1.1), and macroalgae (Figure 1.2). Microorganisms are mainly represented by sessile bacteria, diatoms, microscopic fungi, heterotrophic flagellates, sarcodines, and sessile ciliates. The sessile macroorganisms inhabiting hard surfaces, in turn, serve as a new substrate for colonization by other organisms, including sessile ones. As a result, new sessile organisms of the second, third, and higher orders are involved in the process of successive colonization of the surfaces (Seravin et al., 1985), and

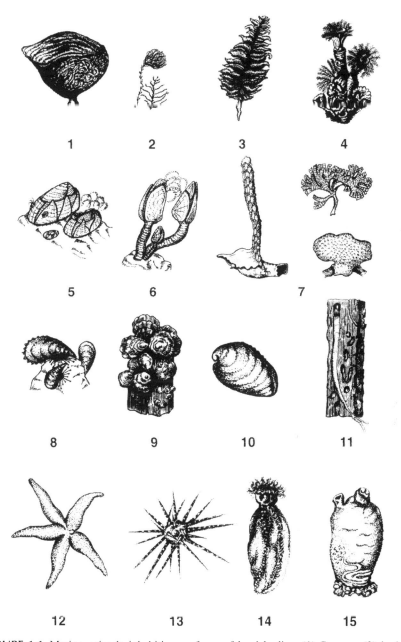

FIGURE 1.1 Marine animals inhabiting surfaces of hard bodies. (1) Sponge; (2) hydroid polyps; (3) coral sea pen; (4) polychaetes of the family Serpulidae; (5–6) cirripedes: acorn barnacles *Balanus* (5) and goose barnacles *Lepas* (6); (7) bryozoans; (8–11) mollusks: mussel *Mytilus* (8), oyster *Ostrea* (9), abalone *Haliotis* (10), shipworm *Teredo navalis* and its tunnels in wood (11); (12–14) echinoderms: starfish *Asterias rubens* (12), sea urchin (13), sea cucumber (14); and (15) ascidian.

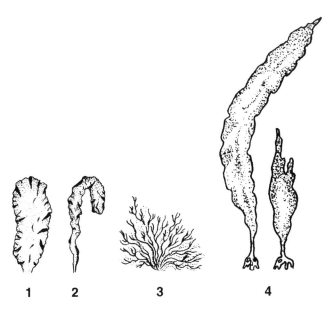

FIGURE 1.2 Marine macroalgae. (1–2) Green algae *Ulva* (1) and *Enteromorpha* (2); (3) red alga *Ahnfeltia*; (4) brown alga *Laminaria*.

thus these communities acquire a characteristic multilayered vertical structure (Partaly, 1980; 2003).

Another life form typical of communities inhabiting hard substrates is composed of the so-called borers, among which, together with sessile animals and macroalgae, vagile animals also occur. Borers demonstrate high specialization and a close physiological and biological connection with the hard substrate (see Section 1.3). The material they inhabit serves not only as shelter for them but also as a source of food. The paradox is that gradually eating the hard substrate (wood, stone, etc.), they may finally destroy it so thoroughly as to deprive themselves of the initial shelter.

Besides the two specialized groups, hard substrates are also inhabited by such vagile invertebrates as turbellarians, nematodes, errant polychaetes, crustaceans, gastropods, echinoderms (starfish and sea urchins), and also vagile microorganisms (mainly diatoms and various protists). The complicated branching, multilayered structure of the community formed by sessile macroorganisms reduces the hydrodynamic action upon vagile forms and serves as a kind of protection for nonattached species. Macroalgae, settling on the hard surface, form a kind of canopy over it, which creates an additional substrate and also shelter for vagile organisms living on and under it. Thus, among sessile organisms, vagile crustaceans, worms, mollusks, and also echinoderms find their abode. It is also highly probable that vagile organisms inhabiting hard substrates, including hard grounds, may possess mechanisms of increased adhesion to the surface on which they move, since even at the bottom, they usually live under the condition of augmented hydrodynamic activity. If they did not possess such mechanisms, they would be easily washed away from the surface.

Sessile, boring, and vagile forms inhabiting hard bodies are characterized by their position on the surface, fast adherence, and typically by being attached to the surface. In marine and fresh waters, the communities inhabiting hard substrates of different nature and origin are represented by similar life forms and may be considered as a single ecological group (Railkin, 1998a). Within its limits, according to the substrate criterion, smaller groups can be distinguished: communities of epibenthos, inhabiting non-living substrates, such as submerged rocks, stones, hard ground, etc. (Savilov, 1961; Khailov et al., 1992; Oshurkov, 1993); communities of epibionts inhabiting the surfaces of underwater animals and plants, sessile and vagile (Wahl, 1989, 1997); fouling communities on man-made structures (Costlow and Tipper, 1984), and some others.

Of course, not all communities possess all the characteristics described above. Any scheme, including the one above, is idealized to some extent. Thus, the multi-layer structure of communities does not attain proper development. Yet such major characteristics as the dominance of sessile species, their edifying role in communities, and finally their surface position on the substrate, are always present.

Relegating of the hard-ground populations to the communities of hard substrates needs further comments. Let us seek them in the detailed study performed by A.I. Savilov (1961). In the Sea of Okhotsk, he distinguished zones of prevalent development of different ecological groups. Among them, the fauna of hard grounds (rocky, gravelly, sandy, and dense sandy-silt ones) is considerably developed in terms of its abundance and biomass. It is characterized by the prevalence of immotile seston feeders, represented by numerous species of sponges, hydroids, soft corals, gorgonarians, cirripedes, some bivalves, brachiopods, bryozoans, and ascidians, i.e., the same groups (Figure 1.1) that inhabit hard substrates beyond the bottom. Communities of hard ground are subject to faster flows than those occurring on soft ground. Similar descriptions of communities inhabiting hard ground are to be found in the works of other authors (e.g., Osman, 1977; Sebens, 1985a, b; Protasov, 1994; Paine, 1994; Osman and Whitlatch, 1998).

In spite of a near 100-year history of studying hard substrates (Seligo, 1905; Zernov, 1914; Hentschel, 1916, 1921, 1923; Duplakoff, 1925; Karsinkin, 1925), there is still disagreement concerning the terms used to represent the communities of microorganisms (Cook, 1956; Sládečková, 1962; Gorbenko, 1977; Weitzel, 1979; and others) and macroorganisms (Tarasov, 1961a, b; Konstantinov, 1979; Braiko, 1985; Iserentant, 1987; Wahl, 1989, 1997; Railkin, 1998a; etc.) inhabiting them. For example, a number of authors (Reznichenko et al., 1976; Braiko, 1985; Hüttinger, 1988; Zvyagintsev and Ivin, 1992; Tkhung, 1994; Clare, 1996; Zvyagintsev, 1999, and others) consider that fouling communities represent a special assemblage of organisms on artificial substrates and man-made structures rather than on natural objects. Other authors (e.g., Mileikovsky, 1972; Zevina, 1994; Grishankov, 1995; Walters et al., 1996; Targett, 1997; Railkin, 1998a; Rittschof, 2000) regard fouling as the process of colonization of any substrate, including natural (living and non-living) ones, and also as the result of this process — the communities formed on various hard substrates.

A.A. Protasov (1982, 1994) analyzed over 350 sources from the 1920s to the early 1980s and found 21 terms for designating those communities. Six of them appeared the most widely used: Aufwuchs (Seligo, 1915), Bewuchs (Hentschel, 1916), periphyton (Behning, 1924), fouling (Visscher, 1928), and two Russian terms *obrastanie* and *perifiton*, translated into English as fouling and periphyton, respectively. These terms were used in 89% of the cases and other terms were employed in 11%.

In view of the common features of communities inhabiting hard substrates in the aquatic medium, considered above, it is possible to unite them into one ecological group. Following the historical tradition of assigning Greek names with the ending -*on* to large ecological groups of water organisms (plankton, nekton, neuston, pleuston), communities on hard substrates can truly be called *periphyton*, from the Greek περιφύω (περι, meaning *around* and φύω, meaning *to grow*, i.e., *to overgrow*). This term was first suggested by Behning (1924, 1929), though in a more narrow sense, to designate fouling of objects introduced into water by man. To designate communities inhabiting hard substrates, I will mainly use the term *hard-substrate communities*. Development of such communities will be referred to as biofouling or simply fouling, and the organisms forming them as foulers.

Unlike organisms inhabiting the surfaces of hard substrates, the typical inhabitants of soft grounds are vagile or sedentary organisms that live mainly within the ground and rarely on its surface. It should be noted that the soft grounds include the sediments (clay, silt, or fine sand) with particles below 1 mm in size. Four life forms of the inhabitants of soft grounds can be distinguished (Zernov, 1949): (1) vagile forms inhabiting the surface, not infrequently partly submerged into the ground (for example, echinoderms, crustaceans); (2) small vagile forms living between ground particles; (3) large vagile burrowing forms; and (4) sedentary forms.

It should be emphasized that sedentary or slow-moving invertebrates inhabiting the soft bottom do not get attached to its particles but are only anchored in it or on it. Therefore they are not attached to the substrate as are the typical inhabitants of hard substrates. A stronger connection with the soft ground is achieved by different means: due to the flattening of the body (e.g., many mollusks, starfishes, some urchins, encrusting bryozoans, calcareous algae), the thickening of the skeleton (a number of polychaetes, brachiopods, mollusks, echinoderms, etc.), forming tubes out of ground particles (polychaetes). Sedentary organisms not infrequently develop special rootlike outgrowths to hold themselves in sand and silt, which they do not possess when they inhabit a hard surface. This can be observed in a number of sponges, soft corals, polychaetes, bryozoans, and ascidians (Savilov, 1961; Zenkevich, 1977; Railkin and Dysina, 1997). In some cases such appendages may be considerably developed. When typical inhabitants of hard grounds colonize soft ones, they usually first get attached to some hard substrate: fragments of shells, mollusks, shelters of other invertebrates, small stones, pebbles, etc. (Savilov, 1961; Zenkevich, 1977). Usually these substrates are not to be seen from the surface of the soft ground since they gradually sink into the ground together with the organisms inhabiting them.

Many invertebrates inhabiting soft ground live under conditions different from those characteristic of the hard substrates. This can be accounted for by the fact that many species live within the ground. Even species constantly existing on the surface of the soft ground live, as a rule, under the condition of poor water exchange which occurs in the near-bottom layer. Soft grounds are inhabited by organisms adapted to life in narrow spaces and able to move within the ground. They are oxygen deficient and have little if any light (Burkovsky, 1992; Valiela, 1995). Marine benthic grounds are also characterized by a low pH and reduction-oxidation potential (eH) values. Specific microorganism activity sometimes results in accumulation of a great quantity of hydrogen sulfide, leading to the phenomenon known as "kill". The toxic effect of sulfides is based on oxygen radicals (see Section 10.5) being formed from reactions with sulfides (Tapley et al., 2003).

All the above allows us to distinguish the communities inhabiting soft ground as a single ecological group, which may be called *emphyton*, from the Greek εμφύω (εμ, meaning *in, inside* and φύω, meaning *to grow*) (Railkin, 1998a).

The same species of macro- and microorganisms may be members of communities inhabiting soft ground and hard substrates, including hard ground (e.g., Savilov, 1961; Oshurkov, 1993). During reproduction periods, they release propagules into the plankton. These propagules can settle on hard substrates and soft ground and participate in the development of associations on them. Thus, there is a regular exchange of dispersal forms between the communities of hard and soft substrates (Figure 1.3). Owing to this process, colonization of new and recruitment of inhabited hard substrates is carried out, the species composition and size structure of the community is maintained, and in case of disruption, their restoration is fast. Exchange is most intensive in the coastal areas, where especially high abundance of organisms is observed on hard substrates and soft ground.

As a result of the exchange of dispersal forms between epibenthic communities inhabiting hard ground and those formed on near-bottom hard substrates, both natural and artificial, a convergent similarity may be observed in both species composition and abundance. Thus, V.V. Oshurkov (1985, 1992) established a high similarity in species composition and abundance in the perennial fouling communities on asbestos cement and fiberglass in the water column with the closely located communities developing on the stone bottom and on a sunken ship. In all cases, the dominant species was the mussel *Mytilus edulis*.

Similar communities may develop at the same stage of succession in the same region only in the presence of similar abiotic conditions, character and properties of substrates. If at least one of those conditions is not met, the species composition and abundance of communities developing on different hard substrates in the same water area may be rather different. This has been noted repeatedly in the literature (Reznichenko et al., 1976; Zvyagintsev and Ivin, 1992; Zvyagintsev, 1999; Kashin et al., 2000). G.B. Zevina (1972) noted that "ship hull fouling differs from that of pipelines or buoys but in principle these differences are neither greater nor less than those between the fouling of ship hulls, seines, and stones or rocks, i.e., between the fouling of natural and artificial objects" (p. 36).

In microorganisms (bacteria, unicellular algae, and protists), the dispersal forms are their vegetative (and sexual) cells, and also spores and cysts, which may be carried by water and air currents to long distances, resulting in their ubiquitous distribution. The dispersal forms of macroalgae are motile or immotile spores, whereas those of invertebrates and ascidians are motile larvae. In their distribution, besides currents, an important role is played by their own motility and selectivity

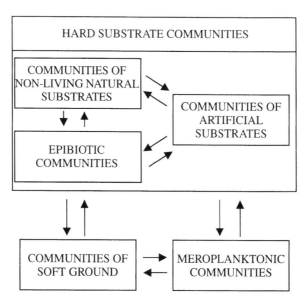

FIGURE 1.3 Exchange of dispersal forms between the communities of hard substrates and soft grounds.

in the choice of substrates. That is why macroalgae, and especially invertebrates, not infrequently occur only on certain substrates and in certain biotopes.

Organisms inhabiting hard substrates and soft ground possess similar adaptations to the habitat, the most important of which are, first, development of specialized structures and behavioral responses to hold on to the dense substrate or to live in it; second, avoiding being buried under detritus or hiding from it in the soft ground. They also possess a common pool of dispersal forms in the meroplankton (see Figure 1.3) and similar colonization cycles (Chapter 8). Based on these and other common characteristics, considered above, the communities of periphyton and emphyton, inhabiting hard and soft substrates, respectively, can be rightly united into one ecological group — benthos (Mileikovsky 1972; Zevina, 1994; Railkin, 1998a), in the same way that plankton and nekton are combined into one higher group, pelagos.

The life cycle of the organisms inhabiting hard (and soft) substrates consists of two parts and three periods: reproduction, dispersion, and growth. The short planktonic part of life is spent in the water column. A number of species, particularly planktotrophic invertebrates, pass there at certain stages of their development as part of the so-called meroplankton (see Chapter 3.1). As plankton, the organisms are insufficiently protected from predators and chance elimination. Their death rate is high. The second (the main) part of life is more prolonged. It proceeds on the hard surface, or, to be more exact, on the hard–liquid interface. Therefore, this part of life may be called periphytonic. It consists of the growth and reproductive periods. Periphytonic life is led by juvenile and adult forms of macroorganisms and also by microorganisms. At this stage of the life cycle they are less subject to the action of eliminating factors. Inhabitants of soft ground have colonization cycles with similar spatial and temporal patterns.

1.2 THE PHENOMENON OF CONCENTRATION OF ORGANISMS ON THE SURFACES OF HARD BODIES

Life in sea (and fresh) water is not distributed uniformly. Near the shores, in the near-surface water layer, and on the bottom, a great concentration of organisms is observed, which the outstanding Russian biogeochemist V.I. Vernadsky (1929, 1998) called the "thickening of living matter." In seas and oceans it is especially great on the water–air interface in the neustal (Zaitsev, 1970; 1997), water–soft ground interface in the bottom benthic communities (Zenkevich, 1956, 1977), water–hard substrates interface on the bottom of reservoirs (including hard grounds), and within the water column (Zevina, 1972, 1994).

V.I. Vernadsky (1929, 1998) was the first to understand and thoroughly and comprehensively analyze the role of living organisms in the change and transformation of the Earth's envelopes: the atmosphere, lithosphere, and hydrosphere. An important place in his theory is occupied by the conceptions of "diffused" life, resulting in a diffuse distribution of chemical substances and elements in the Earth's envelopes, and of concentrated, "thickened" life. He distinguished four static accumulations of life: two films of a vast size, planktonic and bottom ones (benthos) and two huge thickenings, the littoral (marine) and Sargassian, associated with kelp. These accumulations of organisms underlie the exchange of matter and energy in the hydrosphere.

A.V. Lapo (1987), in his book *Traces of Bygone Biospheres*, developing Vernadsky's conception of "thickening of living matter," i.e., of the accumulation of organisms, distinguishes additional upwelling, reef, and rift accumulations, justly relegating the former to plankton and the two latter to benthos. It should be noted that the reef thickening is the thickening of living matter which occurs on the hard substrate, the reef.

From the modern point of view it would be possible to discuss several large ecological groups, actually representing gigantic zones of organism concentrations on hard substrates all over the planet. They are:

1. The population of natural non-living (inert) hard bottom substrates in the coastal and shelf zones around the continents, including the hard grounds;
2. The reef population, which, owing to the vast territories it occupies and the great concentration of living organisms on reefs, deserves being treated as an independent group;
3. The ecological group of epibionts inhabiting the surfaces of living organisms;
4. The population on man-made structures;
5. The population of rifting and anthropogenic oceanic flotsam.

The main part of natural and artificial substrates is situated in the coastal and shelf zones of seas and oceans. The major part of microorganisms and multicellular animals and plants inhabiting the aquatic medium is concentrated on them. According to available estimates (Gromov et al., 1996), which seem to be low, their overall area is comparable with that of soft grounds in shallow waters, constituting about

2.74×10^7 km^2 (Zenkevich, 1956). Calculations show that 99% of the total biomass of the bottom population is concentrated around the continents within an area equal to 25% of the entire ocean floor of the Earth. The shelf appears the most inhabited. The animal biomass in this littoral "thickening of living matter" is 10 to 1000 times as great as in the open ocean (Figure 1.4).

According to the data available (Granéli, 1994), over 98% of marine animal species live on the bottom. At least 127,000 species live on gravelly to rocky bottom substrates, and only 30,000 species inhabit soft grounds.

On the hard grounds of the Sea of Okhotsk (Savilov, 1961), the White Sea, and Sea of Japan (Oshurkov, 1993), the biomass of animals and plants is from several times to several scores great as that on nearby soft grounds. The same distribution of biomass is also observed on the shelves of other seas (Zenkevich, 1956, 1977).

In all, the coral reefs occupy about 7.2×10^6 km^2, i.e., about 2% of the total area of the Earth's oceans, equal to 3.61×10^8 km^2. The biomass of the animal population of the reef itself may reach several kilograms per square meter, which is 10 to 1000 times as much as the biomass of zoobenthos of soft grounds around the reef (Sorokin, 1993). The macroalgae (coralline, thalline, and filamentous ones) attached to the hard surface of the reef are responsible for up to 30–50% of its total autotrophic production. On "algal" reefs, where living corals with their algal symbionts are not well developed, the autotrophic production of organic substances may reach 70–81%. The number of species of animals and plants in such communities is impressive. Thus, on an individual reef there may be more than 50 species of sponges, 100 to 200 species of polychaetes, 100 to 250 species of crustaceans, 150 to 500 species of mollusks, and 50 to 100 species of echinoderms. Here from 130 to 2200 species of fish live and feed. Their biomass is from 3 to 23 tons per hectare, which is the record value for marine biotopes, the average abundance of fish being from 2 to 40 individuals per square meter.

However, the coral reefs of the tropical oceans are not at all exceptional biotopes. In the temperate (boreal) waters, vast biomasses of organisms are also concentrated on hard substrates. Thus, out of more than 400 species of benthic invertebrates in the Solovetsky Bay of the White Sea, 68% inhabit hard natural surfaces: stones, hard ground, macroalgae, invertebrates, and ascidians (Grishankov, 1995).

There is also a considerable concentration of organisms on individual hard natural substrates. For example, about 180 species of invertebrates inhabit *Laminaria saccharina* (Sidorov, 1971), whereas 197 species of animals and plants live on *L. japonica* growing in mariculture (Ivin, 1995).

According to N.N. Marfenin (1993a), the surface area of only one small colony of hydroids exceeds that of the substrate it has colonized by scores of times. Taking into consideration the great abundance of sessile and vagile organisms inhabiting the bottom and also the water column, the area of living substrates appears to exceed that occupied by the inhabitants of soft grounds.

Epibiosis is widespread in seas and oceans and occurs in tens of thousands of species, including almost all phyla of marine animals and plants (Wahl and Mark, 1999). Most macroalgae, all the sponges and barnacles, most cnidarians and bryozoans, and also tube-building polychaetes, mollusks, brachiopods, some echinoderms, and ascidians are involved in the processes of epibiosis.

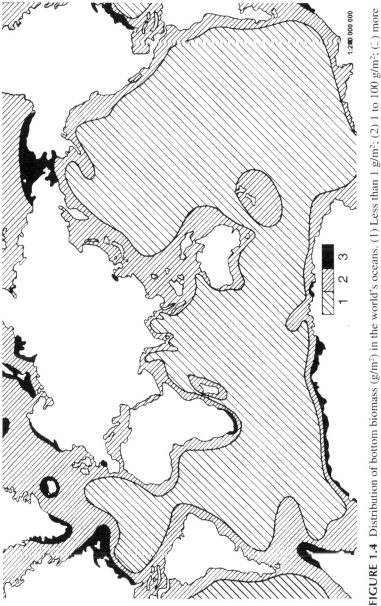

FIGURE 1.4 Distribution of bottom biomass (g/m²) in the world's oceans. (1) Less than 1 g/m²; (2) 1 to 100 g/m²; (3) more than 100 g/m². (After Gromov et al., 1996. With permission of the Central Administrative Board of Navigation and Oceanography, the Russian Federation Ministry of Defence.)

An impressive picture is also presented by microorganisms. The proportion of attached bacteria is usually from 20–30 to 50–60%, sometimes up to 90% of their total number in freshwater reservoirs (Punčochař, 1983; Hoppe, 1984). It is highly probable that in the sea a considerable number of microorganisms (bacteria, diatoms, protists) are also concentrated on hard substrates.

This is supported by the author's own observations (Railkin, 1998b). If marine microorganisms removed from some natural substrates (macroalgae, stones, wood, artificial polymeric materials) were placed as cell suspension into a Petri dish, settlement on the walls and bottom would start within the very first minutes. Some microorganisms (bacteria and diatoms) would colonize the hard surface within 3 to 6 h, others (ciliates), no later than in 24 h. In spite of the fact that many microorganisms swim well in water (bacteria, flagellates, ciliates, some sarcodines), most of them would be concentrated on the hard surface. These observations agree well with the known facts on fast adhesion (good sorption) of bacteria in seawater (ZoBell, 1946; Marshall et al., 1971; Zviagintzev, 1973; Railkin et al., 1993b).

The total surface of various materials and industrial structures used in marine environments is great and continues to increase. According to some estimates (Reznichenko, 1978), their total area is about 5000 km^2. Given the time that has elapsed since then it is reasonable to assume this value has already doubled. In any case, almost a quarter of the area colonized by foulers falls to ship hulls and other vessels. A band 100 m wide composed of marine artificial substrates would belt the whole globe. The total fouling biomass on all the anthropogenic surfaces is more than 6×10^6 tons (Reznichenko, 1978), and the number of species, 4000 (Crisp, 1984).

Many species of invertebrates and macroalgae are concentrated on artificial solid surfaces, their abundance and biomass being several times, and sometimes even several scores of times as great as that on soft grounds (Zevina, 1994). These values are especially great on inert substrates in the surface layer in the coastal zone. For example, in the White Sea, the biomass per 1 m^2 of commercial mussels in suspended mariculture at an age of 4 to 5 years is almost 6 to 10 times as great as that in their natural dense settlements — the so-called mussel banks (Galkina et al., 1982; Kulakowski, 2000). These mollusks are not only a source of food but contain some valuable substances used in medicine, perfumery, and agriculture (Kulakowski, 2000). It should be noted that at the end of the last century, the world trade balance of aquaculture was twice as great as that in the field of microelectronics (Goudet, 1991).

In the climax communities of high boreal waters, the biomass of coastal fouling can be weighed in kilograms, whereas in the subtropical and tropical zones, it is measured in tens and hundreds of kilograms per square meter (Reznichenko et al., 1976; Zevina, 1994). The absolute record is the biomass of *Megabalanus tintinnabulum*, observed in Nachang Bay (South China Sea). It was 301 and 343 kg/m^2 on rocks and piles, respectively (Zevina and Negashev, 1994).

Some other examples, despite being industrial and protected from biofouling, also demonstrate the phenomenon of macroorganism concentration on artificial solid substrates, with high densities of some species and groups of invertebrates. Thus, the average abundance of *Balanus reticulatus* dominating on coastal traffic ships in the South China Sea was 365 individuals/m^2, which is tens and hundreds of times as great as that of other species and groups of macrofoulers (Zevina et al., 1992). The density of polychaetes in fouling of biohydrotechnical structures in the northwestern part of the Sea of Japan was about 1 to 3 thousand individuals/m² (Bagaveeva, 1991). In the White Sea, on buoys fouled by hydroids, the density of *Mytilus edulis* juveniles, preferring to settle on such filamentous structures, reached about 8×10^6 individuals per square meter of the filamentous substrate (Zevina, 1963). On artificial materials (ropes, netting) under the conditions of mariculture, *M. edulis* forms dense clusters of individuals attached with their byssus threads perpendicular to the substrate. Their density in such settlements may reach several thousand individuals/m² within 4 years of growing (Kulakowski and Kunin, 1983; Loo and Rosenberg, 1983; Kulakowski, 2000).

Communities inhabiting hard substrates can attract other organisms. Within the zone of a great accumulation of hard natural substrates, the productivity of plankton is high, which causes great accumulation of fish (Zenkevich, 1977; Sorokin, 1993; Gromov et al., 1996). In regions of the blue mussel mariculture, the biomass of bacterial and algal plankton is 10 to 13 times as high as that in the adjacent water areas, where their mass settlements are absent (Galkina et al., 1982).

The vast abundance and biomass of organisms on hard natural substrates, including hard grounds, determine their important ecological role. Foulers usually form short detritus and grazing food chains, characterized by high efficiency. Communities of foulers, such as oyster and mussel beds, work as real biofilters (see Section 6.5), passing vast volumes of water through themselves, extracting pollutants and pathogenic microorganisms, precipitating suspended particles and thereby purifying and clearing water. The great ecological role of communities inhabiting hard substrates makes them an effective instrument of environment protection, in particular in restoring disrupted ecosystems by means of artificial reefs which are colonized by foulers and accompanying organisms (e.g., Khailov et al., 1994; Sherman et al., 2001; Svane and Petersen, 2001; Alexandrov et al., 2002).

All the above indicate a high accumulating ability of hard substrates, favorable conditions for survival and the possibility of growing on them. The very phenomenon of the "thickening of living matter" (Vernadsky, 1929, 1998) on hard substrates may be designated by the common term "the concentration of organisms on hard substrates." The same phenomenon is observed in freshwater reservoirs as well, but to a lesser extent than in seawater.

It should be noted that species of some macroalgae and attached animals, especially sponges, corals, and ascidia, are fouled rather weakly. As will be shown in Chapter 10, such natural protection is mediated by the release of toxic metabolites on the protected surface or into the water around it. The surfaces of industrial objects may be protected from fouling by special chemicals (Chapter 10). Thus, everything that is not protected from fouling is fouled. The above does not at all contradict the idea of concentration. It only shows that the constant tendency of organisms for concentrating on hard surfaces may be coupled with the action of other factors, preventing this process temporarily or constantly.

Why organisms are concentrated on water–air, water–soft ground, and water–hard surface interfaces and around them is not yet completely understood. It might be suggested that there are both common and specific explanations for this. An interface disrupts a certain (though conditional) homogeneity of the water mass. Living organisms settling on it or around it adapt their environment to their requirements, creating flows of substances and energy between the different media. Consequently, as a result of the organism's activity, chemical and physical gradients appear and are maintained, which contribute to the formation and development of communities (Aizatulin et al., 1979). At the same time, the closest living space controlled by attached organisms becomes biologically (and ecologically) inhabited and maximally adapted to the requirements of organisms and communities, as shown on macroalgae and invertebrates (Khailov et al., 1992, 1995, 1999). Thus, under the influence of the organisms settling on hard surfaces, the space around them becomes structured.

The concrete causes of concentration of organisms reflect the specificity of communities. In the case of neuston, the determining role seems to be played by the nutrients accumulating in the foam of the surface film (Zaitsev, 1970; 1997). They may serve both for feeding and for attracting microorganisms and maybe other inhabitants of the neuston. In its turn, accumulation and reproduction of microorganisms in the film (bacteria, multicellular algae, and protists) create a nutrition base for the development of the higher trophic levels and may attract multicellular organisms there.

Higher abundance and the biomasses of benthic organisms on the bottom in the coastal zone and on the shelf, as compared to the continental slope and deeper areas of the ocean, may be accounted for by a number of reasons. Among them are the diverse conditions of life on the bottom, the presence of a vast number of ecological niches, a considerable number of substrates for the settlement of the propagules (larvae and spores) of macroorganisms (Zenkevich, 1956, 1977; Oshurkov, 1993; Zevina, 1994; Paine, 1994; Kusakin and Lukin, 1995). The best trophic, temperature, and photic conditions in these shallow parts of the ocean contribute to the growth and reproduction of organisms (Valiela, 1984, 1995). It seems to be also important that the coastal currents detain some of the dispersal forms, sometimes a considerable number, in the shelf zone without carrying them into the open sea (see, e.g., Mileikovsky, 1968a; Martin and Foster, 1986; Lefévre, 1990).

The discovery of concrete reasons for the concentration of organisms on the solid body–water interface is practically equivalent to the solution of the problem of why organisms foul solid surfaces of natural and artificial origin. This book attempts to answer this question. As will be shown in subsequent chapters, the solution of this problem allows us to understand not only the reasons for colonization of hard substrates, but also to determine the approaches to increasing productivity of mariculture, and protection from biofouling.

1.3 BIOFOULING AS A SOURCE OF TECHNICAL OBSTACLES

The surfaces of technical objects immersed in seawater differ in configuration, size, texture, and material, and can be protected or unprotected from fouling. The dynamic, gaseous, temperature, and chemical regimes in which they are exploited are also different. They are subject to colonization by dispersal forms of microorganisms, invertebrates, and macroalgae to a different degree. The concentration of organisms on hard substrates, considered in Section 1.2, is the main source of biological problems arising during exploitation of technical objects in sea (and fresh) waters.

Seven types of marine anthropogenic objects are distinguished: vessels and their water conduits, navigational equipment, stationary structures, industrial pipelines, fixed submerged surfaces, and flotsam (Reznichenko, 1978). Ship hulls and flotsam, which is mainly oceanic debris, rank highest in terms of size and extent of fouling (ship hulls account for 24% of the total area and 85.5% of the total biomass, which is about 5×10^6 tons, and flotsam accounts for 70% of the area and only 5.6% of the biomass). The role of other anthropogenic objects in the concentration of bio-foulers is not as great.

In spite of evident differences between ships and flotsam from different points of view, they have one thing in common, which is their significant role in the random dispersion of different species of marine animals and plants to great distances, even between continents (Scheltema, 1971; Kubanin, 1980; Scheltema and Carlton, 1984; Carlton and Hodder, 1995; Zvyagintsev, 1999, 2000, etc.). The dispersion of invertebrates to great distances is briefly considered in Chapter 3 in connection with the coastal and oceanic drift of larvae.

As a result of larval drift, the transport of foulers by ships, and their rafting upon various objects afloat on the sea surface, a number of invertebrates and macroalgae are carried to geographical zones, regions, and biotopes new to them, and in some cases could become naturalized there. This results in the extension of ranges of these species, in their biological progress, and, in some cases, they replace native species. The problem of invasion by species and disturbance of marine (and fresh-water) ecosystems is one of the ecological problems of the twentieth century (Carlton et al., 1990; Carlton and Geller, 1993; Zevina, 1994; Sherratt et al., 1995; Galil, 2000; Zvyagintsev, 2000).

In Russia, a radical change in the fauna and coastal ecosystems of the Caspian Sea took place after the opening of the Volga-Don canal. Within a decade, about 20 invertebrate species migrated to the canal from the Black and Azov Seas and became naturalized (Zevina, 1972, 1994). Their immigration restructured the life in the Caspian Sea. Before this, fouling of ship hulls in the Caspian Sea had been practically negligible. However, the introduction of the barnacles *Balanus eburneus* and *B. improvisus*, the bryozoan *Conopeum seurati*, the polychaete *Mercierella enigmatica*, and other foulers caused the biomass of fouling on man-made structures to increase by 10 to 15 times, reaching up to 20 kg/m² on ship hulls and 40 kg/m² on buoys.

The sedentary polychaete *Mercierella enigmatica* got dispersed to different regions of the world's oceans from the coast of India (Zevina, 1994). This species colonized the shores of the Atlantic and the Pacific oceans within 50 years. The process of invasion became especially pronounced in the 1990s. During this period, invasions of the Japanese alga *Undaria pinnatifida* into Australian waters (Hay, 1990), the European bryozoan *Membranipora membranacea* and the nudibranch *Tritonia plebeia* into the northwestern Atlantic Ocean (Lambert et al., 1992), the Venezuelan bivalve *Perna perna* into Texas (Hick and Tunnel, 1993), and the European zebra mussel *Dreissena* into North America (Effler et al., 1996) were recorded.

The fouling of ship hulls and other vessels is of special practical significance. It depends on the region of operation, the time ratio of anchorage and sailing, speed

FIGURE 1.5 Fouling of propeller and rudder of a vessel. (Photo: S.I. Maslennikov. Used with permission.)

regime, the method of hull coating, and docking frequency. As a rule, high-speed boats spending little time in ports and a lot of time in the open sea and protected from fouling are least susceptible (Zevina, 1994). If the above conditions are not observed, they get fouled intensively. During one docking, up to 200–400 tons of fouling biomass may be removed (Redfield and Ketchum, 1952; Lebedev, 1973).

Fouling of ship hulls results in loss of speed, which may decrease by 40% or more (Redfield and Ketchum, 1952). This results in additional fuel expenditure to maintain the necessary speed. Friction resistance increases with sufficiently strong fouling of the hull by both micro- and macroorganisms (Redfield and Ketchum, 1952). However, the greatest impediment to the movement of the ship is macrofouling. With its development the initially smooth surface becomes rough and sometimes even knobby (Tarasov, 1961b). In Russia, according to technical standards, the hull roughness cannot exceed 0.12 to 0.15 mm at the time of building. In the process of exploitation, it becomes much greater. Increasing the hull roughness by only 0.025 mm raises its friction resistance by 2.5% and results in extra fuel consumption (Gurevich et al., 1989). In some cases, fouling of the propeller blades (Figure 1.5) is a more important cause of fuel waste than fouling of the ship's hull.

The negative effect of biofouling on vessels does not consist solely in decreasing their speed. The dense layer of macroorganisms, e.g., bryozoans, on certain parts of the ship hull may screen the release of toxic substances from antifouling coating and thus reduce its effectiveness.

All ship systems coming in contact with seawater are subject to biofouling. Pipes and water exchangers are especially affected (Yakubenko and Shcherbakova, 1981; Adamson et al., 1984; Yakubenko et al., 1984). The rate of seawater intake into the pipelines is rather high, which makes it easier for the larvae to enter them and also

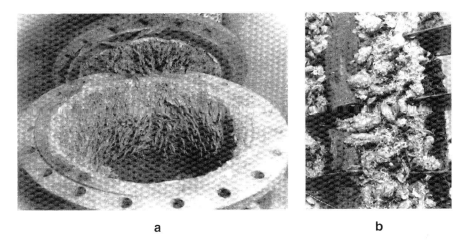

a   b

FIGURE 1.6 (a) Fouling of the inner wall of a pipe by hydroids. (Photo: E.P. Turpaeva. Used with permission.) (b) Fouling of stationary structures.

facilitates the inflow of nutrients to those organisms that have already settled there. Pipelines usually have a small diameter and intensive fouling reduces their carrying capacity, hampering their operation and sometimes (in case of blockages) leading to the break down of the units and mechanisms cooled by water. Common organisms inhabiting the inner walls of piping are bivalves, hydroids (Figure 1.6a), polychaetes, barnacles, bryozoans, and ascidians.

A frequent cause of ship wreckage is engine failure owing to the heavy biofouling of fuel lines (Bowes, 1987). There appears to be enough moisture and organic substances left in the fuel tanks for microorganisms to develop in this damp atmosphere. When sufficiently developed, they may totally block the piping, disrupting the fuel supply and stalling the engine. In the open sea, during a storm, this may lead to shipwreck.

A great danger is presented by the fouling of heat exchangers, in which bacteria play an important role (Adamson et al., 1984; Charaklis et al., 1984). The development of bacteria on the inner walls of heat exchangers stimulates settlement of the larvae of invertebrates, accelerating the process of biofouling (see Sections 5.5 and 5.8). The layer of micro- and macrofoulers, together with sediments and corrosion products, serves as a buffer between service water and water pumped in from the sea. This biological heat insulating layer reduces the effectiveness of heat exchangers, which results in energy losses and premature wear of different machines and mechanisms. Biofouling accelerates corrosion of the metal walls of heat exchangers.

In sea and fresh water, any technical objects are subject to biofouling: pipelines, navigational equipment, offshore oil and gas platforms, and port structures. Stationary structures are especially strongly fouled (Figure 1.6b).

Hydroids, barnacles, mollusks, and bryozoans settle in industrial pipes taking in seawater. Their development and biomass are determined by the parameters of intake sites and the velocity of water flow in the pipes. Water intakes and collectors are most subject to fouling. The biomass of hydroids in them may reach 6 to 10 kg/m^2, that of barnacles and bivalves, 9 kg/m^2, that of bryozoans, 2 kg/m^2,

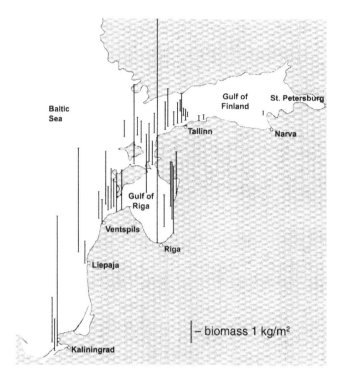

FIGURE 1.7 Fouling of buoys in the Baltic Sea. (After Zevina, 1972. With permission of the Publishing House of the Moscow State University.)

whereas the total values on grates, which are especially heavily fouled, may be as great as 16 kg/m^2 (Starostin and Permitin, 1963). Such intensive fouling seriously impedes the working of industrial enterprises, and in some cases even results in emergency situations.

Biofouling of navigational buoys in the coastal waters of even such a temperate sea as the Baltic sea may reach masses of several kilograms per square meter (Figure 1.7), whereas in subtropical seas, it may be up to 70 kg/m^2 (Yan et al., 1994). Biofouling hampers the operational characteristics of buoys and beacons, and may even cause them to sink. Fouling of oil and gas platforms increases their drag (Zvyagintsev and Ivin, 1995), so that in stormy weather they may be overturned by waves and hurricanes. Fouling is dangerous for plants using seawater in technological cycles (Turpaeva, 1987b) and for tidal power plants (Usachev, 1990).

Marine foulers are capable of destroying objects made of a variety of materials. This process has been especially well studied with regard to metals, concrete, and wood. Biocorrosion is a frequent cause of destruction of materials in the water. It occurs as the result of electrochemical processes, on the one hand, and of the biochemical activity of the organisms, on the other. Mechanical damage of objects and materials by foulers is caused by their growth and feeding. In the latter case, invertebrates are highly specialized. Some of them make tunnels in wood, others destroy concrete, still others damage the shells of commercially important mollusks.

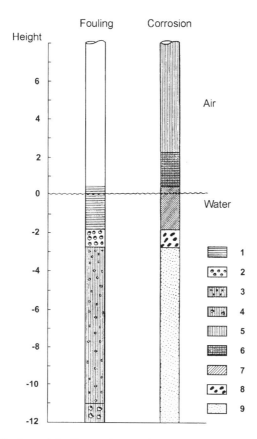

FIGURE 1.8 Distribution of fouling and corrosion on metal piles in the Caspian Sea. (1) Algal zone; (2) *Mytilaster* zone; (3) hydroid zone with small accretions of *Mytilaster*; (4) large accretions of *Mytilaster*; (5) flaky corrosion; (6) very strong uneven corrosion; (7) strong uneven corrosion; (8) large corrosive spots; (9) small corrosive spots under *Mytilaster* accretions. (After Zevina, 1972. With the permission of the Publishing House of Moscow State University.)

The common mechanisms of corrosion are conditioned by the electrochemical heterogeneity of the organisms and the metal surfaces immersed in the electrolyte solution (sea water) and the processes of leveling the potentials of their anode and cathode sites (Lyublinskii, 1980). The contact of organisms with a metal surface usually intensifies its corrosion in the marine medium (Redfield and Ketchum, 1952; Ulanovskii and Gerasimenko, 1963; Terry and Edyvean, 1981; Gerchakov and Udey, 1984; Korovin and Ledenev, 1990; Lukasheva et al., 1992). The specific mechanism of adhesion of the organisms, their metabolism, and distribution on the metal influence the corrosion processes. In some cases, firm adhesion of the organism to the surface may slow down or even prevent corrosion. Under the action of macrofouling, corrosion occurs both in case of dispersed and dense settlement of sessile forms, but if attachment is loose it occurs in the latter case. The correlation between vertical distribution of corrosion zones and biofouling of piles (Figure 1.8) leaves no doubt as to its causes.

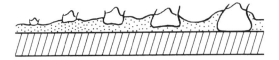

FIGURE 1.9 Destruction of anticorrosion coating by barnacles. (After Tarasov, 1961a. With the permission of *Zoologicheskii Zhurnal.*)

The usual agents of corrosion under aerobic conditions in the marine medium are thiobacteria and heterotrophic bacteria (e.g., Andreyuk et al., 1980; Kwiatkowska and Wichary, 2001; Kaluzhny and Ivanov, 2002). Fast corrosion in the presence of these organisms is associated with the chemical processes in which they are involved, whose products dissolve metals, concrete, and other materials (Sand, 2000). As a result, new heterogeneous microsites keep appearing and corrosion of the surface accelerates. The oxidative activity of the enzymatic apparatus of thiobacteria is extremely high. The rate of iron disulfide oxidation caused by thiobacteria is hundreds of thousands times higher than chemical oxidation. Destruction of iron structures in sea water is associated with sulfur compounds, which are always present in a dissolved state in water or are adsorbed on the surface. The reactions involving oxidative enzymes cause the formation of free sulfur and sulfuric acid, which subsequently becomes one of the leading factors in microbiological corrosion. The mechanism by which thiobacteria damage other materials is essentially the same. The usual bacterial microfoulers of the genera *Achromobacter*, *Bacillus*, *Flavobacterium*, *Micrococcus*, *Pseudomonas*, and *Vibrio* are heterotrophic (Gorbenko, 1977). The main mechanism of the corrosion of materials by these organisms consists in releasing exometabolites, which create an aggressive medium on the surface (Andreyuk et al., 1980; Sand, 2000; Kwiatkowska and Wichary, 2001). These are first of all organic acids, carbon dioxide, hydrogen sulfide, ammonia, peroxides, and enzymes.

It should be noted that besides numerous examples of biologically induced corrosion, there are many indications of the possibility of microorganisms inhibiting corrosion (Little and Ray, 2002). The probable mechanisms of inhibition consist in releasing exometabolites inhibiting corrosion and the formation of a diffusion barrier over the surface, slowing down the removal of corrosion products.

Barnacles may damage paint coating by growing into it (Figure 1.9). Young crustaceans, having shells only several millimeters high, destroy five-layer antifouling coating 0.2 mm thick with the sharp edges of their shells (Tarasov, 1961a).

Wood and stone borers are rather specialized organisms (Turner, 1984; Kleemann, 1990, 1996; Il'in, 1992a; Lebedev, 1992). They inhabit submerged structures made of wood, concrete, and other materials in great numbers and damage them. In the course of several months wood borers are capable of drilling so many tunnels in wooden piles (Figure 1.10) that they become unfit for use. Bivalves of the families Teredinidae and Pholadidae are wood borers (Il'in, 1992a; Nair, 1994). They feed mainly on wood, digesting it by means of cellulase, and also on bacterial plankton (Mann, 1984). They obtain additional nitrogen-containing substances from symbiotic bacteria and also by absorbing dissolved amino acids (Turner, 1984).

FIGURE 1.10 A wooden pile damaged by borers.

Since ancient times, one of the most well-known wood borers has been the mollusk *Teredo navalis* (Figure 1.11), which was called a "shipworm" for its worm-like shape and the damage it does to ships. It inhabits the south seas, e.g., the Black Sea, and drills tunnels in wooden ship hulls, wharf piles, etc. The boring organ of this and other wood-boring mollusks is the shell, which is shifted to the anterior part of the body. It has changed beyond any recognition, has diminished in size and developed sharp dentation along the valve edges, which serve to bore tunnels in the wood. Of great importance, besides the shipworm, are mollusks of the genera *Bankia*, *Xylophaga*, and *Martesia*. In harbors, they damage submerged wooden structures, cellulose-fiber cables, and even concrete structures. The *Martesia* mollusks are especially dangerous because they damage wooden structures protected by a special compound containing creosote. *M. striata* can drill even the lead sheathing of power cables and concrete (Fischer et al., 1984).

Another group of wood borers are crustaceans of the families Limnoriidae and Chelluridae. *Limnoria lignorum* have been found to feed on microscopic fungi and heterotrophic bacteria inhabiting their tunnels (Boyle and Mitchell, 1984). The food of other boring crustaceans seems to be similar. These microorganisms also live on the tegument of the borers, which spread their spores by moving within the wood. In this way, they create as it were fungal and microbial "plantations." Unlike tere-dinids, which can completely destroy the pile base of a wharf within two or three summer months, the speed of crustacean boring is not high. In *Limnoria*, it does not exceed 2 cm per year (Il'in, 1992a). Crustacean wood borers live close to the surface and do not penetrate deep. Yet by destroying wood centimeter by centimeter they may finally reach its center.

Of course, division of organisms into wood and stone borers is somewhat conditional because a number of invertebrates that damage wood are also capable of drilling limestone, stones, and other hard materials. Such an example was given above: the wood boring mollusk *Martesia*, besides wood, damages concrete and even metal. Nevertheless, distinguishing the stone borers as a separate category of marine biodeteriorating organisms may be justified by the fact that in most cases they drill not wood but sedimentary and metamorphic rock.

If only some representatives of mollusks and crustaceans could be assigned to the group of wood borers, stone borers may be found in all large taxa (Lebedev, 1992). Some microorganisms (cyanobacteria) and green algae destroy hard rocks and artificial materials by releasing acids, dissolving them, as well as other metabolites that act in a similar way. In boring sponges, e.g., genera *Cliona* and *Mycale*, specialized amebocytes disintegrate limestone chemically and "scrape it out" with their pseudopodia where it is dissolved (Sorokin, 1993). There are polychaetes releasing acid secretion which dissolves hard rock. They clean out a pit formed on the surface, and subsequently a tunnel, with their parapodial setae. Bryozoans can also use chemical substances for drilling substrates. Stone-boring mollusks and echinoderms (sea urchins) mostly drill mechanically (Lebedev, 1992). Bivalves drill with their shells, which are not infrequently transformed into a bore. Sea urchins, for instance the purple sea urchin *Strongylocentrotus purpuratus*, damage hard materials with the teeth of their Aristotle's lantern (mouth apparatus) and with thick motile spines covering their body.

Stone borers destroy hard metamorphic rocks (granite, marble, basalt), contributing to their erosion, and their presence is conducive to the formation of precipitate rocks, such as limestone and sand. Their role in damaging submerged structures is little studied. However, they may destroy natural materials used in construction, such as limestone, shell rock, stone, and concrete.

Thus, the organisms inhabiting hard substrates, settling and developing into juvenile and adult forms, do harm to different technical objects. They not only hamper exploitation, but also act as biodeteriorating agents.

According to expert estimation, marine biofouling ranks first among problems related to biodeterioration (Reznichenko et al., 1976). Prevention of fouling by means of special coatings and removal of fouling from technical objects require great financial investments. For example, in the U.S. oil industry alone, 16 to 18 billion dollars are spent annually on antifouling protection (Aroujo et al., 1992), while expenditures due to excess fuel consumption in the Navy is about 75 to 150 million dollars annually (Haderlie, 1984; Lewis, 1994). At the beginning of the twentieth century, the American researcher Visscher (1928) estimated annual losses of U.S. shipping companies caused by fouling at 100 million dollars. By the end of the century, this value has grown by an order (Gurevich et al., 1989). The general damage from marine organisms, whose greater portion falls on macro- and microfoulers all over the world, amounts to about 50 billion dollars a year, according to some estimates (Zevina, 1994).

Modern means of protection of submerged technical objects from biofouling and biodeterioration are chemobiocidal, i.e., they are based on using chemical substances to kill dispersal, juvenile, and adult forms of foulers. These substances are not harmless for the marine environment and the organisms inhabiting it. As the result of development and intensive use of the highly effective organotin biocides, many technological problems were solved and ships were continuously protected from macrofouling for up to 5 or more years. Yet many coastal ecosystems, especially those of harbors, were threatened with extinction. The restriction of use of organotin-based ship paints during the 1980s and the 1990s and the ban on their use since 2003 do not completely resolve ecological problems, since other chemobiocidal industrial means of protection (for example, copper-based paints) have the same drawback, though to a lesser degree (see Chapter 9).

Consideration of the processes and mechanisms resulting in concentration of organisms on hard surfaces (Chapters 2 to 6) and their analysis (Chapters 7 and 8) will allow us to establish the general features of colonization and consider ecologically safe approaches to the protection of engineering objects from biofouling (Chapters 10 to 11).

2 Biofouling as a Process

2.1 COLONIZATION

In its extended sense, the word colonization means the process of spreading over some new territory. As regards hard surfaces in a water medium, the terms "colonization," "biofouling," and sometimes "fouling" may be synonymous (Wahl, 1989). Biofouling as a process means biological fouling, as distinct from other forms of fouling, i.e., the accumulation of deposits of different kinds and origin on the surface, such as the products of corrosion, crystallization, chemical reactions, suspended particles, detritus, ice, etc. (Characklis et al., 1984; Bott, 1988). Thus, biological fouling is a special case of colonization of hard surfaces by living organisms in the water medium.

Unfortunately, there are only a few studies in the literature in which a comprehensive and detailed analysis of problems associated with colonization processes has been performed. According to a study by W.G. Characklis (1984), entitled "Development of Biofilm: Analysis of the Process," bacterial colonization includes:

1. Transport of organic molecules and bacteria towards a submerged surface
2. Adsorption of organic molecules, as a result of which the surface becomes conditioned, i.e., more favorable for attachment of bacteria
3. Attachment of bacteria to the conditioned surface
4. Metabolism of attached microorganisms, as a result of which they adhere to the surface faster
5. Growth of bacteria
6. Detachment of part of the bacterial film

In Characklis' classification (1984), emphasis is laid on the mechanisms of these processes. For the purposes of our analysis it should be accepted that neither the second (adsorption of molecules) nor the fourth (bacterial metabolism) stages are processes of hard surface colonization proper, although they facilitate the settlement of bacteria. Detachment of bacteria occurs as a result of the overdevelopment of the biofilm and loss of mechanical strength, and under the action of the current. This specific feature of microfouling films is not inherent in macrofouling.

The processes of hard-surface colonization were comprehensively discussed by M. Wahl (1989, 1997). He considered them in great detail, especially in connection with epibiosis, although his general statements can be applied to the characteristics of colonization of any surface. Like many other authors, he distinguishes: (1) biochemical conditioning (adsorption of macromolecules and ions); (2) bacterial colonization; (3) colonization by unicellular eukaryonts; and (4) colonization by multicellular eukaryonts. This scheme, with certain reservations, is applicable to the

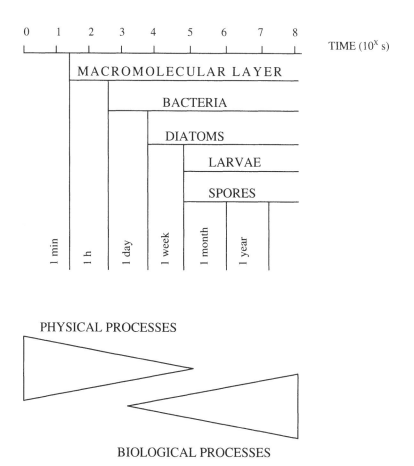

FIGURE 2.1 Chronology of hard surface colonization. (Modified from Wahl, M., *Mar. Ecol. Progr. Ser.*, 58(1–2), 175, 1989. With permission of the author and *Marine Ecology Progress Series.*)

colonization of any surface, not only that of living organisms but also of non-living objects. M. Wahl (1989) gives a highly schematized chronology of colonization processes and estimates the relative role of physical and biological mechanisms involved (Figure 2.1). Bacterial colonization is of a mixed physical and biological nature. However, when settling of diatoms starts, the biological factors begin to prevail over the physical ones.

A discussion of the various views held by investigators on the particular processes that take place during biological fouling would take more space and time than are available and would still not clear up the problem. Therefore, analysis of the causes of colonization of hard surfaces and the concentration of great numbers and biomass of foulers on these surfaces demands special consideration, which is what this book is devoted to.

To understand the mechanisms of colonization and concentration of foulers on hard natural surfaces and man-made structures, analysis of the elementary processes of biofouling is important. Theoretical analysis of protection against biofouling requires identifying the key processes necessary and sufficient for fouling of a surface.

Usually colonization is regarded as a sequence of accumulation and growth (see, e.g., Cairns, 1982a; Caldwell, 1984; Stevenson, 1986). Accumulation is understood to be the buildup of fouling on a hard surface as the result of transport by current, settlement, and attachment of propagules. The transport of foulers to the substrate by currents and their settlement are commonly referred to as immigration (e.g., Stevenson, 1983). In spite of the fact that the mechanisms of hard-surface colonization by microorganisms, spores of macroalgae, and animal larvae are different (see Chapters 4 to 7), phenomenologically they proceed rather similarly and involve the same processes: transport to the surface, settlement, attachment, and growth. It should be noted that transport may be carried out both passively, by means of the current, and due to the locomotor activity of propagules, juveniles, or adults. As for colonization by propagules of macroorganisms, their development and metamorphosis on the hard surface should be taken into account, since without them further growth of juvenile and adult individuals is impossible.

Analysis of the published data (Chapters 3 to 7) given in this book made it possible to reveal those common processes that underlie biofouling of any natural and artificial hard surfaces by any organisms. They are (1) transport, (2) settlement, (3) attachment, (4) development, and (5) growth. These elementary processes replace each other sequentially during surface colonization by micro- and macrofoulers. In some groups of foulers, for example in many microorganisms, the above sequence may be shorter and may not involve development as an independent process. The growth of microorganisms, as concerns colonization, may be interpreted as population growth, i.e., increase in their abundance (and biomass) by cell division. The growth of macroorganisms on a hard surface may take place concurrently with their development (metamorphosis). A marked increase in the abundance of macroorganisms due to growth is usually observed after the end of their development.

As for surface conditioning, considered by a number of authors (e.g., Characklis, 1984; Kent, 1988; Wahl, 1989, 1997) as the initial colonization process, it should be noted that though it is important for the beginning of bacterial fouling, in my opinion it cannot be regarded as a colonization process proper, since the organisms themselves are not involved.

The above scheme is applicable to all foulers with certain exceptions. The phenomenological resemblance of colonization processes in micro- and macroorganisms is conditioned by:

1. The similarity of their life forms where sessile ones dominate (see Section 1.1)
2. A common environment, which determines the conditions of transport by the current, settlement, attachment, nutrition, and growth
3. The presence of a hard surface as a substrate
4. The limited surface area of the hard substrate, due to which its colonization follows the island model (MacArthur and Wilson, 1967)

Although the colonization processes have been called elementary, in reality, they consist of simpler ones. Transport phenomena responsible for dispersion of foulers include both their transport by currents and their own movement, active and passive. Settlement consists of movement towards a hard surface, contact with it, exploration, evaluation, and selection or rejection of the substrate. Attachment also consists of stages controlled by different mechanisms. Development and growth of settled and attached larvae of invertebrates and spores of macroalgae appear to be multi-stage and complex processes.

Yet distinguishing transport, settlement, attachment, development, and growth as independent processes, while analyzing colonization (biofouling) and concentration of organisms on hard substrates, is justified by the fact that from the phenomenological point of view they are really indivisible, and in this sense, elementary processes. Their mechanisms are different in micro- and macroorganisms; still, the visible character of their occurrence and their general biological content are rather similar in both cases, which will be shown in subsequent chapters.

2.2 PRIMARY SUCCESSION

The study of succession of communities inhabiting hard surfaces is important to an understanding of how they are formed and develop, and how and why concentration of organisms on the liquid–solid interface takes place.

Besides establishing the general process of development of communities, the study of succession provides a basis for ecological prediction of biofouling in certain regions of the ocean where human economic activity is carried out (Zevina, 1972, 1994; Reznichenko et al., 1976). Thus, it is directly related to the prevention of marine biofouling. Knowing the ways in which species development creates obstacles for the functioning of technical objects makes it possible to remove them from the biofouling zone and take other measures to reduce and even prevent biofouling. There are cases when visits of sea vessels to brackish-water ports (for instance, those of St. Petersburg and Ho Chi Minh) did delay fouling development (Zevina, 1990).

The peculiarity of succession of hard-substrate community is its two-stage character (Little and Wagner, 1997). Microorganisms (bacteria, unicellular fungi, algae, protists) are the first to colonize surfaces immersed in seawater. This stage of succession is referred to as microfouling. At the next stage of macrofouling, propagules of macroorganisms, spores of macroalgae, and larvae of invertebrates and lower chordates (ascidians) settle on the hard surfaces. Succession of communities of micro- and macrofoulers develops with the participation of phenomenologically similar colonization processes (biofouling) but cannot be reduced only to them (Valiela, 1984, 1995). Succession mechanisms of micro- and macrofouling are similar but there are some differences between them (Connell and Slatyer, 1977; Gorbenko, 1977; Valiela, 1984, 1995; Oshurkov, 1992; Little and Wagner, 1997).

Let us consider development of successions on chemically inert hard surfaces. In this way, it would be possible to concentrate on the general regularities, setting aside the peculiarities associated with different properties of hard substrates which may be not only inert but attractant, repellent, toxic, and biocidal (see Chapters 1, 5, 9, and 10) and thereby exert an active influence on colonization and succession.

Microfouling communities are not infrequently referred to as bacterial-algal films or simply biofilms (e.g., Bryers and Characklis, 1982; Characklis, 1984; Hamilton, 1987; Little and Wagner, 1997). This is associated with the fact that bacteria and diatoms are dominant in them (Avelin, 1997; Avelin and Vitalina, 1997). Communities of microorganisms may be called films because of their physical character and placement on the surface. In the White Sea, for example, in climax communities of microfouling that developed on polymer plates, with the total number of cells equaling 10^7 cells per 1 cm^2 of fouled surface, the ratio bacteria/diatoms/heterotrophic flagellates was 640:4:1 (Chikadze and Railkin, 1992). At the same time, the proportion of the other unicellular organisms (yeast, autotrophic flagellates, sarcodines, and ciliates) was only about 0.15% of the total number of cells. Similar proportions of the main microfouler groups were reported by the American researcher C.E. ZoBell (1946) for the U.S. Pacific Coast communities. Bacteria and diatoms usually constitute the main biomass of microfouling, which may reach 99.9% of the adhered organic materials (Cooksey and Cooksey, 1988).

C.E. ZoBell (ZoBell and Allen, 1935; ZoBell, 1943, 1946), who used glass microscope slides widely in his research, showed that bacteria were the first to settle on them. He repeatedly pointed out that first they were represented by small cells. Subsequently, the slides became fouled by diatoms and protozoa, and later, by multicellular organisms, such as cirripede larvae.

A detailed study of microfouling development was carried out in the Black Sea by Yu.A. Gorbenko (1977). In particular, he established the following sequence of artificial substrate colonization: bacteria, yeast, heterotrophic flagellates, diatoms (almost concurrently with the latter), and finally ciliates. In other seas, microfouling successions are not as well studied. Yet the general order of settlement from plankton appears to be close to the above. For the White Sea the following sequence on hard surface was established: bacteria, diatoms, autotrophic flagellates, heterotrophic flagellates, amoebae, heliozoans, and ciliates (Laius and Kulakowski, 1988; Railkin, 1998b, 2000). A similar sequence was observed off Miami Beach in the Atlantic Ocean (Redfield and Deevy, 1952): bacteria, then, almost concurrently, diatoms, autotrophic and heterotrophic flagellates (algae and flagellates, according to the authors' terminology), and finally ciliates. Several thousand kilometers to the west of Miami Beach, in the Gulf of Mexico, the following order of microorganisms appearing on metal surfaces was observed: bacteria, diatoms, and ciliates (Little, 1984). Wahl (1989), who had summarized the published data on epibiosis and fouling, assumed the following sequence: bacteria, diatoms, protozoa. A similar sequence (bacteria, unicellular algae, protozoa) was noted for metal surfaces in the review by Little and Wagner (1997).

Thus, in accordance with the published data, the following stages are characteristic of succession of marine microfouling: bacterial, autotrophic, and heterotrophic (protozoan). It is of interest that development of microperiphyton in fresh waters passes through the same common stages (Cairns, 1982b). The above generalization holds for marine and fresh waters on hard substrates of different kinds and different surface properties (Little and Wagner, 1997).

Among fouling bacteria, the genera *Pseudomonas*, *Vibrio*, *Micrococcus*, *Achromobacter*, *Flavobacterium* are widespread, and *Bacterium*, *Bacillus*, and *Sarcina* are less frequent (ZoBell, 1946; Gorbenko, 1977; Avelin and Vitalina, 1997). The common diatoms in fouling include the pennate forms *Nitschia*, *Navicula*, *Cocconeis*, *Licmophora*, *Synedra*, *Amphora*, *Achnanthes*, *Bacillaria*, *Biddulphia*, and the centric forms of the genera *Melosira*, *Fragilaria*, *Grammatophora*, *Rhabdonema*, *Berkeleya*, and others (Gorbenko, 1977; Avelin, 1997; Railkin, 1998b, 2000). Among heterotrophic flagellates, there are *Bodo*, *Spumella* (=*Monas*), *Pteridomonas*, *Metromonas*, *Monosiga*, *Codonosiga*, and others (Railkin, 1995c).

Development of microfouling communities proceeds as a biological succession, and it has been shown that in a number of cases the mechanism of facilitation is involved (Connell and Slatyer, 1977), in accordance with which the preceding stage of succession is conducive to the onset of the following one. The first colonists on submerged surfaces are rod-shaped chemoheterotrophic bacteria (see, e.g., Corpe, 1972; Kjelleberg, 1984; Railkin et al., 1993b). They appear on substrates within 1 to 2 h or even earlier and are dwarf copiotrophic bacteria capable of growing only in high concentrations of nutrients, under conditions that are not usually found in marine and oceanic waters. Why does their development appear to be possible in these oligotrophic waters? The fact is that on hard immersed surfaces adsorption of organic and inorganic substances and ions starts from the very first minute, and saturation of the surface with them and dynamic balance with the environment is established within hours (Baier, 1984; Wahl, 1989, 1997). Thus, concentrations of nutrients necessary for the development of copiotrophic bacteria are created. Having exhausted the stores of nutrients, the copiotrophs prepare the nutrient medium favorable for the development of oligotrophs, metabolically active at lower concentrations of food. The above mechanisms of the regular replacement of copiotrophic forms by oligotrophic ones determine the direction of the initial bacterial succession.

Bacteria involved in the first colonization stage may prepare the microconditions for the development of the filamentous and stalked bacteria *Hyphomicrobium* and *Caulobacter* at the final stage of the bacterial succession (Little and Wagner, 1997).

Gorbenko (1977) gave experimental proof of the fact that bacteria stimulate the development of diatoms, thus facilitating the onset of the autotrophic phase of succession. He established that in the developing community of microfoulers there were significant correlations between the abundance of morphologically and metabolically different groups (rod-shaped bacteria, cocci, heterotrophs, oligonitrophilic bacteria, diatoms, etc.) (Gorbenko, 1977; Gorbenko and Kryshev, 1985). These data may serve as indirect testimony of interaction between them. These facts suggest that the development of microfouling proceeds similarly to autogenic succession (Odum, 1983; Begon et al., 1996), that is, it is determined mainly by biological interactions and not by the influence of external, for instance, seasonal factors, as in the case of allogenic succession, though it certainly depends on them.

The growth processes may play the leading role in the alternation of the dominant species (groups) of microfoulers (Avelin, 1997; Avelin and Vitalina, 1997; Little and Wagner, 1997), since it is growth, and not accumulation, that mainly contributes to the abundance of microorganisms on hard substrates (Punčochař, 1983). In accordance with the models of Connell and Slatyer (1977), succession may be based on the facilitation by early-succession species of the reproduction of late-succession species. The early-succession species act by creating conditions favorable for the

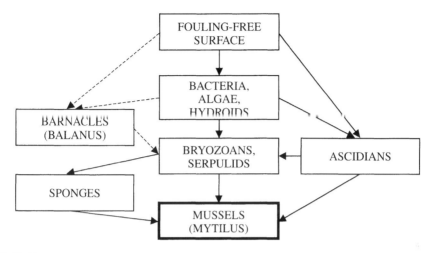

FIGURE 2.2 A classical scheme of succession of a biofouling community. Dashed lines show collateral ways of succession, bold lines show climax communities. (From Scheer, B.T., *Biol. Bull.*, 89, 103, 1945. With permission of the *Biological Bulletin*.)

late-succession species or by inhibiting the reproduction of other early-succession species, without negative influence upon late-succession species (or groups). Of course, conditions more favorable for reproduction of one species than another may be determined not only by the surface properties of hard substrates but also by the environment. Therefore, mechanisms governing succession should be the subject of special study in each case. Yet, it should be admitted that there are very few works where a particular succession mechanism has been rigorously demonstrated.

In the surface water layer, microbial films develop on hard surfaces fairly rapidly: in a fortnight in the boreal-arctic waters during the warm season (Laius and Kula-kowski, 1988; Railkin, 1998b), and in a week in the subtropical zone (Redfield and Deevy, 1952). Therefore, many dispersal forms of macrofoulers (larvae and spores) settle on the already formed biofilms (see reviews by Scheltema, 1974; Mitchell and Kirchman, 1984; Seravin et al., 1985; Morse, 1990; Pawlik, 1992; Slattery, 1997). Analysis of the published data shows that microfouling does not simply precede macrofouling chronologically. In many cases, it induces and stimulates macrofouling (see Section 5.5). This suggests that microfouling is an independent stage of auto-genic succession.

Many studies (see reviews by Sutherland, 1974; Harris and Irons, 1982; Osman, 1982; Schoener, 1982; Smedes, 1984; Braiko, 1985; Oshurkov, 1993; Paine, 1994; Zevina, 1994; Valiela, 1995) have been devoted to the study of macrofouling suc-cession. According to the classical work of Scheer (1945) carried out in the harbor of Newport on the Pacific Coast (California State, U.S.), the fouling of a surface free from organisms starts with the formation of a microbial film on it. At the first stage of macrofouling succession it is colonized by fast-growing, and at the second stage, by slow-growing organisms (Figure 2.2). The succession ends with a short-term climax stage. In Scheer's experiments, the stage of fast-growing organisms was represented by hydroids, serpulids, bryozoans, or the ascidian *Ciona*. The development of the community was completed by slow-growing mollusks of the genus *Mytilus*, which are dominant at the climax stage. There were some other ways of community development. For example, settlement of cyprid larvae of *Balanus* was determined mainly by the presence of conspecific individuals and considerably less dependent on the presence of a microfouling film. Thus, the climax state characterized by the dominance of bivalves on the substrate was achieved within a few years, that is, much sooner than in terrestrial ecosystems (Connell and Slatyer, 1977).

Scheer's conception of succession (1945) was supported and developed in many subsequent studies (Redfield and Deevy, 1952; Zevina, 1972; Rudyakova, 1981; Chalmer, 1982; Harms and Anger, 1983; Smedes, 1984; Braiko, 1985; Oshurkov, 1985; Hirata, 1987; Khalaman, 1989, 2001b; etc.), reviewed by G.B. Zevina (1994). According to these works, the first stage of succession, from several days to 2–3 weeks long, is associated with the development of a microbial film on the substrate. At the first stage of macrofouling it is colonized by fast-growing, more often colonial foulers, such as hydroids, bryozoans, and also solitary sea anemones and polychaetes. Its duration is from 2–3 weeks to 1–2 years. The second stage of macrofouling is represented by slow-growing invertebrates (mollusks, sponges) and ascidians. It is usually interrupted by the action of physical factors and is therefore short-lived, lasting for several years, rarely longer.

Most studies on succession have been carried out near sea coasts, in shallow waters, that is, in unstable environments. In accordance with this, some scientists are doubtful of the possibility of identifying general processes of macrofouling succession, pointing to its considerable variability (Osman, 1982). Facts are cited (Pisano and Boyer, 1985), that cast doubt on the dependence of each stage of succession upon the preceding one. Under the conditions of frequent disturbances of the physical medium the stable climax state may not be achieved at all in a number of cases, which gives some authors reason to doubt the very existence of climax in macrofouling (Sutherland, 1974, 1984; Sutherland and Karlson, 1977).

In spite of these difficulties with the succession theory, many researchers (see above) still think that succession of macrofouling is quite a natural process, and its final (climax) stage characterized by a stable abundance and number of species is achieved within one or several years, depending on the climatic zone and depth, that is, many times faster than land succession (Connell and Slatyer, 1977).

If artificial substrates are regarded as "islands" colonized by propagules of foulers (e.g., Cairns and Henebry, 1982; Osman, 1982), biofouling can be considered within the framework of the theory of island biogeography (MacArthur and Wilson, 1967). According to this theory, the number of species colonizing an island is determined by their immigration and extinction. The balance state is characterized by the equality of the rates of these processes.

According to the theory of island biogeography, succession sequence would depend on what particular season it is when the substrates are introduced into the sea and at what depth, how close they are to the mass settlements of foulers, and how many propagules can be found in the plankton.

In any case, the first stage of succession would seem to be microfouling, which proceeds especially fast on all submerged substrates. At the same time, a three-stage development of macrofouling may be different from Scheer's scheme (1945) for various reasons. For example, in the blue mussel culture in the White Sea, the artificial substrates (ropes and sheets of fishing net) are set in the upper 3-m layer only 1 week before the mass settlement of the larvae of these bivalves (Kulakowski and Kunin, 1983; Kulakowski, 2000). This provides optimal conditions for the blue mussels to "monopolize" the substrates. It is natural that the stage of fast-growing macroorganisms is omitted in this process (Khalaman, 1989).

In the biofouling of artificial materials in the water column and on the bottom, the succession sequence of macrofoulers appears to be really dependent on the season and the depth at which the experimental panels were set (e.g., Oshurkov and Seravin, 1983; Oshurkov, 1985, 1986). Yet the general sequence remains unchanged. The film of microfouling is consistently colonized by fast-growing species. As the result of competition, they are subsequently replaced by slow-growing species, which occupy the dominant position in the climax community.

V.V. Oshurkov (1985, 1986, 1992, 1993) carried out long-term investigations in the same biotopes and on identical substrates in Kandalaksha Bay (the White Sea) and in Avachinsk Bay (the Pacific). As a result, he found that the succession variant similar to that studied by Scheer (1945) for macrofoulers and also observed by many other authors (see review by Zevina, 1994) was the most frequently observed, but not the only variant possible. It is typical of unstable biotopes, e.g., coastal areas, shallow waters, and estuaries. The dominance of bivalves (e.g., blue mussels and oysters) is one of the stages of succession in such communities but not necessarily the final one. Monopolization of substrates by mollusks is a comparatively short process, lasting only one or several years. The disturbances commonly occurring in such biotopes as the result of storms and predation (e.g., by starfish), can damage and destroy communities, throwing them back to earlier stages of succession, where fast-growing organisms dominate. Such cyclic, practically unfinished successions occur repeatedly and create an illusion of high-rate succession.

Under more stable conditions (at greater depths) in the White Sea, bivalves can be replaced by brown algae, whereas in the Pacific they are replaced by encrusting coralline algae (Oshurkov, 1992, 1993). The age of algal communities is scores of years and that of corallines up to 80–100 years, which is quite comparable to the duration of land successions (Connell and Slatyer, 1977). The general scheme that may be constructed on the basis of Oshurkov's studies (1985–1993) is given in Figure 2.3. As can be seen, the climax (final) stage is represented by communities in which the dominant forms are sea anemones, cirripedes, ascidians, sponges, or algae. The community of the bivalve *Mytilus edulis* may be either a final or an intermediate stage of succession. In the latter case it is replaced by communities with prevalence of the ascidian *Styela rustica*, algae, or other bivalves (Oshurkov, 1992, 1993; Khalaman, 2001a, 2001b). The course of development depends on the conditions under which biofouling takes place: the geographical region and the location of the water area, the degree of the physical stability of the environment, the set of species in adjacent communities, the season in which the substrate was submerged, the duration of its submersion in the water, and the distance from the bottom.

Analysis of the literature on the development of hard-substrate communities under relatively stable conditions shows that their succession is a directed and, in

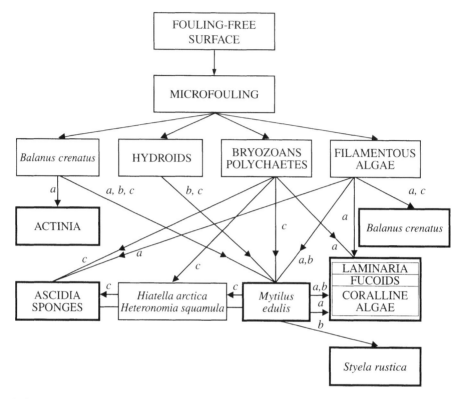

FIGURE 2.3 A generalized scheme of the main directions of succession of fouling communities on chemically inert substrates (based on data compiled by V.V. Oshurkov, 1985–1993). Letters indicate different position of the substrates: on the bottom (a), above the bottom (b), and in the water column (c). Bold lines show climax communities. Some links between the boxes "fouling-free surface," "microfouling," etc. are not shown.

this sense, predictable process. The primary succession includes four consecutive stages: microfouling, the stage of fast- and slow-growing macrofoulers, and the climax.

As for the mechanisms of macrofouling succession, there are some data testifying to the possibility of the preceding stages facilitating the subsequent ones. Thus, Scheer (1945) showed that hydroids prepared the microenvironment for bryozoans to colonize substrates. On the coast of Delaware the mussel *Mytilus edulis* settled better on the substrates occupied by ascidians and hydroids (Dean and Hurd, 1980).

Yet other authors favor the inhibition mechanism over the facilitation mechanism (Connell and Slatyer, 1977). The study of the succession of algal communities on hard substrates (rocks and boulders) on the southern coast of California (Sousa, 1979a, 1979b, 1980) has shown that the pioneer species *Ulva* spp., which are good interference competitors under stable conditions, restrained the colonization of substrates by other species. In the case of desiccation and grazing, to which *Ulva* is sensitive, it was replaced by later arrivals. When those species were removed artificially, *Ulva* prevailed temporarily, because it possessed a higher growth rate. The red algae *Gigartina canaliculata* and *Gelidium coulteri* seem to inhibit the settlement and growth of other species, as their removal results in a greater number of species in the vacant space. Sutherland (1977) has shown that after removal of the bryozoan *Schizoporella unicornis* from experimental panels the abundance of other species grew: the sponge *Mycale cecila*, barnacles *Balanus* spp., the oyster *Ostrea equestris*, and the ascidian *Styela plicata*.

Under unstable conditions the succession process may be often interrupted, which in extreme cases can cause the formation of bare patches, partially or completely free from fouling. It is natural that in such cases the succession returns to its earlier stages. On such bare patches, the succession occurs not simultaneously but in accordance with the real degree of disturbance, the biotic environment, and the colonization potential of those species that are recruited to regenerate the disturbed patches (e.g., Paine, 1994). Succession in such a disturbed community occupying a series of patches must be nonlinear, and occur according to a more complicated scheme (Valiela, 1995). A relatively constant colonization potential of macrofoulers, under the local control of recruitment by predators, may produce the effect of long-term stability, which has been shown for hard-substrate communities in southern New England, U.S. (Osman and Whitlatch, 1998). Besides predation, grazing and competition for resources (space and sources of food) may exert an important influence both on the structure of communities (see Section 6.5) and the course of succession (e.g., Sebens, 1985a; Paine, 1994).

2.3 RECOVERY SUCCESSIONS. SELF-ASSEMBLY OF COMMUNITIES

A natural or artificial disruption of the succession process may inhibit further development of macrofouling for a time, returning it to an earlier stage (see Section 2.2). Yet this does not solve the problem of protecting industrial objects from biofouling.

The abundance and the species structure of the community may for some time return to the initial state owing to the recovery successions, the biomass and the species diversity of the community turning out to be even higher than the initial ones (see below). This may be the case when succession is controlled by facilitation mechanisms (Connell and Slatyer, 1977). In the case of inhibition mechanisms, the dominant position may be occupied by species of later succession stages (see Section 2.2).

Underwater mechanical cleaning of a ship hull from macrofouling is not always effective and may result in still faster fouling (Lebedev, 1973; Zevina, 1990). This is associated mainly with the fact that such strongly adhering organisms as barnacles may be removed only together with the layer of antifouling coating. As a result, the cleaned surface turns out to be quite unprotected and is quickly fouled again. Underwater cleaning followed by painting of the ship hull can help extend the effective life of intact copper-based coating up to 5 years (Preiser et al., 1984).

The process of recovery of mechanically disturbed communities in natural waters was studied in greater detail using fresh-water microperiphyton as an example, using the method of experimental plates, which is quite often referred to as a method of fouling plates or artificial substrates (e.g., Sládečková, 1962; Bamforth, 1982; Protasov, 1987). As was established in the study of freshwater communities, the abundance and species structure are restored within 3–9 days of disruption, while in some cases it takes longer (Peterson et al., 1990, 1994; Stevenson, 1990; Peterson and Stevenson, 1992). It should be noted that the number of species in regenerated communities may be even higher than in the initial (undisturbed) communities (Cairns et al., 1971; Peterson and Stevenson, 1992). This may be associated with the appearance of bare patches, where new organisms can settle, and with substances stimulating their development, which are released during partial destruction of microfouling films.

Similar processes of recovery also take place in marine communities. Temporary disturbance of their development cannot disrupt the general trend of the succession to a climax. Thus, colonization of clean (sterile) sand, placed at a chemically polluted benthic site, by marine psammic ciliates was even faster than at the control (non-polluted) site, and the community was recovered within a fortnight (Burkovsky and Kashunin, 1995). Complete mechanical removal of microfoulers should be repeated under marine conditions at least every third day to be really effective (Caron and Sieburth, 1981).

My research (Railkin, 1998b) has shown that marine and fresh-water microfouling communities, carefully removed from the substrate in such a way as not to damage the microfoulers, assemble themselves from the resulting suspension on the horizontal surface. The sequence of recovery of the initial species composition and abundance and the spatial structure corresponds to the general succession sequence of the main microfouler groups. At the same time, the rate of the formation of a new (recovered) community appears to be higher than that of succession and does not depend on its stage. The phenomenon observed can be referred to as the self-assembly of communities.

It should be noted that earlier (Lima-de-Faria, 1988) this term was used to denote a mechanism resulting in a cohabitation of individuals of one or several species, that is, its original meaning was close to that assumed by me. As community factors, this author considered signal molecules and also some physical factors (light, color, sound) recognized by individuals and supporting the constitution and integrity of macroorganism communities.

In order to observe the self-assembly of a community of microorganisms (Railkin, 1998b), it is necessary to take a fouled surface (a glass slide, a macroalga, a stone, etc.) and carefully remove the microfouling with a hard brush or a rubber spatula. Microscopic control has shown that over 95% of the visible cells is removed practically undamaged from a smooth fouled surface. Fouling should be washed off into sterile water. It should cover the bottom of a crystallizer or another vessel with a layer not more than 5 cm thick. Exceeding this level will not allow all the slow-settling cells (e.g., bacteria) to reach the bottom and adhere to it during self-assembly (12 to 24 h). If the quantity of detritus in the microfouling suspension is very great it may completely cover the bottom, which will make adherence of microorganisms difficult or even impossible.

To meet the above requirements in the simplest way, the area of the substrate from which microfouling is washed off should be equal to that of the vessel bottom.

Otherwise, the necessary dilutions should be performed; the proper concentration of microbial cells and detritus should be specially calculated and controlled microscopically (Railkin, 1998b). All the preparatory procedures and the self-assembly of communities itself should be performed at a sufficiently low temperature (no higher than 5 to 10°C) in order to slow down the reproduction of microorganisms and not to change the relative abundance of different groups. Self-assembly should be performed in vessels with a flat bottom

How self-assembly proceeds can be seen when observing microfouling, which develops on artificial inert substrates in the coastal area of the White Sea (Railkin, 1998b). Having carried out all the above manipulations and used the bottom of the Petri dish as an experimental surface, it is possible to observe vagile bacteria settling on the bottom within 5 to 15 min. This is the first (bacterial) stage. Following the bacteria, after approximately 15 min, the first algae, namely the passively settling diatoms, become involved in the process. Autotrophic flagellates appear as part of the fouling community after 30 min. They complete the autotrophic stage of self-assembly. At the same time (more often a little later), heterotrophic flagellates move from the water column to the bottom. Further sequence of the community structure regeneration is as follows: naked amoebae, heliozoans, and ciliates. As a result, no later than 12 to 24 h from the beginning of the experiment a community is formed, which is not dissimilar from the initial one in its species composition and abundance structure. The abundance of individual groups may be restored even earlier (Figure 2.4). The sequence of self-assembly corresponds to the general succession sequence described above for the White Sea (see Section 2.2): bacteria, diatoms, autotrophic and heterotrophic flagellates, sarcodines, and ciliates. Thus, self-assembly, as well as microfouling succession, consists of a bacterial, an autotrophic, and a heterotrophic stage.

The development of the four-layer structure of the microfouling community (Railkin, 1998b) in the process of self-assembly starts with the formation of the bacterial layer, also containing motile diatoms and other algae. The second layer is formed by vagile protozoans (heterotrophic flagellates, amoebae, ciliates). The third layer comprises sessile protozoans (heterotrophic flagellates, ciliates) and solitary attached diatoms. The increasing complexity of the spatial structure is associated with the appearance of large colonial diatoms on the substrate, which form the fourth layer above it.

The self-assembly of microfouling communities is not a mere mechanical process of sedimentation of microorganisms. Indeed, the order of appearance of different groups of microorganisms in the community, corresponding to their rate of immigration, is, on the whole, opposed to their rates of sedimentation or active movement. If the groups were ranged by the rate of immigration, one end of the range would be occupied by bacteria, the other by ciliates. On the contrary, according to the rate of swimming (see Section 3.2), the ciliates would be first, and the bacteria last.

Self-assembly seems to be an emergent property of microorganism communities inhabiting the hard surfaces of natural and artificial origin. The groups of microorganisms do not possess this property separately, it is inherent in the community as a whole.

Self-assembly differs from succession in having a higher rate. The duration of the community development up to the climax stage in the surface layer of the White

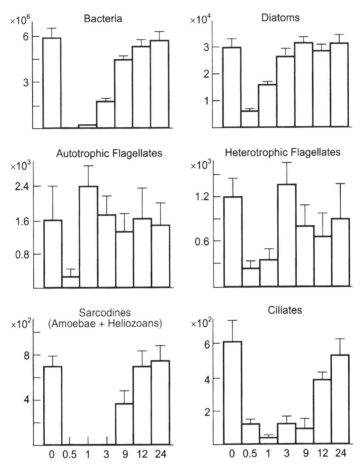

FIGURE 2.4 Recovery of the main groups of microfouling in the process of the community self-assembly (based on data from Railkin, 1998b). Abscissa: time from the beginning of self-assembly, h; ordinate: abundance, cells/cm^2 (for ciliates, cells/dm^2).

Sea is 3 to 5 weeks in summer and autumn, whereas in the laboratory it does not last more than 24 h. Self-assembly under laboratory conditions lasts 12 to 24 h independently of what stage of development the natural community is at. It should also be pointed out that, unlike succession, during self-assembly the exact species sequence of substrate colonization is not observed. Yet full coincidence of the group sequence of colonizing of vacant substrate takes place.

When comparing the patterns of recovery successions of microfouling communities in the field and in the laboratory, it should be noted that there are other differences between them. After mechanical disturbance of the communities in the field, their abundance and species composition is mostly disturbed, whereas in the laboratory, it is their spatial structure that is disrupted. Under field conditions, the source of microorganisms for regeneration is plankton, while in the laboratory experiment, it is the suspension of microorganisms removed from hard substrate ("laboratory plankton"). Both in the field and in the laboratory, recovery was complete and included such colonization processes as immigration and accumulation. At the same time, a considerable role in the field belonged to reproduction of microorganisms on hard substrates. Community self-assembly was impeded on vertical surfaces but especially strongly in the current, even a weak one (1 cm/s). In the latter case the abundance of some groups was not completely restored even within 24 h.

After destruction, which may be caused, for example, by storms or human activity, benthos communities of multicellular organisms on soft and hard grounds may also recover. After removing the fauna from small experimental areas (10×10 cm) in Old Tampa Bay, Florida (U.S.), recovery proceeded quickly: within 7.5 h in the case of infaunal communities and 1.8 d in the case of epifaunal communities (Bell and Delvin, 1983). Such a high rate of recovery was determined by the dominant participation of adults in this process. Finer mechanisms of recovery (self-assembly) of hard-substrate communities of macroorganisms seem to invite additional study.

The above examples show that temporary interruption of biofouling by itself does not solve the problem of protection from it, as disrupted communities recover after some time or continue their development in another direction, which involves the processes of accumulation and growth. Yet in accordance with the degree of disturbances in the community structure the ways and rates of recovery successions may be different. To make protection from biofouling by means of disrupting the succession process effective, it is necessary to repeat this disruption with a frequency exceeding the frequency of accumulation of organisms on hard substrates, that is, continuously or almost continuously. An absolutely new approach seems to be necessary for designing protection from biofouling. It should be based on the analysis of key (the most sensitive) colonization processes whose complete inhibition would impede colonization and development of communities on hard substrates.

3 Temporary Planktonic Existence

3.1 RELEASE OF PROPAGULES INTO PLANKTON

The sources of colonization of living and non-living (inert) surfaces in the marine environment are communities inhabiting hard substrates of natural and artificial origin, and also soft grounds (see Chapter 1). They release dispersal forms, usually called propagules, into water: microorganisms, animal larvae, and macroalgal spores, which are potential colonists (foulers).

The contribution of different hard-substrate communities to the colonization (biofouling) process is not the same. It may depend on the intensity of propagule production in communities, their species composition, the distance from fouled bodies, the pattern of the currents in the region under consideration, the season, and a number of other factors. Close to the coast the main contribution seems to be made by the bottom communities, whereas in the open ocean, owing to distance from the shores, the role of fouling communities developing on floating objects and oceanic debris becomes greater.

In order to become part of the plankton and be carried by the current to the appropriate substrates, microorganisms must be washed off the surface or detach themselves from it. Let us consider the ways sessile and motile forms enter the plankton.

In case of excessive development, the gelatinous matrix of the biofilm which is inhabited by microorganisms and covers the submerged objects becomes insufficiently durable, and scraps of it are detached and carried away by water (McIntire, 1968). This process is especially manifest when the current is strong. It occurs both on natural (macroalgae and stones) and artificial substrates.

Rather a common mode of microorganisms entering the plankton is resuspension of marine bottom sediment, for instance, sand or silt grounds. The flow rate of 10 cm/s, typical of the littoral zone, is quite sufficient to raise from the bottom and to carry away fine grains of sand, silt, and detritus, together with the organisms inhabiting them (de Jonge and van den Bergs, 1987). Some, such as diatoms, are not attached to the particles firmly enough and are washed away into the water. Resuspension may claim up to 45% of the phytobenthic cells in the upper 0.5-cm sediment layer (Delgado et al., 1991). Sediments raised from the bottom release other microorganisms and also small multicellular organisms.

A great role in the passive release of microorganisms into the plankton is played by detritus, both in the bottom sediments and in the water column (Gorbenko, 1990). According to my observations, the abundance of diatoms, heterotrophic flagellates, and ciliates on detritus particles 0.05–0.10 to 1–2 mm in size is quite comparable to that on hard inert surfaces in climax microfouling communities, whereas that of bacteria is still greater.

Microorganisms, motile when suspended, may leave the surface, including detritus particles, on their own. Their emigration can be observed in still water under laboratory conditions. To demonstrate this, it is necessary to place a clean glass slide on the bottom of a Petri dish filled with sterile sea water, and above it, at a close distance (0.1 to 0.5 mm), a glass slide containing microfouling on its lower surface. This can be achieved by placing pieces of safety razor blade or a coverslip between the two slides. Placing both slides under the microscope, within 12 h one can observe that the character of fouling on the slides is very similar. This is caused mainly by the emigration of motile organisms from the upper to the lower slide.

Larvae of invertebrates and ascidians are released actively, by swimming. On the contrary, spores of red and a number of green algae, which have no locomotor flagella, are ejected from sporangia and carried by currents (South and Whittick, 1987). Release of motile spores of brown and green algae also occurs under pressure arising in the sporangium.

The time of release of benthic animals and foulers into the plankton may be synchronized with natural periodical processes: light and dark, tidal and lunar cycles (see reviews by Giese, 1959; Neumann, 1978; DeCoursey, 1983; Morgan, 1995). Release of the larvae of many species studied may be synchronized with a certain phase of the cycle: with the dark or the light time of the day, with the high tide, or with the full moon. Such a strategy of reproduction reduces the probability of encountering predators, decreases the death rate of larvae during hatching, and, as a consequence, increases the reproductive success of the species (Giese and Pearse, 1974; Christy, 1982; Morgan, 1990, 1995). In the littoral hydroid species living as epibionts on macroalgae, the release of larvae may take place at low tide, which determines their prevalent settlement close to parental colonies on vacant areas of algae (Orlov and Marfenin, 1993; Orlov, 1996b; Belorustseva and Marfenin, 2002).

The propagules of macrofoulers are part of the plankton only until they settle, and therefore are referred to as meroplankton or temporary plankton (Ehrhardt and Seguin, 1978). Instead of this term, Mileikovsky (1972), not without reason, suggested the term "pelagic larvaton." Further on the former term is used, as it is the one more widely used in the literature on the problem of fouling.

Four ways of development are known (Mileikovsky, 1971): viviparity, direct development, and development of the pelagic or demersal larva. Of these, only pelagic development may provide recruitment of populations and communities on hard surfaces, both at the bottom and in the water column. Indeed, on the basis of viviparity (hatching of juvenile individuals) and direct development (proceeding externally under the cover of egg shells), colonization of hard substrates at the bottom is possible, though at a fairly limited distance from their birthplace; but the colonization of substrates in the water column is unlikely. Consequently, with such mechanisms of development the connection between bottom communities and those inhabiting hard substrates in the water column is limited and may be totally disrupted. Demersal development (Mileikovsky, 1971), proceeding close to the bottom or on the bottom, imparts a somewhat greater dispersal potential to the larvae than viviparity and direct development. It has been described in some species of polychaetes and echinoderms.

At the same time, larvae with pelagic development provide recruitment of all the communities on hard surfaces independently of their position in the reservoir, and also a continuous connection between them. In the opinion of Mileikovsky (1971), pelagic development gives considerable advantages for dispersal to great distances and, consequently, for the expansion of the species range, affords opportunities for colonization of new substrates and biotopes, and also for a fast recovery of disturbed populations and communities owing to recruitment. It should be added that pelagic development may be the only means of larval development that causes biofouling of natural and artificial hard bodies in the water column. Therefore, it is quite possible that pelagic development is the main mechanism of development in species inhabiting hard substrate. It occurs in approximately 70% of the species of benthic invertebrates (Thorson, 1950; Lefèvre, 1990).

Larvae with the pelagic type of development may be feeding (planktotrophic) or non-feeding (lecithotrophic). Planktotrophic larvae are the most widespread. They are observed in 90% of species with pelagic larvae (Thorson, 1950).

The length of the period during which larvae belong to the plankton until their settlement is mainly determined by the peculiarities of their nutrition and development. The planktonic existence of pelagic lecithotrophic larvae is rather short. It is limited by the stores of yolk and usually does not exceed several hours or days. The larvae of sponges, cnidarians, polychaetes of the family Spirorbidae, archeogastropods, most common encrusting bryozoans, and ascidians are lecithotrophic.

Most polychaetes and mollusks, crustaceans, heilostomate bryozoans of the genera *Membranipora*, *Electra*, *Conopeum*, and echinoderms have planktotrophic larvae. The life span of planktotrophic larvae is weeks or months. It should be noted that the first dispersal larval form of cirripeds, the nauplius, feeds, whereas the other dispersal and settling form, the cypris, does not. The cyprid larva is considered to be able to stay in the plankton for up to 2 months or more until it finds a substrate suitable for settlement. In any case, in experiments the cyprid larvae of *Balanus balanoides* did swim without settling during such a long period (Kamshylov, 1958; Holland and Walker, 1975). It should be borne in mind that as the result of this they lost the ability to settle under experimental conditions.

3.2 BUOYANCY AND LOCOMOTION OF PROPAGULES

An important condition of meroplanktonic life, that is, temporary life as part of plankton, is maintenance of positive buoyancy by the propagules. The strategies of the dispersal forms of micro- and macrofoulers, in spite of their different level of organization, size, and energy resources, are generally similar. They are aimed at attaining a maximum viability and dispersion across as great a territory as possible. The same purpose is served by behavioral responses of larvae to light, gravity, and pressure, which will be considered below (see Section 3.3).

Dispersal forms are usually somewhat heavier than water, but many of them have a near-neutral buoyancy. This is characteristic of almost all microfoulers (with the exception of diatoms). Spores of macroalgae, larvae of sponges, cnidarians, echinoderms, a number of bryozoans, ascidians, and early larvae of polychaetes and mollusks do not have heavy protective covers or shells and therefore they are only a little heavier than water. In larvae of other invertebrates or in later larval stages possessing shells or chitinous skeletons, buoyancy is lower. These are nauplii and cyprids of cirripedes, clad in a chitinous shell, late polychaete larvae with their chitinous setae, late mussel larvae possessing rudimentary calcareous shells, and also larvae of a number of bryozoans, possessing thin bivalve shells. Larvae of these invertebrates are good swimmers and in this way are capable of compensating for their negative buoyancy (Chia et al., 1984).

Much heavier than water are diatoms, which are encased in silica frustules. They are motionless when suspended and can move actively only when in contact with a surface. Therefore, they sink in water, which decreases the length of their existence in the plankton, and consequently reduces the probability of settling on the proper substrate. However, the high abundance of diatoms easily compensates for this.

On the whole, microorganisms, larvae, and spores have a negative buoyancy. Regardless of whether they move in the water or not they are always affected by gravity. The rate of sinking of propagules (both motile and non-motile) is determined by the effect of gravity, Archimedes' buoyancy force, and the force of water resistance. Stokes law, based on this relation, reads that the settling (or free falling) velocity V of a spherical body with radius r and density p in fluid with density p_0 ($p \geq p_0$) and dynamic viscosity μ is determined by the equation:

$$V = \frac{2}{9} r^2 (p - p_0) g / \mu \qquad (3.1)$$

where g is gravitational acceleration. In particular, it follows that the rate of passive settlement is directly proportional to the square of the linear size of the body, and also to the difference between the density of the body and that of the fluid. The latter reflects the balance between gravity and Archimedes force. Experimental studies (Rudyakov, 1986) have shown that Stokes law adequately describes sinking not only of tiny organisms, such as diatoms, but also of larger multicellular organisms, such as larval and adult crustaceans.

Equation 3.1 can be reduced to:

$$V = a L^b \qquad (3.2)$$

where a and b are coefficients, and L is the linear size of the body. According to the data of Yu.A. Rudyakov and V.B. Tseitlin (1980), who analyzed their own and published data on pelagic fishes, crustaceans, including larval forms, chaetognaths, and phytoplankton, pelagic organisms with negative buoyancy within the range of 0.1 mm to several centimeters, tend to sink at a rate approximately equal to one body length per second. This relation is more precisely expressed by the empirical formula:

$$V = 1.11 L \qquad (3.3)$$

where V is measured in mm/s, and L in mm. This formula holds at +20°C. A correction must be applied for other temperatures, since both density and viscosity of water change with temperature.

For organisms with body density considerably different from that of water, Equation 3.3 is not accurate enough. For example, for diatoms, whose density is 2.6 g/cm^3 (South and Whittick, 1987), calculations made using this formula yield values of sinking rate several times lower than the real ones. According to the results of laboratory experiments with different pennate and centric diatom species from freshwater plankton, their rate of sinking is estimated to be from 0.0005 to 0.004 cm/s and is almost independent of the morphology of the cells (Smayda and Boleyn, 1966a, 1966b). Similar average values (0.0002 to 0.003 cm/s) were recorded in the marine environment (Bienfang and Harrison, 1984; Riebesell, 1989).

During the period of "algal bloom" (intensive cell division), a great quantity of mucus is released on the surface of the diatom frustules. This mucus glues the cells into huge aggregates referred to as "marine snow." Their rate of sedimentation is hundreds of times as high as that of individual cells and constitutes about 0.1 cm/s (Stemacek, 1985).

Microorganisms motionless in the water column (aflagellate forms and spores of bacteria, diatoms, etc.) may compensate to some extent for their negative buoyancy, for instance, by accumulating gases in vacuoles or reserve lipids. This is known for some protists (Dogiel et al., 1962). Being lighter than water, lipids reduce their body density. Lipids represent one of the products of photosynthesis in diatoms, besides carbohydrates (Raymont, 1980). Calculations show that when the lipid content in marine diatoms rises from 9 to 40%, their density is reduced from 2.60 to 1.15 g/cm^3, and the rate of sinking drops by 25% (Smayda, 1970). Diatoms may possess some other mechanisms ensuring a near-neutral buoyancy, which is discussed in the literature (Raymont, 1980). These mechanisms are known to be energy dependent (Waite et al., 1992). Their source of energy is light, in its absence it is respiration.

Though sedimentation of motionless microorganisms and some propagules of potential macrofoulers is a real ecological phenomenon, leading to the deposition of silicon, organic and other substances on the sea bottom, their rate of sinking is not great, even in diatoms. The transfer of water masses together with planktonic organisms in seas and oceans occurs not only horizontally, but also vertically, especially upwards (Bowden, 1983), which contributes to the maintenance of passively sinking organisms in the water column. The rate of ascending and descending flow is usually a fraction of a centimeter per second. In the coastal waters, where most meroplankton is concentrated, a significant role is played by the turbulent mixing of water masses because of the irregularity of bottom relief and small depths (Ozmidov, 1968; Bowden, 1983). By force of the above (lower sinking rates, ascending water currents), many motionless microorganisms may stay in the meroplankton for quite a long time, until they settle on some substrate.

Motile suspended propagules can, to a certain extent, regulate their vertical position and the duration of their staying in the plankton. Even bacteria, the slowest swimmers, do not behave like inert particles. The speed of sinking in seawater, calculated by formula (3.1) for bacteria of size 1 to 2 μm is 0.00003 to 0.0001 cm/s. Yet the speed of swimming of bacteria, in Berg's estimation (1985), is not lower than 0.001 cm/s, whereas the greatest theoretically possible value does not exceed 0.1 cm/s. The speed of chemotactic movement in bacteria, measured experimentally, varied from 0.005 to 0.05 cm/s (Blackburn and Fenchel, 1999). Thus, bacteria can obviously resist gravity as the speed of their own movement is an order of magnitude or more higher than that of sinking. The same holds even more for motile propagules of a larger size, whose swimming speed is higher than in bacteria.

Larvae of invertebrates and motile spores of macroalgae possess special adaptations for regulating their vertical position. They are: substances reducing the body density, appendages reducing the sinking rate, and finally swimming.

As with many other organisms constantly or temporarily existing in the plankton, larvae possess certain amounts of lipids increasing their buoyancy (Chia et al., 1984). Lipids are the main energy resource in the larvae of marine benthic invertebrates. Especially rich in lipids are pelagic lecithotrophic larvae (Raymont, 1983).

Another passive way of maintaining buoyancy is development of all kinds of appendages and other external structures, acting as parachutes. According to Stokes law considered above, the appendages reduce the sinking speed because they increase the cross-section area of the body and, correspondingly, the resistance to sinking. Such structures are, for instance, tufts of setae on the anterior end of many larvae and on parapodia of nectochaetes, arms of echinoderm larvae, appendages of cyprid larvae of barnacles, and other such structures (Figure 3.1).

The main mechanism by which larvae of invertebrates, ascidians, and zoospores of macroalgae maintain their vertical position is their motor activity. It is manifested not only in locomotion (active movement in space) but in their behavioral reactions to light, gravity and hydrostatic pressure (Crisp, 1984), and also in vertical migrations (Rudyakov, 1986). These problems will be considered in greater detail below (see Section 3.3).

Important reviews on larval locomotion, which are still topical, were written by M.I. Konstantinova (1966, 1969), S.A. Mileikovsky (1973), F.-S. Chia et al. (1984), and A. Metaxas (2001). Different swimming mechanisms are known. Ciliary movement as a more ancient way of locomotion is common among ciliated amphiblastules and parenchymulae of sponges, planules of scyphoids and hydroids, actinules of corals. Movement by beating of crown cilia is observed in cyphonautes of bryozoans. It is characteristic of early larval stages of polychaetes, trochophores bearing two ciliate belts of different length (prototrochs) above and below the mouth opening. At the metatrochophore stage, body segmentation starts and ciliary zones and tufts of locomotory setae develop on each segment. In late larvae, nectochaetes, transition from a mainly ciliary to a mainly muscular form of locomotion by means of rowing parapodial movements is completed. Anyway, at this stage polychaetes pass over to the near-bottom way of life. Swimming of bivalves and gastropods passing through the stages of trochophore and veliger in their development is maintained only by ciliary beating until settlement. Even pediveliger, with its well-developed foot adapted to crawling on the substrate, swims excellently using its velum. The combination of ciliary

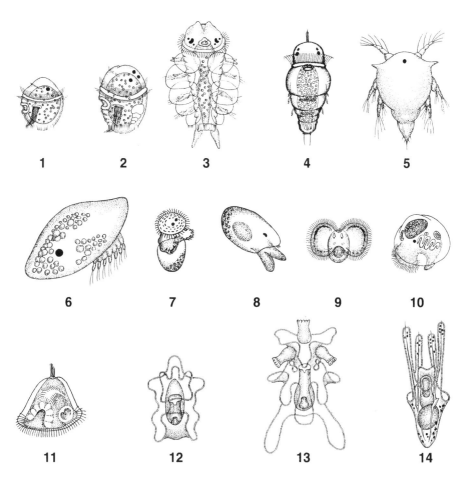

FIGURE 3.1 Larvae of sessile and vagile invertebrates. Polychaete larvae: (1) trochophore, (2) metatrochophore and (3) nectochaete of *Harmatoë imbricata*, (4) nectochaete of *Circeis spirillum*; cirripede larvae: (5) nauplius and (6) cypris of *Semibalanus balanoides*; mollusk larvae: (7) veliger and (8) pediveliger of the limpet *Testudinalia tessellata*, (9) veliger of *Littorina littorea*, (10) veliconcha of *Mytilus edulis*; bryozoan larva: (11) cyphonautes of *Electra pilosa*; echinoderm larvae: (12) bipinnaria and (13) brachiolaria of the starfish *Asterias rubens*, (14) pluteus of the sea urchin *Strongylocentrotus droëbachiensis*. (Unpublished drawings by M.B. Shilin. With permission.)

and muscular movement is characteristic of variously shaped echinoderm larvae, though they mainly use cilia organized into circumoral bands. Nauplii and cyprids of cirripedes move only by means of sharp and frequent strokes of their legs, i.e., by means of muscles. In them, as well as in larvae of other crustaceans, the periods of locomotion alternate with those of short-term rest. The same is also observed in larvae of bivalves and ascidians. Alternation between the periods of motor activity and rest accompanied by the sinking of larvae is considered as an adaptation directed at reducing energy expenditure while maintaining a certain position in the water column (Chia et al., 1984; Rudyakov, 1986; Metaxas, 2001).

The usual velocities of tidal flow in the littoral fringe are tens of centimeters per second; they may reach several meters per second in narrow straits and fjords, whereas in the open seas and oceans the values are smaller (Ozmidov, 1968; Bowden, 1983). The swimming rate of most larvae of invertebrates and motile macroalgae spores is considerably lower than the above values. Therefore, it is possible to suggest that horizontally, they are mainly carried by currents (e.g., Abelson and Denny, 1997), which does characterize them as planktonic, or, to be more exact, as meroplanktonic forms.

Yet it does not follow that larvae can be considered as passively transported particles, as some scientists do (e.g., Hannan, 1981, 1984) regarding polychaetes. Active vertical movement and behavioral reactions to environmental factors allow them to choose their habitat. It has been mentioned above that the swimming velocities of propagules are considerably higher than their rate of passive sedimentation under the action of gravity. Experimental data (Lefèvre, 1990) show that the number of pelagic larvae transported by the current turns out to be smaller in potential places of sedimentation than it would be if they were transported as passive particles.

Velocities of swimming in larvae of various invertebrates, including foulers, differ considerably and embrace a great range of values, from approximately 0.003 cm/s in amphiblastules of some calcareous sponges (Konstantinova, 1966) to almost 4 to 5 cm/s in cyprids of *Balanus crenatus* and *Balanus (Semibalanus) balanoides* (Crisp, 1955). Table 3.1, based on well known reviews (Mileikovsky, 1973; Chia et al., 1984) supplemented with other data, presents information on typical swimming rates of the larvae of foulers.

Generalizations that can be made using this information are as follows. As a rule, the velocity of ciliary movement is lower than that of muscular movement. In primitive larvae of sponges and cnidarians, swimming is mainly rather slow, although in some species it may be rather fast. Muscular movement is more effective, such as that in cirripedes and ascidians.

Maintenance of propagules in the water column, whose mechanisms have been considered above, is prerequisite for their being carried by the currents, i.e., drifting. Duration of planktonic life and the distance to which larvae may be transported together with water masses are rather different in different species and groups of invertebrates. Some of them are philopatric, whereas others may be carried by currents to considerable distances from their parent habitations.

3.3 TAXES AND VERTICAL DISTRIBUTION OF LARVAE

Biologists' conceptions on taxes and vertical distribution of the larvae of benthic animals were formed under the influence of G. Thorson (1964), who had collated a vast literature as well as his own material. On the basis of data on 141 species he divided them into three groups according to their response to light during the early period of planktonic life. The first and the most numerous group (82% of all the species) was photopositive, the second (12%), indifferent, and the third (6%), photonegative. The first group, including 116 species, was made up of hydroids; the polychaetes *Hydroides dianthus*, *Ophelia bicornis*, *Polydora ciliata*, *Spirorbis* spp.; the cirripedes *Balanus amphitrite*, *B. crenatus*, *B. eburneus*, *Elminius modestus*, *Semibalanus balanoides*; mollusks — the

TABLE 3.1
Swimming Velocity of Larvae in the Main Groups of Marine Foulers

Species	Type of Larva	Maximum Swimming Velocity, cm/s ↑	↓	→	?	Reference
		Sponges				
Haliclona sp.	Amphiblastula			1.00		Bergquist et al., 1970
H. tubifera	Amphiblastula			0.36		Woollacott, 1993
		Hydroids				
Dynamena pumila	Planula			0.03		Author's data
Gonothyraea loveni	Planula			0.08		Railkin, 1995b
		Polychaetes				
Harmothoë imbricata	Trochophore				0.11	Konstantinova, 1966
Polydora ciliata	Nectochaete				0.11	Konstantinova, 1969
Scoloplos armiger	Trochophore				0.08	Konstantinova, 1969
		Cirripedes				
Balanus crenatus	Cypris			3.90		Crisp, 1955
Lepas pectinata	Nauplius				0.40	Moyse, 1984
Semibalanus balanoides	Nauplius				0.43	Singarajah, 1969
S. balanoides	Cypris			4.80		Crisp, 1955
		Bivalves				
Mercenaria mercenaria	Veliger	0.13				Carriker, 1961
Mytilus edulis	Veliger				0.11	Konstantinova, 1966
M. edulis	Veliconcha	0.40				Bayne, 1976
Pecten maximus	Late veliger	0.14				Cragg, 1980
Teredo pedicellatus	Veliger				0.75	Isham and Tierney, 1953
		Bryozoans				
Membranipora sp.	Cyphonautes				0.19	Konstantinova, 1966
		Echinoderms				
Asterias rubens	Bipinnaria				0.03	Konstantinova, 1966
Ophiopholus aculeata	Ophiopluteus				0.01	Konstantinova, 1966
		Ascidians				
Ascidia mentula	Tadpole larva				0.30	Berrill, 1931
Botryllus gigas	Tadpole larva				2.00	Berrill, 1931
Ciona intestinalis	Tadpole larva				0.40	Berrill, 1931
Ecteinascidia turbinata	Tadpole larva		1.24			Bingham and Young, 1991
Styelopsis grossularia	Tadpole larva				1.0	Berrill, 1931

Note: Arrows indicate the direction of movement of the larvae; ? indicates that the direction was not specified.

oysters *Crassostrea virginica*, *Ostrea edulis* and the shipworm *Teredo* spp.; the bryozoans *Bowerbankia* spp., *Bugula* spp., *Celleporella hyalina*; the ascidians *Botryllus schlosseri*, *Ciona intestinalis*; and representatives of some other groups. All the animals listed above inhabit hard substrates and some of them also soft grounds. This group includes species not only with planktotrophic but also with lecithotrophic larvae. The larvae of most invertebrates and all ascidians are known to possess photoreceptors, whereas sponge and hydroid larvae have no eyes but nevertheless respond to light (for instance, Ivanova-Kazas, 1975, 1977, 1995; Crisp, 1984).

Further observations of the larvae of sponges (Uriz, 1982; Wapstra and Soest, 1987; Wielsputz and Saller, 1990), hydroids and scyphoids (Williams, 1965; Chia and Bickell, 1978; Otto, 1978; Railkin, 1995b; Orlov, 1996b), polychaetes (Wilson, 1968; Evans, 1971; Eckelbarger, 1978; Marsden, 1991, Dirnberger, 1993), cirripedes (Rzepischevsky et al., 1967; Lewis, 1978; Elfimov et al., 1995), mollusks (Bayne, 1976; Heslinga, 1981; Kasyanov, 1984a), bryozoans (Ryland, 1976; Mihm et al., 1981; Brancato and Woollacott, 1982; Woollacott, 1984), echinoderms (Kasyanov, 1984b), and ascidians (Millar, 1971; Hurlbut, 1993; Railkin and Dysina, 1997), showed that the behavior of many species studied under laboratory conditions may be generalized in a scheme that agrees with Thorson's conceptions (1964) on the vertical distribution of photopositive larvae in the sea. During the first period after hatching, called the swimming stage, the larvae keep to the upper part of the aquarium, exhibiting positive phototaxis and not infrequently also negative geotaxis. Yet their response is soon reversed. The larvae sink to the lower horizon where they at first swim close to the bottom. Later their movement slows down and they pass over to the so-called crawling stage and then settle. It should be noted that the same taxes are also characteristic of planules of the colonial hydroid *Clava multicornis*, lacking cilia and unable to swim (Orlov and Marfenin, 1993). They crawl on the substrate, moving like the larva of a geometrid moth. Positive phototaxis and negative geotaxis allow larvae of this species to ascend sloping surfaces and selectively inhabit thalli of the brown alga *Ascophyllum nodosum*.

Owing to the behavioral activity described above, planktotrophic and many lecithotrophic larvae are mainly concentrated in the surface waters of seas and oceans (Zenkevitsch, 1956, 1977). For instance, in the offshore regions of the White Sea under stratification conditions, over 99% of all meroplanktonic larvae is concentrated in the upper level 10 m thick (Shilin et al., 1987; Shilin, 1989; Maximovich and Shilin, 1993). Similar vertical distribution is also observed in other seas.

On the shelf and in the open sea meroplankton distribution is similar to that described above. The difference is only quantitative. The upper 50- to 300-m layer is the richest in larvae (e.g., Mileikovsky, 1968b). It contains dispersal forms of polychaetes, mollusks, cirripedes, bryozoans, and echinoderms.

Some pelagic larvae may spend part of their development at the water–air interface in the hyponeuston, in the layer 0 to 5 cm thick, and may be observed as deep as 0.5 to 1.0 m, for instance in the Black Sea (Zaitsev, 1970; Alexandrov, 1986). The proportion of such larvae is different in different seas: up to 50–60% in the White Sea and only 5–6% in the Barents Sea (Shuvalov, 1978). The early stages of development of polychaetes, bryozoans, gastropods (Shuvalov, 1978), and cirripedes (Alexandrov, 1986) may proceed in the hyponeuston.

The above scheme, according to which larvae swim towards light and stay in the surface layer almost until they settle, should be especially applicable to planktotrophic larvae whose settling horizon is lower than that in which their dispersal stages occur. It can be expected that larvae with a prolonged planktonic period will show a well-expressed swimming stage under laboratory conditions as well. This stage will be delimited in time from the subsequent crawling stage.

Possible deviations from the above scheme are mainly characteristic of lecithotrophic larvae. They rise to the surface only for a short time, if at all, the main part of the day being spent in the near-bottom layers. In the laboratory, their swimming stage is also not well expressed and is not always revealed at all. Not all larvae possess motor response to light in the form of taxes. In some of them, positive photokinesis (Fraenkel and Gunn, 1961) is observed. It is manifested in reduced swimming velocity or an elevated number of rotations when approaching light. In both cases this results in an accumulation of larvae in the zone of optimal illumination. This reaction was observed in the planules of the hydroid *Gonothyraea loveni*, usually within the first hour after hatching, and was not observed later, though in some experiments, in the absence of appropriate substrates, the planules settled only in two weeks.

Not all larvae are photopositive during the initial period after hatching. There are species with a negative or neutral response to light, but their number is not great. According to Thorson's review (1964), of typical inhabitants of hard substrates, larvae of the corals *Acropora brüggemanni*, *Galaxea aspera*, and *Pachycerianthus multiplicatus* and the polychaetes *Scoloplos armiger* and *Spirorbis rupestris* are photonegative at the initial stage of swimming. Neutral and almost unexpressed response to light at the initial period of swimming is manifested by the hydroid *Tubularia larynx*, the polychaete *Hydroides norvegica*, the bivalves *Ostrea* spp., *Mercenaria mercenaria*, the bryozoan *Alcyonidium polyoum*, and the ascidians *Molgula citrina*, *Dendrodoa grossularia*, and *Styela partita*.

In larvae of invertebrates, together with photoreceptors, statocysts and baroreceptors are described (Prosser and Brown, 1961; Ivanova-Kazas, 1977, 1995; Crisp, 1984; Kasyanov, 1984a, 1984b; Zevina, 1994; Elfimov et al., 1995). One of the reasons for larvae maintaining a certain vertical position is their response to hydrostatic pressure (Knight-Jones and Morgan, 1966; Crisp, 1984). Its influence upon mollusk larvae is combined with the action of gravity (Rice, 1964). When a veliger descends passively, water pressure on it increases. This serves as a signal for upward movement. As a result, the larva returns to its former position. The pressure acting on it lowers and stops being a factor, intensifying the vertical component of its movement. The veliger ceases movement and sinks passively. The alternation of the phases of active swimming and sinking maintains the vertical distribution of veligers. A similar mechanism is present in the cyprid larvae of cirripedes (Knight-Jones and Morgan, 1966) and larvae of some polychaetes (Evans, 1971). Another type of reaction is shown by cirripede nauplii (Knight-Jones and Morgan, 1966). The action of pressure on them is combined with the influence of the intensity and direction of light. If pressure increases they exhibit a photopositive reaction and swim upward; if pressure lowers they exhibit a photonegative reaction. Such behavior makes it possible for them to keep to a certain horizon. At the presettlement stage larvae, as a rule, stop responding to hydrostatic pressure but stay sensitive to light and not infrequently to gravitation.

In the tropical, temperate and cold waters, zooplankton is observed to ascend to the upper layers in the evening and night, and descend in the morning (Raymont, 1983). This is conducive to the reduction of predation press and better employment of trophic resources (Metaxas, 2001). Phototaxis underlies the daily migrations of zooplankton (Ringelberg, 1995). The distribution of phytoplankton also does not stay constant during the day. When it is light motile algae congregate near the surface, whereas they descend when it is dark (South and Whittick, 1987). The distribution of meroplanktonic larvae during the day also changes owing to their vertical movement to considerable distances (Rudyakov, 1986). It is especially expressed in crustacean larvae, which possess a developed muscle system and powerful locomotory appendages. The I–III stage nauplii and cyprids of the barnacle *Balanus improvisus* exhibit daily vertical migrations in the Black Sea (Alexandrov, 1986). Nauplii, being a main dispersal form of barnacles, are concentrated near the surface for the best part of the day, whereas cyprids, as a settling form, stay close to the bottom. Research carried out in mesocosms 9 m deep, modeling stratification of the surface sea waters, has shown that vertical migrations are also typical of mollusk larvae (Gallager et al., 1996).

It is possible that one of the important factors determining the vertical distribution of planktotrophic larvae may be the distribution of their food objects. In any case, the planktonic distribution of larvae of the bivalve *Mytilus edulis* correlates well with that of microalgae that serve them as the main source of food (Dobretsov and Railkin, 2000; Dobretsov and Miron, 2001).

The above vertical distribution of larvae is certain to be an adaptation to temporary planktonic life. Inhabiting the upper, well-aerated, and well-heated photic layer, where microalgae, bacteria, and protists are concentrated, on which planktotrophic larvae feed, creates favorable conditions for their nutrition and development. Here the velocity of currents is greater than at the depth. Therefore, being close to the surface, larvae may be carried by currents to considerable distances and settle far from parent habitats. This is conducive to their dispersion and reduces competition for space and resources as compared to the situation when they would settle in the biotopes where their parents live. For example, change in the vertical position of late barnacle larvae within the range of several centimeters corresponds to several kilometers' difference in the position of their settlement sites (Bousfield, 1955). Descending to the lower horizons before settlement reduces the probability of their being carried into the open sea and creates prerequisites for settlement in places most favorable for development and life of adult forms. Thus, it is quite reasonable to speak of dispersal and presettlement stages in plankthotrophic larvae of species inhabiting hard substrates and soft grounds. The dispersal stage in most species occurs in the upper water layer the presettlement one is usually associated with, descending to the deeper or near-bottom layers.

3.4 OFFSHORE AND OCEANIC DRIFT

Motile macroalgal spores (Reed et al., 1992) and lecithotrophic larvae (Crisp, 1984) have a short swimming period. Therefore, their dispersion from hatching places may vary from several centimeters (Belorustseva and Marfenin, 2002) to tens or hundreds of meters, and only rarely reach several kilometers (Graham and Sebens, 1996). The dispersion potential of long-living planktotrophic larvae is much greater. In the open ocean, they may be carried out by currents over distances of up to hundreds or thousands of kilometers (see review by Scheltema, 1986b).

In the interior sea waters (partly closed gulfs, bays, and fjords) or at individual semi-isolated sites for instance, in river estuaries, where ports are often situated, macrofoulers are either recruited from the local populations or migrate from neighboring ones, depending on the hydrological conditions in those water areas. The system of currents is not infrequently organized in such a way that it prevents the larvae from being carried into open regions (e.g., Panov et al., 1988; Lagadenc et al., 1990). This may be promoted by the adaptive behavior of larvae considered in Chapters 3 and 4, and additional physical factors, such as internal waves, periodical alternation of currents, thermocline, halocline, etc. (e.g., Lefèvre, 1990; Roegner, 2000). In such cases planktotrophic larvae may settle not only in the territories occupied by the neighboring populations of the species but also in the habitat of parental populations, which should limit the genetic exchange between populations and influence the microevolutionary processes.

In the open sea areas within the shelf zone, larvae are mainly dispersed along the shoreline by local tidal currents, whose speed is not great and is from one to several tens of centimeters per second. Yet even such moderate currents carry them to considerable distances, sometimes to tens or even hundreds of kilometers (e.g., Crisp, 1958; Mileikovsky, 1968a), since larvae (especially planktotrophic ones) can stay in the plankton for a long time until they find a substrate suitable for settlement. It should be emphasized that it is along the shoreline that the main stationary hydrotechnical constructions are situated. Local freight and passenger transport by ship is also carried out there. Therefore, the main stream of propagules is practically directed at engineering objects situated on the shelf, which increases the probability of their fouling and biodamage and impedes protection measures.

Because the offshore drift and the subsequent settlement of larvae are repeated from year to year they result in the gradual shifting of the borders of the ranges of some species and their expansion to new territories. This is how S.A. Mileikovsky (1971) describes this process (p. 201):

> Such spreading into the new areas occurs in repeated successions: pelagic larvae drift with local currents, settle and establish a new reproductive bottom generation; then the new larvae, produced by this generation drift further and, in turn, establish new territories which are then occupied anew by fresh generations of bottom-dwelling adults of the species. Such successions, many times repeated, led for example to the wide spreading in the marine waters of North-Western Europe … of the Australasian barnacle *Elminius modestus* … This process has evidently played an important role in the practically worldwide distribution of the polychaete *Mercierella enigmatica* … An additional example is mass development of the polychaete *Polydora ciliata* ssp. *limicola* in one of the lymans (little bays) of the north-west part of the Black Sea.

The barnacle *E. modestus*, which migrated from the shores of Western Europe to the British Isles expanded along the Welsh shore at a speed of 20 to 30 km per

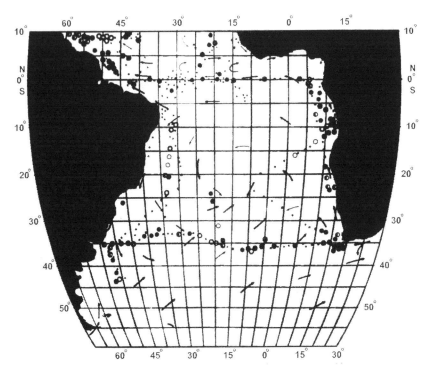

FIGURE 3.2 Transatlantic drift of cirripede larvae. The arrows show the direction of the current and the dots, places where larvae were found. (After Scheltema, R.S. and Carlton, J.T., *Marine Biodeterioration: An Interdisciplinary Study*, Costlow, J.D. and Tipper, R.C., Eds., Naval Institute Press, Annapolis, MD, 1984. Used with the permission of the United States Naval Institute.)

year! The above examples demonstrate the great role played by the benthic populations and all the communities inhabiting hard substrates on the bottom and in the water column, in dispersion of mass species of foulers across large water areas.

Although the scale of the above phenomena is sufficiently imposing, the transoceanic drift of larvae is still more so. Our idea of it was formed under the influence of the works of G. Thorson (1950), S.A. Mileikovsky (1968a, 1968b, 1971), R.S. Scheltema, and other researchers (Scheltema, 1971, 1986a, 1986b; Scheltema and Carlton, 1984; Carlton and Hodder, 1995). Larvae appear to be able to be transported by oceanic currents to hundreds or even thousands of kilometers. As a result, new territories are colonized, genetic exchange between distant populations is carried out, and the borders of specific habitats are expanded (Strathmann, 1986). All this is conducive to the biological progress of species inhabiting hard substrates and soft grounds. The general picture of the drift and dispersion of larvae in the open ocean is striking and the known examples are simply amazing. In Figure 3.2 the transatlantic drift of gooseneck barnacles between the African and the American continent is shown. Comparison of their positions as regards currents suggests that they are first carried from the south to the north along the western shore of Africa, by the Bengel current and then to the west, towards the coast of South America, by the

South-Trade-Wind current (Scheltema and Carlton, 1984). There may be a return drift (Figure 3.2). It may proceed under the influence of the South-Trade-Wind current along the American continent both toward the north and the south, with the participation of the West-Wind current, crossing the Atlantic Ocean from west to east. Similar data were also obtained for the shipworm *Teredo* (Scheltema, 1971).

The larval drift and transport of juvenile and adult foulers (hydroids, polychaetes, barnacles, mussels, bryozoans) with the oceanic flotsam and ships, including the ballast water brought by tankers around the world, create prerequisites for the foulers to invade new geographical regions. In a number of cases, this has resulted in elimination of aboriginal species and radical changes in the ecological situation, with far-reaching consequences (see Section 1.3). For example, colonization of the Great Lakes by the bivalves *Dreissena* (*D. polymorpha* and *D. bugensis*) inhabiting European rivers created a serious national problem, which involved heavy financial and labor inputs (Effler et al., 1996).

4 Settlement of Larvae

4.1 THE REASONS FOR PASSING TO PERIPHYTONIC EXISTENCE

Having passed through the planktonic stage (prolonged in planktotrophic and short in lecithotrophic larvae), the larvae of invertebrates get ready for settlement and further development. The transition from planktonic (pelagic) to periphytonic existence (see Section 1.1) is the key moment in the life cycle of organisms inhabiting hard surfaces of natural and artificial origins. Indeed, in many species, the metamorphosis into an adult is possible only on a hard surface. In some cases, in the larvae of those species whose adults are attached, metamorphosis may also end during the free-swimming stage, according to laboratory observations (Berrill, 1931; Chia and Bickell, 1978; Cloney, 1978). In these cases, the organisms do not establish contact with the substrate, and, if they were in the sea milieu, they surely would be quickly eliminated.

The settlement of propagules (larvae and spores) is an active behavioral process; however, it is also influenced by environmental factors and depends on the properties of the hard surface. The larvae may settle repeatedly, exploring different surfaces and becoming part of the plankton, until they find a favorable substrate; that is to say, they actively choose their permanent habitat.

Settlement may be considered as one of the colonization processes, which include passing from the plankton to the hard surface (settlement proper) and crawling along it until the beginning of attachment in sessile species (metamorphosis in motile species) or returning to the plankton. As the propagules move along the surface they explore it and determine whether that surface should be chosen or rejected as a substrate. For individuals that have already passed through metamorphosis during their planktonic life, settlement is considered to be only a sinking to the surface, accompanied by the exploration and choice (or rejection) of a substrate. Some other workers (e.g., Scheltema, 1974; Chia, 1978; Burke, 1983) hold similar views.

In the literature there is another approach to the definition of the term "settlement" that is widely accepted. Most authors (see the reviews in Crisp, 1984; Lindner, 1984; Hadfield, 1986; Pawlik, 1992; Elfimov et al., 1995; Slattery, 1997) distinguish between the following stages of settlement: exploration of the substrate, temporary attachment (adhesion) of a larva, and final attachment. This approach is justified by the fact that settlement and movement along the surface always involve some form of temporary attachment. Therefore, their terminological distinction is really difficult. However, in my opinion, it is not reasonable to equate attachment to settlement on this ground, as attachment is a separate colonization process (see Section 2.1). Equally unreasonable is including the process of metamorphosis in that of settlement, as some authors do (e.g., Davis, 1987; Orlov, 1996a, 1996b). We will return to this debate again in Chapter 6, in connection with defining the terms "adhesion" and "attachment."

What makes a larva leave the plankton for a hard surface, beginning its transition from planktonic to periphytonic existence? The answer to this question is very important, both for understanding the mechanisms of colonization of natural and artificial hard substrates and for determining how to protect man-made materials and constructions from biofouling. In my opinion, there are three main reasons for this. First, a larva settles only when it is ready (competent) to transform into a juvenile, i.e., at a certain stage of development. It is during this period that the larva is especially sensitive to those environmental factors that will cause its settlement. The presence of these factors, which are usually called "cues" or "signal factors," is the second reason for larval settlement. The third reason is the collision of the larva with the hard surface. This is decisive, though rather often it depends on random events.

The above reasons are not equally important. The intrinsic process of development and the physiological state of the larva are most significant. They are what determine the onset and sequence of events during the settlement and metamorphosis of a larva. The environmental cues (mainly chemical) may induce settlement; however, in many cases their influence is limited to increasing or decreasing the amount of propagules settling on natural or artificial substrates. The presence of a hard surface for settlement and metamorphosis is certainly the decisive factor in passing over to the periphytonic way of life. Though some species that normally inhabit hard surfaces may also exist on soft ground, for most of them, living on the hard surfaces is not only a typical but a necessary condition of survival. In such cases the attached species act as "edificators" in hard-substrate communities.

The larval state called *competence* is usually regarded as a physiological state in which it is capable of metamorphosis under certain environmental conditions (Crisp, 1984; Pawlik, 1992). The history of the problem and the modern conception of "metamorphic competence" are discussed by Hadfield (1998). Some authors (e.g., Burke, 1983) relegate larval attachment to metamorphosis while others (Pawlik, 1992; Orlov, 1996a, 1996b) consider metamorphosis to be part of settlement. This can hardly be justified. Settlement and attachment usually precede metamorphosis or occur concurrently with it (Crisp, 1984) and are not, strictly speaking, metamorphosis. They do not characterize the transformation of the larval form into a juvenile, though they are conducive to the onset and progress of this process. In my opinion, the conception of larval competence should include its readiness for settlement and, for sessile species, also readiness for attachment, since, being competent for metamorphosis, the larvae also should be competent for both settlement and attachment. Thus, the competence of macrofouler larvae (spores) may be regarded as a physiological state achieved at a certain stage of development in the plankton that characterizes their capacity for settlement, attachment, and metamorphosis (development).

There are a number of reviews in which the signal significance of abiotic and biotic environmental factors for settlement is considered (Scheltema, 1974; Ryland, 1976; Crisp, 1984; Cameron, 1986; Hadfield, 1986; Rittschof and Bonaventura, 1986; Svane and Young, 1989; Morse, 1990; Pawlik, 1992; Abelson and Denny, 1997; Rittschof et al., 1998; Clare and Matsumura, 2000). They will be discussed in much more detail in Chapters 4 and 5. For now we will simply summarize them and note the following. Light and gravity are usually noted among the main abiotic factors that orient larvae during the settlement period, because they direct the larvae to a certain water layer in which their settlement is to be accomplished. For a number of species, for instance, for cirripedes (Crisp, 1955; Mullineaux and Butman, 1990), this orienting factor may be the current. During settlement on macroalgae, in the process of establishing epibiotic relations, signal function may be performed by chemical substances that are present on the metabolically active surfaces of plants or released by them into the water. Some polychaetes, cirripedes, and mollusks induce the settlement of larvae of only their own species. The latter use chemoreception to recognize the settlement inductor present on the surface of an adult animal. The stimulating influence of the chemical factors of bacterial and algal films on the settling larvae have been observed in many cases; such films cover all hard surfaces exposed to the marine environment. Some of them also act as inductors, causing not only settlement but the subsequent attachment and metamorphosis. All of the above factors, both biotic and abiotic, in the long run ensure finding a suitable substrate for settlement. The ultimate choice of habitat occurs on the hard surface. The larvae not only evaluate the extent to which the surface is favorable, but they often find a specific site on it where they will settle. Thus, the conditions of the sea milieu and hard surface influence the settlement of competent larvae and macroalgal spores.

Passing over to periphytonic stage is affected not only by the chemical factors of the surface; in many cases, the physical factors of the hard surface also play an important role. Indeed, propagules of macrofoulers may settle on chemically inert surfaces of experimental plates. They may settle on various engineering objects. Finally, larvae and spores also may colonize toxic antifouling coatings. These and many other facts suggest that merely making contact with a hard body is sufficient for larvae and spores to settle on it. Consequently, the settlement of macroorganisms on uninhabited surfaces may be induced by their physical contact with them.

4.2 TAXES AND DISTRIBUTION OF LARVAE DURING SETTLEMENT

During the presettling period and at the settlement stage, larvae are not sensitive (or slightly sensitive) to hydrostatic pressure. At the final stage of their planktonic existence they become especially sensitive to light and gravity. This allows some larvae to descend to the near-bottom layers, whereas others, on the contrary, are able to ascend and settle close to the surface.

There are only three vectors in the sea — direction of light, gravity, and current — with regard to which the larvae and spores can orient themselves in space and with respect to hard surfaces (Crisp, 1984). Propagules of most species possess a swimming velocity that is considerably lower than that of the current (e.g., Butman, 1987; Abelson and Denny, 1997). Therefore, many of them are carried away by the current and cannot use it as an environmental orienting factor to reach a hard surface. Cyprid larvae are an important exception (Crisp, 1955; Mullineaux and Butman, 1990); they are good swimmers, and the current appears to help them find a surface for settlement. These larvae are rather sensitive to the current, and, under experimental conditions, they do not settle in still water at all (Crisp, 1955). The function of perception of water movement near the substrate surface may be performed by setae surrounding their frontal horn pores (Elfimov et al., 1995). A certain gradient of current velocity near the surface, which is necessary for settlement, is different for different species (Crisp, 1955). For example, in an experiment, *Semibalanus balanoides* settled and adhered at greater current velocities than *Balanus crenatus*, which is well accounted for by the conditions under which they live in the sea. The former species is littoral, whereas the latter is sublittoral. *Elminius modestus* exhibits maximal settlement at lower velocities of the current than *S. balanoides* (Crisp, 1955). This agrees with the specific features of their ecology: the former species settles in the sea, in calmer water (usually in sheltered bays and estuaries), whereas the latter does it in highly turbulent waters on the exposed shore.

One of the possible mechanisms of dispersion of lecithotrophic larvae of hydroids, whose colonies develop on algae, is their crawling from the maternal colony to the nearby free surfaces of algae at low tide, when the water is still, i.e., without using currents (Orlov and Marfenin, 1993; Orlov, 1996b; Belorustseva and Marfenin, 2002). In any case, partial drying for 1–3 h stimulated the settlement of *Clava multicornis*, *Dynamena pumila*, and *Laomedea flexuosa*. Such an additional strategy of dispersion and settlement during drying at low tide may be more widespread in littoral animals. This interesting question awaits further study.

As mentioned above, the main abiotic factors orienting larvae during settlement are light and gravity. They determine the general direction of movement and not infrequently the place where larvae would most probably settle.

G. Thorson (1964) summarized the data on larval phototaxes in 141 species of marine bottom invertebrates, both during their planktonic life and during settlement. Using the results of his analysis of phototaxis reversal, two groups of dispersal forms of foulers can be distinguished. The first, and the most numerous one, contains larvae changing a positive reaction to light into a negative one during settlement. It includes many polychaetes (*Ophelia bicornis*, *Polydora* spp., *Spirorbis* spp., etc.), bryozoans (*Bowerbankia pustulosa*, *Bugula* spp., *Celleporella hyalina*, *Watersipora cucullata*, etc.), ascidians (*Ciona intestinalis*, *Ascidia nigra*, *Botrillus schlosseri*, etc.), and cirripedes (*Balanus eburneus*, *B. improvisus*, *B. perforatus*, *Pollicipes spinosus*). The second group contains larvae that remain photopositive during settlement. Most of these species belong to cirripedes (*Balanus amphitrite*, *B. crenatus*, *Megabalanus tintinnabulum*, *Chthamalus stellatus*, and *Semibalanus balanoides*), but there are some polychaetes (*Polydora antennata*, *Pygospio elegans*, *Spirorbis* spp.) and representatives of other groups. These data are important for analyzing and explaining the regularities of settlement and distribution of invertebrates and ascidians on differently oriented substrates.

The larvae of many species of calcareous sponges and demosponges prefer to settle on the lower sides of experimental plates (Vacelét, 1981). According to some field observations (Oshurkov and Oksov, 1983; Shilin et al., 1987) cyprids of the barnacle *Semibalanus balanoides* and cyphonautes of the bryozoans *Callopora craticula*, *Disporella hispida*, and *Cribrillina annulata* settle mostly on the lower side of horizontal plates and on vertical surfaces; the nectochaetes of *Circeis spirillum* settle mainly on the lower side, as do sponge larvae, whereas the planulae of the hydroids *Obelia longissima* and *Gonothyraea loveni* settle predominantly on the upper side and on vertical surfaces. Other workers have also noted the preferred settlement of cirripede larvae on the lower side of horizontal surfaces (Korn, 1990) and of bryozoan larvae in shaded places (Ryland, 1976; McKinney and McKinney, 1993). The compound ascidians *Aplidium stellatum* settle better on vertical surfaces than on horizontal ones (Gotelli, 1987). Swimming pediveligers of the bivalve *Mytilus edulis* show negative phototaxis and positive geotaxis during settlement (Bayne, 1976); i.e., the vector of their settlement is directed downward. The above behavioral features make it possible to explain why they settle mainly on the upper side of horizontally oriented experimental plates, which has been observed repeatedly by this author in the White Sea and also by other scientists (Oshurkov, 1985; Shilin et al., 1987). A similar distribution pattern is shown by another bivalve, *Hiatella arctica*. These species settle less actively on the lower side of horizontal plates and on vertical surfaces.

Larvae of the solitary ascidian *Molgula complanata* almost always settle on the lower side of the plates, whereas the colonial species *Diplosoma listerianum* and *Botryllus schlosseri* almost equally colonize on the lower side of horizontal plates and both sides of vertical surfaces (Schmidt, 1982), and *Didemnum candidum* prefers the lower side of the substrate (Hurlbut, 1993). Two types of responses to light are typical of ascidian larvae: accumulation before settlement in places with low illumination and the so-called "shadow response," which is manifested in increased motor activity by motionless larvae on shading or an abrupt increasing of illumination (Svane and Young, 1989). Both types of responses allow the larvae to select suitable light conditions; they usually choose to settle in a poorly illuminated place.

Algae prefer to foul horizontal and slightly inclined surfaces, which is an important adaptation to photosynthesis (Vandermeulen and de Wreede, 1982; Konno, 1986; Whorff et al., 1995).

Thus, different patterns of spatial distribution of animals and algae on vertical, horizontal, and inclined surfaces are observed. The settlement pattern of invertebrates and ascidians on differently oriented surfaces seems to be based mainly on larval response to light and gravity, and the same reasons may account for the peculiarities of algal spore settlement. Settlement on vertical surfaces probably testifies to the presence of horizontal movement of larvae near the substrates, which seems to be associated with the absence of response to hydrostatic pressure during the settlement period. A number of important experimental works present additional data on the settlement of invertebrates (Vandermeulen and de Wreede, 1982; Konno, 1986) as well as of spores of macroalgae (Konno, 1986; Whorff et al., 1995) on differently oriented surfaces.

Information on the peculiarities of settlement on differently oriented surfaces may be used to explain and predict the distribution of macrofoulers on the surfaces of technical objects, such as ships. It is necessary to take into account not only light and gravity but also other factors, especially the current, turbulence, and vertical water exchange.

When discussing the behavior of settling larvae, it should be noted that they may change their vertical distribution just before settlement on the substrate. For example, the planktonic distribution of larvae of the mussel *Mytilus edulis* during the presettlement period was found to be different from the distribution on the substrates just after settlement (Dobretsov and Railkin, 2000; Dobretsov and Miron,

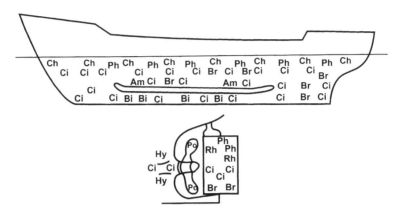

FIGURE 4.1 Distribution of foulers on the ship hull, propeller, and rudder. Macroalgae: Ch – green, Ph – brown, Rh – red. Invertebrates: Hy – hydroids, Po – polychaetes, Ci – cirripedes, Am – amphipods, Bi – bivalves, Br – bryozoans. (After Zvyagintsev and Mikhailov, 1980. With permission of Kasyanov, V.L., Director of Institute of Marine Biology, Vladivostok, Russia.)

2001). Before settlement, larvae become concentrated vertically on the horizon of filamentous substrates on which they initially settle. The biological meaning of such an adaptive strategy is that, owing to a change in the vertical distribution of larvae in the plankton immediately before settlement, the zone where their settlement occurs becomes narrower, increasing the probability of successive selection of the biotope favorable for the life and development of juvenile and adult mollusks. A similar mechanism of vertical redistribution of larvae before settlement was also described in the barnacle *Semibalanus balanoides* (Dobretsov, 1998). Such a mechanism of habitat choice appears not only to occur in these species, but to be more widepread.

Let us consider the peculiarities of distribution of foulers on the submerged part of a ship's hull. Macroalgae settle on the best illuminated parts of the hull. Therefore, they are distributed along the vertical boards below the waterline and are almost never observed on the bottom (Gurevich et al., 1989). It is characteristic that here, as in the sea, a certain vertical zonality is observed: green algae mainly settle closer to the water surface, brown algae, lower, and red algae occupy a still lower position. Such a distribution is determined predominantly by the different responses of algae to the intensity and spectral composition of light. Below algal fouling, and also within it, animal foulers — hydroids, sedentary polychaetes, cirripedes, mollusks, and bryozoans — are distributed in a regular pattern (Figure 4.1).

The bottom of the ship's hull is fouled by bryozoans, hydroids, and cirripedes. It should be noted that the distribution of organisms on the hull also depends on the ability of some species to stay on the hard surface when the current velocity near the hull is high. Therefore, in ship fouling, besides the above-mentioned vertical zonality, a non-uniform distribution of the species and group composition from the bow to the stern is observed (Figure 4.1). Barnacles, which are capable of staying on the surface at high current speeds, inhabit the bow and middle parts whereas polychaetes, which are less adapted to high speeds but well adapted to turbulent current, settle on the stern part. N.I. Tarasov (1961b) justly wrote: "It is hydrodynamics around the moving ship, together with illumination, that determines the peculiarities of distribution of different foulers along its submerged surface" (p. 6).

4.3 SENSORY SYSTEMS PARTICIPATING IN SUBSTRATE SELECTION

The sensory systems of larvae have not yet been sufficiently studied. Nevertheless, data on the morphology of these systems and larval behavior throw light on their function in representatives of various groups of invertebrates and also ascidians. In sponge larvae (amphiblastulae and parenchymulae), no structures have yet been found that could suggest a possible sensory function. Yet they certainly respond to light, showing a negative phototaxis (Maldonado and Young, 1996), and often settle in sheltered and shaded places (Vacelét, 1981). Four pigment cells with light-refracting bodies situated among the flagellate epithelial cells may be responsible for photoreception (Ivanova-Kazas, 1975). Sponge larvae also perceive the microrelief of the substrate, settling mainly on the rugous surfaces (Uriz, 1982). The mechanisms of these reactions (photoresponses and possibly rugophily) are still unknown. Sponges respond to external stimuli slowly since they have low velocity of excitation conduction, which is only 0.02 cm/s (Koshtoyants, 1957).

In cnidarians, at the stages of planula (in hydroid polyps) and actinula (in corals), cells of the ecto- and entoderm become differentiated into sensory, nerve, muscle, cnidocytes, glandular, etc. The development of the nervous system of larvae is still little studied. In adults, it is represented by a nerve plexus with true synapses and usually a small number of synapselike non-polar contacts (Prosser and Brown, 1961; Svidersky, 1979). An important characteristic of their sensory system is epithelial conductance. The rate of excitation conduction is not high (less than 0.5 cm/s), but it is still higher than in sponges (Prosser and Brown, 1961). Certain elements of the nervous system and probably epithelial conductance are undoubtedly developed already in late planulae, and thus the larvae competent to settlement may differentiate between habitats rather well. According to my observations, attachment in the planulae of the hydroid *Gonothyraea loveni* occurs not in 1–2 days but much later, and sometimes is delayed for as long as 1–2 weeks in the absence of favorable conditions. After this, the larvae may settle normally, attach, undergo metamorphosis, and give rise to a new colony. Receptors responsible for the choice of substrate in hydroids may be scattered over the whole body surface of the larvae. The results of experiments with planulae of the polyp *Hydractinia echinata* speak in favor of this (Müller and Spindler, 1972). When the larva was cut transversely, its anterior and posterior halves proved to be approximately equally sensitive to the chemical (bacterial) settlement cue. Yet there seems to be a somewhat greater number of receptors on the anterior end, since the contact chemical induction of settlement and the attachment proper are in most cases carried out with the direct participation of the anterior body end. As in other invertebrates, their receptors are ciliated cells associated with neurons. Chemoreception plays an important role in substrate selection

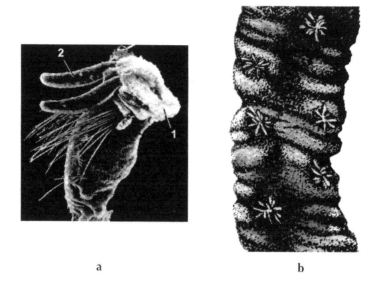

a   b

FIGURE 4.2 Larva of the polychaete *Phragmatopoma lapidosa* just prior to settlement.
(a) Habitus, (b) sensory organs on tentacles; (1) prototroch, (2) tentacle. (After Eckelbarger
and Chia, 1976. With permission of *Canadian Journal of Zoology* and NRC Research Press.)

by the larvae of hydroids (Berking, 1991; Orlov and Marfenin, 1993; Orlov, 1996a),
scyphoids (Neumann et al., 1980), and corals (Morse and Morse, 1991).

Nectochaetes of the family Sabellariidae, having settled on a substrate, explore
it by means of sensory organs (Eckelbarger, 1978) situated on the surface of their
body. Yet these organs are not distributed randomly. In the larvae of *Phragmatopoma
lapidosa*, they are especially numerous on the ventral surface, head, and posterior end
(Figure 4.2). The sensory organs are formed by ciliary cells grouped closely together.
The greatest concentration of these seemingly chemosensory structures is observed on
the tentacles by means of which the polychaetes explore the substrate and assess its
suitability for final settlement. When the larvae find individuals of their own species
living in sand tubes, the substance contained in the tubes serves as a cue for final
settlement (stopping of locomotion) and the beginning of metamorphosis.

Other sabellariid polychaetes display similar behavior (Eckelbarger, 1978).
Mechanoreception is well developed in the larvae of *Polydora ciliata* (Kisseleva,
1967b). The larvae of polychaetes from other families also possess sensory organs
for the selection of substrates according to their chemical or physical properties
(Ivanova-Kazas, 1975). They have sensory ciliary cells that are connected to neuronal
processes. The latter, in their turn, carry the information on the nature of the substrate
to the ganglia of the nervous system (Eckelbarger, 1978).

The cyprids of barnacles possess both chemoreceptors and mechanoreceptors
(Elfimov et al., 1995), located on the antennulae, carapax, and caudal appendages.
Their density is especially high on the attachment disc, located on the underside of
the third segment of the antennulae (Figure 4.3), where several sensory systems can
be distinguished: axial, pre- and postaxial, and radial (Nott and Foster, 1969). All

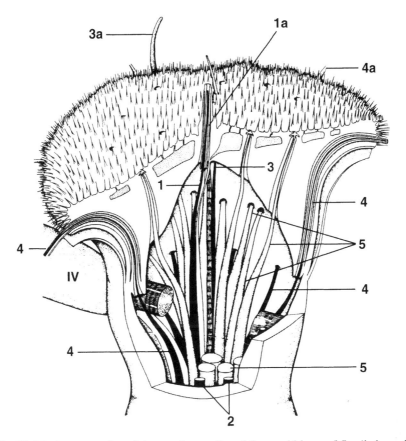

FIGURE 4.3 Reconstruction of the attachment disc of the cyprid larva of *Semibalanus balanoides*: section through the preaxial side. The sensory organs are shown in black: (1) axial, (2) preaxial, (3) postaxial, (4) radial; sensory setae: (1a) axial, (2a) preaxial, (3a) postaxial, (4a) radial, (5) antennular glands; IV – fourth segment of antennula. (After Nott and Foster, 1969. With permission of the Royal Society of London and Prof. J. Nott.)

of them are sensory setae connected to the dendrites of neurons. The axial sensory organ and one of the three radial ones seem to function as chemoreceptors or as chemoreceptors and mechanoreceptors simultaneously. At the same time, the cilia of the pre- and postaxial and radial organs most probably are mechanoreceptors (Nott and Foster, 1969). The same function is performed by sensory cilia located on the fourth segment of the antennulae (Elfimov et al., 1995).

Moving along the surface, the cyprid larvae press their antennulae to it from time to time. This allows them to assess the suitability of the surface by means of mechano- and chemoreceptors. On finding an acceptable site, the larva stops and adheres by means of the attachment disc. Most cirripedes live in groups, and their larvae can distinguish their own species from others when in contact with the specific arthropodin molecules at the bases of their shells (Crisp and Meadows, 1962). Cyprids possess other sensory structures as well, which appear to be used during settlement and substrate selection. These are the nauplius eye and two compound

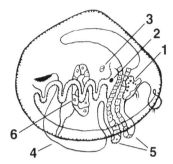

FIGURE 4.4 Sensory and locomotory organs of the pediveliger of the mussel *Mytilus edulis*. (1) Apical plate, (2) eye, (3) statocyst, (4) foot, (5) velum, (6) byssus gland with its duct. (After Kasyanov, 1984b. With permission of the *Biologiya Morya*.)

eyes, the setae of the caudal appendages, sensory organs positioned on the carapax surface, and some others (Elfimov et al., 1995).

Pediveligers of bivalves and gastropods swim before settlement owing to the coordinated ciliary beating of their velum. They possess well-developed sense organs: the apical plate, eyes, and a statocyst — the organ of balance (Figure 4.4) — obviously playing an important role in habitat selection. The sensory systems that take part in the selection of the substrate are mainly represented by sensory ciliate cells on the ventral surface of the foot, rudiments of the osphradium, which assesses the quality of the water entering the mantle cavity, and the eyes (Kasyanov, 1984a, 1989). Having settled on the substrate, the mollusks crawl over it using the foot. In the pediveliger, the elements of the nervous system responsible for the movement of the foot achieve considerable development. They connect the ciliary cells of the foot with the pedal ganglia. The receptor system of the foot, represented by long motile cilia at its end and in the groove, seems to perform a chemosensory but possibly also a mechanosensory function. Having chosen a suitable substrate, the larvae of bivalves get attached to it by strong adhesive threads that are released from the byssus glands (Figure 4.4). The nudibranch *Phestilla sibogae*, which feeds only on corals, possesses an organ located between the lobes of the velum that consists of three types of cells (Bonar, 1978). Of these, flask-shaped ciliated cells are directly connected with the larval nervous system. They are supposed to perform a chemosensory function.

Some bryozoans have a primitive planktotrophic larva with a well-developed digestive system — a cyphonautes. Larvae of other species resemble a cyphonautes but differ from it in the underdeveloped or completely absent digestive canal (Ivanova-Kazas, 1977). The larvae of more primitive forms have a bivalve shell. The swimming cyphonautes of *Electra pilosa* (order Cheilostomata) and *Alcyonidium* spp. (order Ctenostomata) possess an organ of locomotion, the ciliary crown, and a sensory aboral organ (Figure 4.5). The main sensory system used during the substrate selection is the pyriform organ. It is connected to the aboral organ by a bundle of nerve and muscle fibers. On its external surface the pyriform organ has a ciliary field, consisting of sensory cells with long cilia, and on the inner surface there is a glandular field. Thus, besides its main function of surface reception, this organ also performs an additional function of temporary attachment (Reed, 1978). The settled cyphonautes crawls on the surface with the pyriform organ protruded. Its long cilia

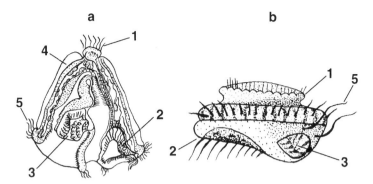

FIGURE 4.5 Sensory, attachment, and locomotory organs of bryozoan cyphonautes. (a) *Electra pilosa*, (b) *Alcyonidium* sp. (1) Aboral organ, (2) pyriform organ, (3) internal sac, (4) shell, (5) crown. (After Ivanova-Kazas, 1977. With permission of Publishing House Nauka, Moscow.)

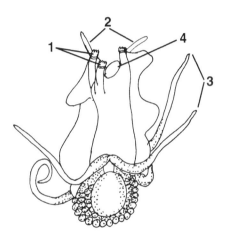

FIGURE 4.6 Sensory, attachment and locomotory organs of the brachiolaria of the starfish *Asterias rubens*. (1) brachiolae, (2) papillae, (3) arms, (4) attachment disc. (Modified after Kasyanov, 1984a. With permission of the *Biologiya Morya*.)

touch the surface and appear to assess its suitability for habitation. The pyriform organ appears to perform a mainly chemoreceptory function (Ivanova-Kazas, 1977). Having found a substrate suitable for metamorphosis and favorable for an adult, the larva turns its inner sac inside out and gets firmly fixed in the chosen place by the sac and the abundantly secreted mucus.

Unlike the bipinnaria developing in the pelagial, the next stage of development of the starfish, the brachiolaria leads a swimming–crawling life and possesses specialized structures for selecting the substrate (Kasyanov, 1984b, 1989). It may cling to a surface with its long arms. The substrate is explored by means of papillae and brachiolae (Dautov and Nezlin, 1985; Byrne and Barker, 1991), located on the anterior end of the larva (Figure 4.6). The sensory ciliary cells present there are connected basally to the subepidermal plexus formed by axons of the nervous system cells. The process of exploration is described in the literature as follows:

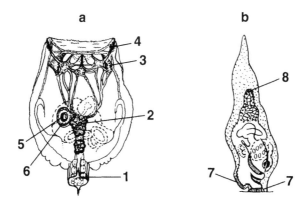

FIGURE 4.7 Sensory, attachment, and locomotory organs of ascidian larvae. (a) Nervous system of *Botryllus niger*, (b) attachment and metamorphosis of *Clavelina lepadiformis*. (1) caudal ganglion, (2) cerebral ganglion, (3) ganglion of the attachment apparatus, (4) papillae of the attachment apparatus, (5) ocellus, (6) statocyte, (7) suckerlike apparatus, (8) retractable tail. (After Ivanova-Kazas, 1995. With permission of St. Petersburg University Publishing House.)

> The brachiolaria orients itself with its ventral side to the substrate, compresses its arms with the ciliate band, and adheres to the substrate with one or two brachiolae, keeping contact with it by means of crowns of papillae. The brachiolaria, so to speak, walks on the substrate by attaching and detaching its brachiolae. The duration of this "walk" is from several seconds to one hour. If the substrate is not suitable for settlement, the larva straightens its arms and swims away. If the substrate is satisfactory, it stops "walking", presses its brachiolae close to the substrate one by one, then straightens them maximally, so that the attachment disc comes into contact with the substrate. (Kasyanov, 1984b, p. 9)

The sea urchin larvae ready for settlement possess rudiments of primary podia bearing sensory ciliary cells (Strathmann, 1978; Burke, 1983). Their outgrowths are connected to the nerve plexus. The sensory cells of the best-studied species, *Dendraster excentricus*, seem to be mainly performing a chemoreceptory function. Contact chemoreception by primary podia is also supposed in *Paracentrotus lividius* (Flammang et al., 1998).

The larvae of ascidians, often referred to as "tadpoles," possess a well-developed nervous system that controls their locomotion and behavior in selecting their habitat. Their organ of locomotion is a tail with a well-developed muscle apparatus. The main sense organs, the statocyte (the organ of balance) and the larval ocellus, lie in the cerebral vesicle (Figure 4.7). In species of the family Botryllidae, the statocyte is associated with the cerebral ganglion and is practically merged with the ocellus, forming a photolith (Ivanova-Kazas, 1995). The mechanoreceptors are located in the caudal part of the body. In a number of species, peculiar coronet cells (the Dilly organ), probably performing the function of a hydrostatic pressure receptor (Svane and Young, 1989), were also described.

As in other animals, after settlement, ascidian larvae start exploring the substrate. The choice of substrate is performed mainly by the receptors of the papillae of the attachment apparatus, which are located at the anterior end of the body (Figure 4.7). Thus, the apparatus simultaneously performs both a sensory and an attachment function. In solitary ascidians, the attachment apparatus consists of eversible and non-eversible, usually sticky, papillae. In compound ascidians it is a sucker releasing a large amount of sticky secretion (Svane and Young, 1989). It is in the compound ascidians that the attachment apparatus is controlled by the nervous system, which allows the larva to attach and detach itself repeatedly, leaving an unsuitable site. The substrate selection does not last for a long time, and soon the larva is finally fixed in its place and starts metamorphosis, retracting its tail within several minutes.

4.4 SELECTIVITY DURING SETTLEMENT

In natural biotopes, the larvae of invertebrates and the spores of macroalgae select their habitats during settlement. According to observations and experiments, habitat selection by the barnacle *Semibalanus balanoides* (Chabot and Bourget, 1988; Le Tourneux and Bourget, 1988; Miron et al., 1999; Olivier et al., 2000) and the mussel *Mytilus edulis* (Dobretsov and Railkin, 2000; Dobretsov and Miron, 2001) progresses in several stages, with the search area gradually being reduced. In particular, during the primary settlement of *M. edulis* in the White Sea, the larva first selects the horizon, then the microbiotope (littoral filamentous algae), and finally the substrate — *Cladophora rupestris*.

Many larvae show a certain selectivity when settling on natural and, under experimental conditions, artificial substrates. This is described in a number of reviews (Meadows and Campbell, 1972; Scheltema, 1974; Crisp, 1974, 1984; Butman, 1987; Morse, 1990; Pawlik, 1992). The selection by larvae (and spores) of a favorable habitat for the life of adults has an adaptive significance. As a rule, the survival rate of larvae is higher on more preferred substrates, as was demonstrated, in particular, for sponges (Maldonado and Young, 1996), polychaetes (Kisseleva, 1967b), barnacles (Miron et al., 1999), mollusks (Moreno, 1995), and ascidians (Davis, 1987; Hurlbut, 1993). Some data on macroalgae (Harlin and Lindbergh, 1977; Figueiredo et al., 1997) suggest that their survival is also connected to substrate selectivity, although the problem does not seem to have been studied experimentally.

Some larvae are very specific in their choice of substrates while others will settle on a wide range of objects. Therefore, by analogy with adult animals, which are divided into specialists and generalists depending on their biotopic selectivity (Begon et al., 1996), larvae can be divided into two similar categories. We shall regard as specialists the larvae that inhabit a limited number of substrates or only one type of substrate (narrow specialization), and those that can colonize various natural substrates will be referred to as generalists.

Sponge larvae are supposed to lack expressed selectivity and be able to settle on any natural or artificial substrates (Bergquist, 1978; Pansini and Pronsato, 1981; Uriz, 1982; Ilan and Loya, 1990). Nevertheless, some species still reveal a certain selectivity. For example, *Halichondria panicea* in the Baltic Sea is usually observed on the red algae *Phycodris* sp. and *Phyllophora* sp. (Barthel, 1986). The problem of sponge larvae selectivity toward hard substrates awaits further experimental study. In any case, the preliminary data suggest (Railkin et al., *in press*) that the settlement and metamorphosis of *Halisarca dujardini* larvae is controlled by some chemical factors of the alga *Fucus vesiculosus* and the microfouling films on its surface.

As a rule, both specialists and generalists can be found within a large taxon. For instance, planulae of the hydroid polyp *Hydractinia echinata* settle only on shells of the hermit crab *Eupagurus* (Müller and Spindler, 1972). Their settlement and attachment are induced by the bacterium *Alteromonas espejina* that lives there (Leitz and Wagner, 1993). Conversely, actinulae of the hydroid *Tubularia larynx* do not distinguish between algal and other substrates (Orlov, 1996a) and can undoubtedly be regarded as generalists. Polychaetes of the genus *Phragmatopoma* settle only on sand tubes of their own species (Eckelbarger, 1978; Pawlik, 1990). At the same time, such motile polychaetes as *Nereis zonata* and *Platinereis dumerilii* settle on sand, silt, shells, and macroalgae under experimental conditions (Kisseleva, 1967a). The larvae of cirripedes, including barnacles, are specialists. Many of them settle only on or near conspecific adults (see, e.g., Knight-Jones, 1953b; Crisp and Meadows, 1962, 1963), yet cyprids of *Semibalanus balanoides* may settle among individuals of several other barnacle species (Moyse and Hui, 1981). In the laboratory, cyprids of *Balanus spongicola* settled and metamorphosed only on shells of living scallops, *Aequipecten opercularis,* on which *B. spongicola* was often found in the ocean, whereas cyprids of *Solidobalanus fallax* settled on different substrates, including non-living ones (Elfimov, 1996).

Bear in mind that the above examples illustrate only extreme cases; many species occupy an intermediate position and may be considered either moderate specialists or limited generalists. J. S. Ryland (1959) studied in detail the distribution of several bryozoan species off Wales (Great Britain). He also investigated the selectivity of the same species under laboratory conditions. He established that the selectivity demonstrated by the bryozoans *Alcyonidium hirsutum, A. polyoum, Frustrellidra hispida,* and *Celleporella hyalina* in the sea on the whole corresponds to the selection of algal substrates by their larvae in the laboratory. The first three species occur in the coastal zone, mainly on fucoids, especially *Fucus serratus*. The same preference was revealed in selection experiments, when different species of algae were simultaneously present in a Petri dish (Figure 4.8). In the sea, *C. hyalina* is found mainly on the alga *Laminaria saccharina*, especially on its rhizoids. At the same time, in selection experiments, it settled well not only on this species but also on the brown algae *Chondrus crispus* and *F. serratus* and the red alga *Gigartina stellata*. Thus, the degree of selectivity of all four of the bryozoan species is not very high, and they can be regarded as moderate specialists.

Solitary ascidians of the family Molguloidea, which inhabit different grounds, macroalgae, and other natural substrates, until recently were not considered to possess any selectivity at all (Svane and Young, 1989). A study carried out on *Molgula citrina* (Railkin and Dysina, 1997) has revealed that its larvae are capable of selecting soft grounds, macroalgae (*Phyllophora brodiae, Laminaria saccharina*), and ascidians (*Styela rustica, Molgula citrina*). It should be noted that the substrate selectivity pattern was close to that observed in the sea, whereas the preferred substrates accelerated the settlement and metamorphosis of the larvae.

An interesting study was performed by A. V. Grishankov (1995). In the Solovetsky Bay of the White Sea, he studied the distribution of 85 species of sessile

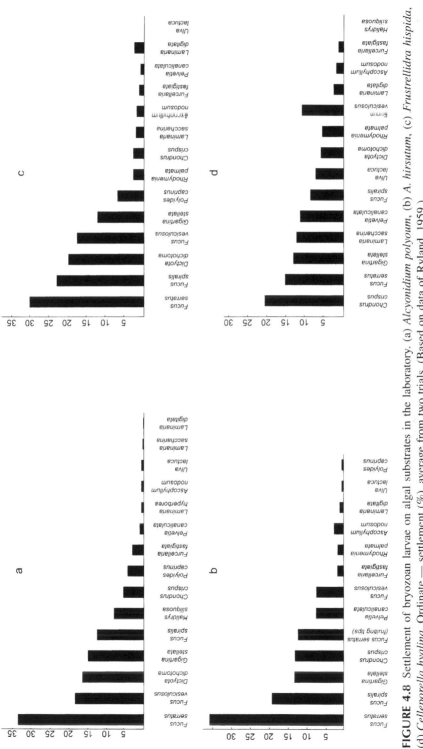

FIGURE 4.8 Settlement of bryozoan larvae on algal substrates in the laboratory. (a) *Alcyonidium polyoum*, (b) *A. hirsutum*, (c) *Frustrellidra hispida*, (d) *Celleporella hyalina*. Ordinate — settlement (%), average from two trials. (Based on data of Ryland, 1959.)

invertebrates on 24 types of hard natural objects, including rocks, empty mollusk shells, macroalgae, and invertebrates. Most of the animals were found to use a more or less limited range of substrates as habitats, i.e., to be to some degree stenotopic and consequently assigned to the group of moderate specialists; 32% of the species were found only on living organisms. These data suggest that specialization is rather common among epibionts. They also demonstrate (based on the material from Solovetsky Bay) that the main structural unit of hard-substrate communities is a consortium or pseudoconsortium whose center is a living or an inert (non-living) object, respectively.

However, it is necessary to note that other workers (Wahl and Mark, 1999), having analyzed observations and experimental studies of more than 2000 cases of epibiosis, concluded that this phenomenon is mostly a facultative one. The discrepancy in the opinions of the cited authors may be caused by different interpretations of the facultative or obligatory nature of epibiosis, and also by the fact that epibiotic species may belong to generalists, specialists, or an intermediate group (see above).

In search of a substrate for further development and life as an adult, larvae can approach a surface, explore it, return to the plankton, and settle again on the same or some different substrate. This goes on until the larva finds a suitable habitat, settles finally, and gets attached (if the adult form is sessile). In many cases, the invertebrate larvae are specialized for settlement on quite definite substrates. If they do not encounter any such substrates, they may remain in the plankton for a period several times as long as the usual duration of their planktonic life. This ability of many species is justly considered to be the manifestation of selectivity during settlement. The delay of development is evident in the planktotrophic larvae, for instance, of polychaetes (Knight-Jones, 1953a; Wilson, 1968), cirripedes (Rzepischevsky et al., 1967; Lewis, 1978), mollusks (Heslinga, 1981), echinoderms (Strathmann, 1978; Highsmith and Emlet, 1986), and other invertebrates (see the review in Pechenick, 1990). It is also observed in lecithotrophic larvae, which usually stay in the plankton for a short time, for instance, in echinoderms (Kasyanov, 1984b), oviparous ascidians (Svane and Young, 1989), and some mollusks (Heslinga, 1981).

The delay of larvae in the plankton may often result in reduced selectivity (Knight-Jones, 1953a, 1953b; Williams, 1965; Strathmann, 1978; Rittschof et al., 1984). They begin to settle on a wider range of substrates, become less selective about their preferred surfaces, or lose the ability to settle and develop, as was demonstrated for the barnacle *Semibalanus balanoides* (Holland and Walker, 1975). In the sea urchins, and the sand dollars *Dendraster excentricus* and *Echinarachnius parma*, juveniles that developed from larvae after they had become competent tended to have a high growth rate, while, conversely, individuals that metamorphosed after a prolonged delay had a low growth rate (Highsmith and Emlet, 1986). The above examples do not cover the whole range of possible variants. Quite a few cases were recorded when prolonged delays in the plankton did not result in reduced selectivity in larvae (see Morse and Morse, 1991). It should be pointed out that, in this case, the larvae as a rule stayed at the stage of development achieved before the delay started.

A prolonged stay of larvae in the plankton, in particular owing to a delay of settlement, must result in their mass elimination (mainly from predators and developmental aberrations) and a drastically reduced settlement potential. As estimated by G. Thorson (1963), less than 0.1% of the total number of larvae released annually into plankton reaches the settlement stage, and only 0.1 to 1.0% of individuals reach maturity. Nevertheless, this abundance is sufficient for the recruitment of bottom populations, since the pool of larvae in the plankton is large enough to compensate for their loss (Mileikovsky, 1971). As a rule, species with a planktotrophic larva that stays in the plankton for a long time produce more larvae than those with a leci thotrophic larva, thus compensating for the loss of larvae during the long period of planktonic life. This can be viewed as the manifestation of the general biological law of a great number of eggs.

The selection by larvae of substrates favorable for their development and the subsequent life of adults is one of the general mechanisms facilitating the concentration of organisms on hard surfaces.

5 Induction and Stimulation of Settlement by a Hard Surface

5.1 TYPES OF INDUCTION AND STIMULATION OF SETTLEMENT

The transition to life on a hard surface, i.e., periphytonic existence (see Section 1.1) is induced and stimulated by certain factors of the surface. Let us classify the types of induction (and stimulation) of settlement, taking the following circumstances into account. In the literature, the surface factors are usually divided into physical and biological. The latter helps to draw attention to the fact that they belong to the biological objects: macroalgae, invertebrate (or vertebrate) animals, or microfouling film. It should be noted that the so-called "biological factors" are such by origin. The concrete nature of their action may be, for instance, chemical or physical. Hereafter, the term "biological factors" will be preferred only for those whose origin is biological and whose mechanism is either unclear or unessential. Such biological factors may be microfouling films, surfaces of adult individuals of some species, etc. Settlement may be induced not only by purely physical and chemical surface factors; for instance, physico-chemical factors may interact or their conjoint influence may differ quantitatively from the simple sum total of the action of these factors. In such cases we shall speak of the combined action of factors. We will hold biological factors with an unidentified mechanism of action to have the same status as combined surface factors.

Planktonic larvae can choose a hard surface from a distance or assess its suitability for final settlement and attachment while in contact with it. Thus, it is possible to speak of distant and contact induction of settlement. Conspecific and heterospecific induction should be also distinguished, i.e., cases when induction is carried out by individuals of the same or another species.

Certainly, in some cases settlement may also take place as the result of non-oriented locomotor activity, i.e., relatively accidentally. Yet typically the choice of substrate and transition to the periphytonic state is obligatory. It is stimulated and induced by specific chemical and physical surface factors. Therefore, the larvae of many species do not settle and start metamorphosis until they find a surface that is suitable as a habitat (see Section 4 4)

Consideration of different settlement cases makes it possible to distinguish between the main types of induction (and stimulation) by physical, chemical, combined (physico-chemical), or biological factors acting in contact or distantly, conspecifically or heterospecifically. Thus, using the above characters for classification,

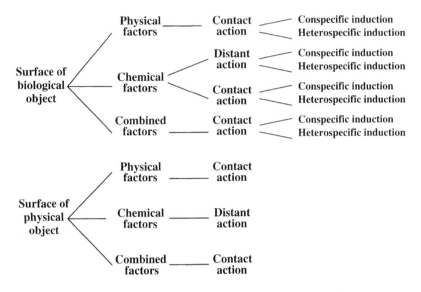

FIGURE 5.1 Classification of types of induction and stimulation of settlement.

we can distinguish between 12 main types of biological induction and stimulation and 6 physical ones, i.e., 18 types altogether, of which only 11 have been described (Figure 5.1). A more detailed classification that takes into account the nature of a biological object (macroalga, animal, microfouling film) or a physical body (natural or artificial) on which settlement occurs would make it possible to consider up to 48 types of induction.

The phenomenon of some species settling preferentially or exclusively on others is usually designated by the term "associative settlement," which was introduced by D.J. Crisp (1974). This general term comprises different cases resulting in the formation of symbiotic (Zann, 1980), parasitic (Pearse et al., 1987), and also grazing and predatory (Pawlik, 1992) associations; epibiotic associations are especially important when discussing the induction of settlement of free living organisms (Wahl, 1989, 1997; Wahl and Mark, 1999). The terms "conspecific" and "heterospecific" induction (stimulation) are convenient for the purposes of our classification because they show whether the larvae (macroalgal spores) and the forms causing their settlement (adult, juvenile, or larval) belong to the same or to different species.

It should be mentioned that physical stimulation and induction almost always occur when contact between larvae and a hard surface takes place. Chemical induction (distant or contact) is conditioned by the properties of a biological or physical object to release or accumulate chemical substances on its surface. There are a number of reviews in which the problems of settlement induction are considered in different aspects (Meadows and Campbell, 1972; Crisp, 1974, 1976, 1984; Scheltema, 1974; Guerin, 1982; Burke, 1986; Hadfield, 1986; Morse, 1990; Pawlik, 1992; Rittschof, 1993; Rodriguez et al., 1993; Slattery, 1997; Rittschof et al., 1998). Here, however, our emphasis will be on analyzing the reasons why benthic organisms concentrate on hard surfaces. First we will consider the phenomenology and mechanisms of settlement on attractive surfaces. The chemical nature of settlement inductors will be discussed in Section 6.3, and the inhibition of settlement by chemical and physical factors will be considered in Chapters 9 and 10.

5.2 DISTANT CHEMICAL INDUCTION

Distant induction under the influence of invertebrates and macroalgae has been found in a few species of hydroids, polychaetes, mollusks, and echinoderms. We are also aware of a limited number of examples of microfouling films causing larval settlement (see Section 5.5). This may be the reason for the impression that distant induction is in general less widespread than contact induction. In spite of the limited number of invertebrate species in which it has been found, there is reason to believe that in reality it occurs more frequently than is known so far.

Settlement by distant chemical induction has been found to occur in hydroids. Some of them, e.g., species of the genera *Sertularella* and *Coryne*, are ship foulers (Chaplygina, 1980). *Sertularella miurensis* and *Coryne uchidai* are found in the ocean, mainly on *Sargassum tortile*, and their larvae are attracted by these brown algae under laboratory conditions (Nishihira, 1967, 1968, cited after Orlov, 1996a). The settlement of planulae is induced by extracts from a sargassum. The substance that causes settlement, as well as attachment and metamorphosis (see Section 6.3), is a terpene compound (Kato et al., 1975).

In the serpulid polychaete *Hydroides dianthus*, adults distantly attract the larvae of the same species (Toonen and Pawlik, 1996). The attractant, which is an unidentified substance released into the water, is responsible for the gregarious settlement of competent larvae.

Settlement by distant chemical induction occurs in several species of mollusks. The tropical nudibranch *Phestilla sibogae*, which lives near the shores of Hawaii, is a predator that feeds on coral polyps (Hadfield, 1978). The coral releases a substance that attracts veligers of the mollusk. Homogenates prepared from the tissues of the prey cause not only settlement but also metamorphosis of the predator (Hadfield and Scheuer, 1985).

Macroalgae, especially *Cystoseira barbata*, may distantly attract the larvae of the motile bivalve *Brachyodontes lineatus*, which forms mass settlements on these algae and near them in the littoral zone of the Black Sea (Kisseleva, 1966, 1967a). When extract of this alga is added to the medium, the veligers swim toward the higher concentration. The attractants are still unidentified substances, soluble in alcohol, and probably low-molecular ones.

Larvae of the oyster *Crassostrea virginica*, when placed in a circular aquarium with clear water in which current is imitated, swim in almost straight paths (Tamburri et al., 1996). However, their behavior changes drastically when water is added from a vessel in which adult mollusks have been kept. The paths of the veliger movements become curved, and they sink to the bottom and settle there. These experiments demonstrate the distant nature of conspecific settlement induction in *C. virginica*.

The most important thing about the above experiments (Tamburri et al., 1996) is that they show the possibility of distant chemical induction of dispersal-form settlement in the natural sea medium in the presence of a current. The rate of diffusion of a chemical substance beyond the boundaries of a hard surface is known to decrease as the water flow over the surface increases (Dodds, 1990; Abelson and Denny, 1997). On the other hand, the velocity of larval locomotion is lower than that of the current, even at a distance equal to the body length of the larva; this is regarded as a serious obstacle for settlement induction by substances that are present some distance away from the surface (Butman, 1986). In the above experiments, as well as in the natural environment, an important role belongs to turbulent mixing, owing to which larvae are able to find a chemical inductor at some distance from its source.

The larvae of the so-called shipworm, the bivalve borer *Teredo*, are attracted to wood from a distance (Harington, 1921; Culliney, 1973). Although other wood-boring mollusks have been less studied in this respect, it is highly probable that their larvae can be chemotactically attracted to wooden constructions. Some of the substances released from the wood may stimulate their settlement.

Sandy-bottom biotopes on the Pacific coast are inhabited by the sea urchin *Dendraster excentricus*, called a sand dollar for its flattened shape. This species often forms large aggregations, consisting of up to several hundreds of animals per 1 m² (Highsmith, 1982). Their formation is associated with a low-molecular substance released by adults that distantly attracts the larvae. The inductor causes both the settlement and metamorphosis of *D. excentricus* (Highsmith, 1982).

When assessing the role of distant chemical induction on the settlement of larvae, the following should be mentioned. A greater number of larvae can be attracted to the surface from a distance than as a result of immediate contact with it. Therefore, it is to be expected that finding a substrate favorable for settlement and development from a distance has certain advantages over coming in direct contact with the surface and is more conducive to the realization of the biological potential of the species. Thus, the mechanism of distant chemoreception and the choice of substrate based on the behavioral reactions of larvae (chemotaxis and chemokinesis) is obviously more advanced from an evolutionary point of view and may be sufficiently widespread in invertebrates.

From the above it is clear that there is a fairly limited number of studies in which it was definitively proved that larvae are distantly attracted to substrates on which they settle. Future studies may supplement the known instances of this kind. For example, the hydroid *Gonothyraea loveni* in the White Sea (Chupa Inlet, the Kandalaksha Bay) settle on the brown algae *Fucus vesiculosus* and *Ascophyllum nodosum*. Under laboratory conditions, planulae of *G. loveni* settle selectively on them (Dobretsov, 1999b). In experiments using chemotactic chambers, homogenates of these algae attracted planulae from a distance. Pediveligers of the blue mussel *Mytilus edulis* in the chemotactic chamber experiments were distantly attracted by washouts of the green alga *Cladophora rupestris* (Dobretsov, 1999a), on whose filaments they settle in the White Sea (Dobretsov and Wahl, 2001). Homogenates of the mantle and adductor muscles of the scallop *Patinopecten yessoensis* attract its larvae from a distance and stimulate their settlement (Zhuk, 1983). Aqueous extracts from the tunics of adult ascidians *Molgula citrina* and some other species cause the settlement and metamorphosis of their larvae (Durante, 1991; Railkin and

Dysina, 1997). These facts may indicate that settlement by distant induction is a more widespread phenomenon than is presently known.

5.3 CONTACT HETEROSPECIFIC CHEMICAL INDUCTION

Contact chemical induction and stimulation of settlement of larvae are quite common. They are represented by three different types. In the first case, exometabolites of the basibiont, which are released and bound on its surface, induce settlement and not infrequently attachment and metamorphosis of the larva of the epibiont of another species, coming into direct contact with the inductor, thus establishing epibiotic relations. In the second case, the larva settles when it comes in contact with adult individuals or larvae of the same species. Such a mechanism of settlement results in the formation of large aggregations of animals, which are of great biological significance. Finally, the third type of contact chemical induction, which is the most widespread, is conditioned by the presence of microfouling films on natural and artificial objects immersed in water.

Larval settlement while in contact with the surfaces of other species of animals or macroalgae has been described for sponges (Bergquist, 1978; Barthel, 1986; Railkin et al., *in press*), cnidarians (Chia and Bickell, 1978; Morse and Morse, 1991), polychaetes (Pawlik, 1990), some cirripedes (Moyse and Hui, 1981), mollusks (Kisseleva, 1967a; Morse, 1992), bryozoans (Crisp and Williams, 1960), and ascidians (Davis, 1987; Durante, 1991; Railkin and Dysina, 1997). This phenomenon is reflected in several reviews (Meadows and Campbell, 1972; Crisp, 1974, 1976, 1984; Scheltema, 1974; Morse, 1990; Pawlik, 1992; Slattery, 1997; Wahl and Mark, 1999).

As a rule, coralline algae induce settlement and metamorphosis in motile herbivorous and predaceous invertebrates (polychaetes, mollusks, echinoderms), which feed on epibionts and thus reduce fouling on the surface of these algae (Johnson, 1995). At the same time, they do not induce settlement of sessile polychaetes, cirripedes, bryozoans, and ascidians on their surface. Yet, when the planulae of the corals *Agaricia humilis* and *A. tenuifolia* come into contact with the encrusting red coralline alga *Hydrolithon boergesenii*, they settle on it and undergo metamorphosis (Morse et al., 1988). They do not occur on other algae commonly found in the same biotopes.

Larvae of the polychaete *Spirorbis spirorbis* settle selectively on fucoids and avoid a number of other brown and red algae. Plates with microfouling films that had been soaked in extracts of *Fucus serratus* became populated by this polychaete 20 times more intensely than the surfaces wetted with water (Williams, 1964). A similar result was obtained in analogous experiments on the bryozoan *Alcyonidium polyoum* attraction with the same species of algae (Crisp and Williams, 1960).

A group settlement of individuals of one species is a characteristic feature of the distribution of cirripedes of the Balanidae (Crisp and Meadows, 1962) and Lepadidae families, especially the genus *Lepas* (Il'in, 1992b), on hard substrates. However, goose barnacles of the genus *Conchoderma* are less specialized and can settle on individuals not only of their own but also of other species (Reznichenko and Tsikhon-Lukanina, 1992). The same is known of the barnacles *Semibalanus balanoides* (Moyse and Hui, 1981). Cirripedes can also settle on sea turtles, sea snakes, and whales (Crisp, 1974; Zann, 1980). These and other factors give evidence to the possibility of the settlement of cirripedes by heterospecific induction.

A number of mollusks settle selectively on red coralline algae. These are, for example, the 13 species of gastropods of the genus *Haliotis* (Morse, 1992) and the chiton *Katharina tunicata* (Rumrill and Cameron, 1983). In both cases, the natural substance that induces settlement and metamorphosis is γ-aminobutyric acid, bound with a protein in the alga wall (Morse and Morse, 1984). Contact with the coralline alga *Porolithon* sp. reduces the time necessary for settlement and metamorphosis of veligers of the gastropod *Trochus niloticus* several times over (Heslinga, 1981).

The gastropods *Rissoa splendida* and *Bittium reticulatum*, which inhabit the brown alga *Cystoseira barbata* in the Black Sea, were shown to select the alga on which they normally occur in nature in the choice experiments using three species of algae, sand, and mollusk shells (Kisseleva, 1967a). The larvae of these species settle better on the alga than on its preparation obtained by alcohol extraction. Soaking foam plastic in an extract of this alga made it more attractive for the larvae. At the same time, they did not respond to changes in the concentration of the algal metabolites. These and other facts suggest that the settlement of these gastropods is most probably induced when they come in contact with *Cystoseira*.

The above examples characterize contact chemical induction of settlement by heterospecific adults. They are especially important for understanding the way in which epibiotic relationships are formed between macroalgae and the animals inhabiting them. The data presented here show that the macrofouling that has already developed may induce and stimulate the settlement of other animal species, possibly determining and accelerating this process. It should be noted that, in a number of cases, the settlement of some species on others was not an object of special investigation. Therefore it is possible that some of them hereafter will be relegated to distant and not to contact induction.

5.4 CONSPECIFIC CHEMICAL INDUCTION AND AGGREGATIONS

The settlement of marine organisms on hard substrates by large groups of individuals of the same species is to be found both in animals and in macroalgae. According to the above classification, such a pattern of distribution may be conditioned by conspecific contact or distant chemical induction of settlement. Reviews are available (Meadows and Campbell, 1972; Burke, 1986; Pawlik, 1992) in which conspecific induction of settlement is considered, not infrequently referred to as "gregarious settlement" in the literature.

Conspecific induction was first discovered in the oyster *Ostrea edulis* (Cole and Knight-Jones, 1949, cited in Crisp, 1984). The settlement of larvae on plates with and without settled mollusks was compared under mesocosm conditions. Based on the experimental data, the authors concluded that young oysters facilitated the settlement of larvae of their own species.

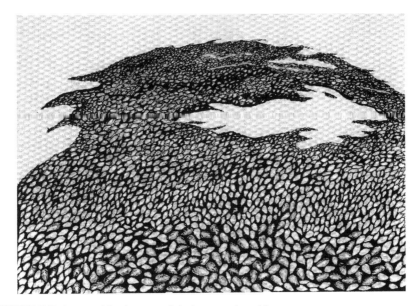

FIGURE 5.2 A mussel bank exposed during very low tide.

Large settlements of littoral and sublittoral cirripedes are well known. Mussels and oysters form vast aggregations of closely packed sessile individuals that are attached to the stony bottom (Kulakowski, 2000). These so-called banks extend for tens and hundreds of meters and may include many millions of individuals (Figure 5.2). Along the coastline, there is a wide band of brown, green, and red algae, many of which form extended thickets represented by individuals of only one species, such as, for instance, the sublittoral settlements of the brown alga *Laminaria hyperborea* near the British coast (Kain, 1979) or the red alga *Ahnfeltia tobuchiensis* in the Sea of Japan (Kudryashov, 1980). *Ahnfeltia* forms several layers whose area reaches hundreds of hectares and whose thickness is several tens of centimeters. Mussels, oysters, and the algae *Laminaria* and *Ahnfeltia* are important objects of fishery and aquaculture. They are also abundant in the fouling of different technical objects (Zevina, 1994).

In the modern English-language literature devoted to fouling, the term "gregariousness" is commonly used to designate a monospecific settlement. Yet in the Russian-language works and in translations from the English such terms as "aggregation," "group settlement," or simply "groups" are often used.

According to W. Allee's classification (1931), there are two types of aggregations. The first group consists of individuals that are strongly connected by physical contact; in the second group, contact is not a common rule. Aggregations of organisms inhabiting hard substrates mainly belong to the first type. They are characteristic of species whose individuals are attached, though they have also been observed in motile organisms.

Monospecific aggregations have been described in larval and adult sponges (Borojević, 1969), hydroids (Williams, 1976; Oshurkov and Oksov, 1983; Orlov, 1996b), larvae of scyphoids (Otto, 1978), corals (Duerden, 1902), polychaetes

(Knight-Jones, 1951; Wilson, 1968; Eckelbarger, 1978; Marsden, 1991; Toonen and Pawlik, 1996; Bryan et al., 1997; Chan and Walker, 1998), cirripedes (Knight-Jones, 1953b; Crisp, 1961; Crisp and Meadows, 1962; Lewis, 1978; Oshurkov and Oksov, 1983; Rittschof et al., 1984; Hills et al., 1998), mollusks (Chipperfield, 1953; Bayne, 1964, 1976; Kulakowski and Kunin, 1983; Oshurkov and Oksov, 1983; Kulakowski, 2000), bryozoans (Mihm et al., 1981; Brancato and Woollacott, 1982; Woollacott, 1984; Svane and Young, 1989), echinoderms (Strathmann, 1978; Highsmith, 1982; Kusakin and Lukin, 1995), ascidians (Schmidt, 1982; Svane et al., 1987; Bingham and Young, 1991; Hurlbut, 1993), and macroalgae (e.g., Kain, 1979; Kusakin and Lukin, 1995). Thus, there is no one large taxonomic group inhabiting hard substrates for which aggregate monospecific settlement should not be known. In most of the cases, such settlements developed as a result of the contact chemical induction of a larval settlement by individuals of the same species. However, distant induction may underlie the group settlement of the oyster *Crassostrea virginica* and the primary settlement of the mussel *Mytilus edulis* (see Section 5.2).

The common occurrence of monospecific aggregations seems to be conditioned by the biological advantages of living in groups (Pawlik, 1992). Indeed, it is easier for animal larvae and algal spores to find mass aggregations of adults of their own species and thus to select their habitat. This is obviously facilitated by the large size of an aggregation and the high total concentration of inductors released by it. The close proximity of individuals facilitates cross-fertilization. Defense from predators is more efficient in a group settlement, since the chemical and mechanical means of protection employed by several or many individuals are directed against one common enemy. Some other advantages are not as evident. The aggregated growth of laminaria reduces the action of waves and the current on individual thalli (Bashmachnikov et al., 2002). The sand dollars *Dendraster excentricus*, living in large groups, process and trench the sand and thus protect their juveniles from predation by the crustacean *Leptochelia dubia* (Highsmith, 1982).

Let us consider the mechanism of the formation of aggregates using a well-known example of the contact chemical induction of settlement in cirripedes. Their larvae settle close to adult individuals of the same species (Lewis, 1978) and avoid immediate contact with individuals of other cirripede species (Crisp, 1961). This reduces interspecific competition.

When a cyprid larva meets a conspecific individual, its movement slows down (Knight-Jones and Crisp, 1953) while the frequency of random turns increases. As a result of such behavior, referred to as kinesis (Fraenkel and Gunn, 1961), the larva continues moving within a restricted area and finally settles close to an adult individual. Aggregate settlements have been described both in true barnacles (the Balanidae, Chthamalidae, and Verrucidae families) and in goose barnacles (the Lepadidae and Scalpellidae families). The aggregate behavior of cyprids is based on contact chemoreception (Crisp and Meadows, 1962).

For example, in *Semibalanus balanoides*, the best-studied species in this respect, the epicuticule of the basis of the calcareous shell was shown to contain a glycoproteid whose properties and structure have been studied extensively (Larman et al., 1982). Similar substances, referred to as arthropodins, are also present in other barnacles showing aggregated settlement. If they are applied to some surface, the

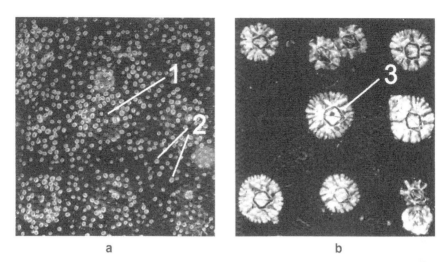

a   b

FIGURE 5.3 Experimental demonstration of contact conspecific induction of settlement in barnacles. (a) Settlement of cyprids around shell bases of adult barnacles; (b) arrangement of adult barnacles. (1) Bases of removed barnacles, (2) settled juveniles, (3) adult barnacles. (After Crisp, 1961. With permission of the *Journal of Experimental Biology* and the Company of Biologists Ltd.)

larvae start settling on it, which does not happen with a clean substrate. This was shown, in particular, for *S. balanoides* and *Elminius modestus* (Crisp and Meadows, 1962; Larman and Gabbot, 1975). Other experiments also have been fairly demonstrative. If young attached barnacles are carefully removed from a hard surface, the larvae will not settle randomly but mainly around the places where adults have been sitting (Figure 5.3). The settlement of cyprids of *S. balanoides* near pits is much more intense when the pits have been treated previously by a settlement factor (Hills et al., 1998). At the same time, cyprids of *Balanus amphitrite* (Rittschof et al., 1984) and *E. modestus* (Clare and Matsumura, 2000) also settle if the glycoproteid is dissolved in water rather than adsorbed on the surface.

Settlement can be induced experimentally by arthropodins of different cirripede species, yet the degree of their influence is different, which seems to be associated with the different structures of the molecules. Extensive comparative studies have been carried out on the barnacle *Semibalanus balanoides*, whose larvae settled on experimental plates impregnated with extracts of animals and plants (Crisp and Meadows, 1962). The inducing effect of extracts of different cirripedes that was determined by these authors is from 66 to 100% when related to that for the extract of *S. balanoides*. In decreasing order of their effects, the extracts form the following series: *S. balanoides, Balanus balanus, Elminius modestus, Lepas hilli*, and *Chthamalus stellatus*. Extracts of other arthropods were 1.5 to 2 times less effective. Extracts of some taxonomically remote organisms, such as the sponge *Ophlitaspongia seriata* and the fish *Blennius pholis*, demonstrate a relatively strong inducing effect (61 and 76%, respectively). Studies performed on other barnacle species confirm that conspecific extracts exert the strongest influence on the settlement of cyprid larvae (Lewis, 1978; Raimondi, 1988).

However, there are data (Wethey, 1984) that cast doubt on the role of contact chemoreception (arthropodins) in the formation of aggregations in cirripedes. Studies of *Semibalanus balanoides* in areas where adult conspecific settlements have been destroyed by storms showed that the larvae did not demonstrate any selectivity toward the bases of the empty shells. These observations gave D. Wethey (1984) reason to suggest that chemical molecules causing aggregate settlement of barnacles were short-lived and not significant for the selection of settlement sites. In my opinion, the data of this scientist do not contradict the investigations of other authors on the same species. On the contrary, they show that group settlement of barnacles is possible only in the presence of arthropodin; in its absence, the cyprid larvae settle individually. This is the very case when the exception only proves the rule.

An additional and possibly even the main influence on the group pattern of barnacle settlement may be played not by the inductor released by the adults but by that released by the larvae. This suggestion is based on experimental data. It was found that the larvae of the barnacles *Semibalanus balanoides* (Walker and Yule, 1984) and *Balanus amphitrite amphitrite* (Clare et al., 1994), while exploring the surface with their antennulae, leave imprints (traces) of their attachment organs on it. Histochemical tests have shown that these traces contain proteinaceous material, which may be an attachment inductor. In any case, several times as many larvae may become attached to a surface with such imprints than to a clean substrate (Clare et al., 1994).

In some cases, the settlement of larvae close to conspecific adult populations may be accounted for by local hydrodynamic conditions as well as by the larval behavior (motor responses and vertical distribution) at the dispersion and settlement stages. This was observed, for example, in the polychaete *Pectinaria koreni* (Thièbaut et al., 1998).

The formation of monospecific thickets of macroalgae has been little studied. Laboratory observations of zoospores of the brown alga *Laminaria saccharina* revealed a group pattern of their settlement (Railkin et al., 1985). If a suspension of laminaria zoospores is placed in a Petri dish, they will move in the water randomly. When they get close to the bottom, they very seldom settle on clean glass; they largely swim away, back into the water column. Yet much more often the mobile spores will settle on already attached germinating embryospores or resting spores or in close proximity to them, and also close to diatoms and particles of plant detritus. As a result of this, groups of two or three, but sometimes ten or more, adjacent spores are formed. If a slide is placed in the spore suspension for several hours and then transferred into clean water, the attached spores will start to germinate and form germinative tubes in two days. Such embryospores are especially attractive to the swimming zoospores. In a parallel experiment, when slides with such embryospores were previously UV-treated and then carefully washed in water, the number of settling zoospores was reduced more than threefold (Table 5.1). These and some other results suggest that the group pattern of settlement and attachment of *Laminaria* zoospores may be conditioned by chemoreception.

Calculations based on my own experience of obtaining zoospores of *L. saccharina* and the existing morphological data (Kain, 1979) show that up to several tens of millions of spores can be released from 1 cm^2 of sporangium surface.

TABLE 5.1
Selective Settlement of Zoospores of the Brown
Alga *Laminaria saccharina* on Natural and
Artificial Substrates

Substrate	Abundance (spores/mm^2)
Glass	354.1 ± 21.5
Plexiglas smooth	288.3 ± 15.6
Plexiglas cellular	411.7 ± 24.4
Microfouling	445.9 ± 34.5
Microfouling UV-treated	295.5 ± 20.9
Glass with germinating spores	2162.6 ± 88.2
Glass with germinating spores UV-treated	684.7 ± 68.8

After Railkin et al., 1985. With permission of the *Biologiya Morya*.

This gives us reason to suggest that, in the ocean, zoospores settle on surfaces already inhabited by settled zoospores. Therefore, the mechanism of group settlement observed in the laboratory is also quite possible in nature. The group pattern of spore settlement seems to determine the growth of the groups of gametophytes and sporophytes, which usually develop directly on the gametophytes and in laminaria reach several meters in length. All of this taken together creates favorable conditions for forming thickets of *L. saccharina*.

The adaptive significance of group settlement of laminaria is that spores settled in groups are more resistant to bacterial lysis, and a greater percentage of them survives in the process of development (Railkin et al., 1985). In the laminaria thickets, the hydrodynamic stress on individual thalli is reduced, owing in particular to these algae smoothing the near-bottom turbulent pulsations (Bashmachnikov et al., 2002).

5.5 STIMULATION OF SETTLEMENT, ATTACHMENT, AND METAMORPHOSIS BY MICROFOULING

It has been mentioned previously that microfouling communities, also referred to as bacterial–algal films or biofilms (see Section 2.2), are mainly represented by bacteria and diatoms. They develop in the ocean on any natural substrates and surfaces of artificial objects, including technical ones. The speed of microfouling development is sufficiently high. In as little as 1 h, settled and attached bacteria can be observed on objects immersed in water (ZoBell, 1946; Costerton et al., 1995). Usually in 1 or 2 weeks in warm (Redfield and Deevy, 1952; Gorbenko, 1977) and temperate waters (Railkin, 1998b), respectively, a noticeable layer of diatoms is formed on inert substrates. In the climax community, bacteria together with diatoms may constitute over 99% of the total number of microorganisms (Chikadze and Railkin, 1992).

The developed microfouling film partly determines the properties of the hard surface that it covers. Some data illustrate this well. For example, under laboratory

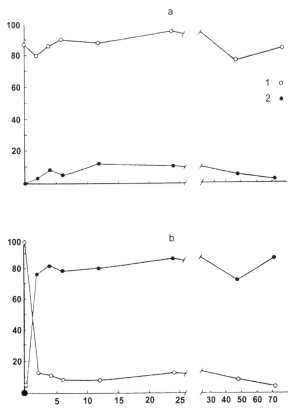

FIGURE 5.4 Settlement of larvae of the bryozoan *Bugula neritina* on polystyrene and glass. (a) Materials soaked in sterile water, (b) materials soaked in aquarium water, rich in micro-organisms. (1) Polysterene, (2) glass. Abscissa: duration of soaking, h; ordinate: settlement,%. (After Mihm et al., 1981. With permission of the *Journal of Experimental Marine Biology and Ecology.*)

conditions, the abundance of the larvae of the bryozoan *Bugula neritina* settling on hydrophobic polystyrene comprises 80 to 90%, and the abundance of those settling on hydrophilic glass is no more than 10% (Mihm et al., 1981). When these materials are covered with bacterial–algal film, the results appear to be opposite (Figure 5.4). It is known that hydrophilic materials become more hydrophobic after being placed in seawater, and, conversely, water-repellent materials improve their "wettability" (Little and Wagner, 1997). Changes in hydrophilic and hydrophobic properties, together with the microfouling film, may play a fairly important role in the effects observed.

Taking into account the fact that under natural conditions dispersal forms usually settle on the developed biofilm, it is reasonable to suggest that the biofilm should not inhibit the colonization process. Indeed, in many studies (see below) microfouling was shown to stimulate and in some cases even induce settlement, which makes it possible to consider microfouling as a stage of succession preceding macrofouling (see Section

2.2). The main data on stimulation and induction of fouling, attachment, and metamorphosis of larvae by microorganisms are given in Table 5.2. Additional information may be found in reviews by J.R. Pawlik (1992), M. Slattery (1997), N. Fusetani (1998), and S.K. Wieczorek and C.D. Todd (1998).

Let us consider some examples in greater detail. The induction of settlement and metamorphosis in the planulae of the solitary hydroid *Hydractinia echinata* is well studied (Müller and Spindler, 1972; Berking, 1991; Leitz, 1998). These processes are caused by the bacterium *Alteromonas espejina* (Leitz and Wagner, 1993). The induction of metamorphosis proved to be contact and to be associated with some so-far unknown lipophilic substances (Berking, 1991). The above mechanism of induction of settlement and metamorphosis ensures the specificity of the epibiotic association between the hydroid and the hermit crab *Eupagurus*, since the bacterium *A. espejina* lives on the surface of its shell.

Colonies of the hydroid *Laomedea flexuosa* in the White Sea occur on the brown algae *Ascophyllum nodosum* and *Fucus vesiculosus*. In the laboratory, the microfouling transferred from the surfaces of six species of macroalgae onto the bottom of Petri dishes owing to its capacity for self-assembly (Railkin, 1998b; see also Section 2.3), noticeably stimulated the settlement of planulae of this hydroid (Orlov et al., 1994). The percentage of settled larvae increased in the series: *Rhodymenia palmata* (20%), *F. inflatus* (31%), *F. serratus* (36%), *Laminaria saccharina* (43%), *F. vesiculosus* (44%), and *A. nodosum* (45%). The planulae of *Gonothyraea loveni*, according to the results of sampling in nature and laboratory experiments, show lesser selectivity with regard to both the macroalgae and the biofilms isolated from them.

Yet, according to other data (Chikadze and Railkin, 1992), microfouling films exert a strong inducing action on the settlement of the larvae of *G. loveni*. Probably owing to distant chemical induction, the planulae demonstrate a selective attitude to biofilms from the surface of *A. nodosum* and *F. vesiculosus*, on which adults are mainly to be found in the sea (Dobretsov, 1999b).

It is quite natural that the main groups affecting larval settlement are bacteria and diatoms, which are common in biofouling. The larvae of the polychaete *Ophelia bicornis* settled on sand soaked in sea water that had been taken from the habitats of this species (Wilson, 1955). The attracting factors are so-far unidentified bacteria and possibly diatoms. Similar data were obtained with regard to another polychaete, *Protodrilus symbioticus* (Gray, 1966). Bacteria from the surface of the green alga *Ulva lobata*, on which adults of the polychaete *Neodexiospira (Janua) brasiliensis* occur, induce settlement and metamorphosis of the polychaete larvae much more efficiently than do diatoms inhabiting the alga (Maki and Mitchell, 1985). In laboratory experiments, the larvae of the polychaete *Spirorbis borealis* settle on the panels covered with films of the green unicellular alga *Dunaliella galbana*, the diatom *Navicula* sp., or several other species of diatoms (Meadows and Williams, 1963). The polychaetes settle seven to nine times more intensively on diatoms at comparable concentrations of the above algae cultures, in which the panels were soaked. More than 90% of the larvae of the same species underwent metamorphosis on algal films obtained in the culture of *Chlamydomonas* or *Synechococcus* (Knight-Jones, 1951).

TABLE 5.2
Facilitation and Induction of Settlement, Attachment, and Metamorphosis of Larvae by Biofilms

Species of Larvae	Place	Source of Biofilm	Effect of Biofilm	Reference
Spongia				
Halichondria panicea	Lab	MF	F s	Zhuravleva and Ivanova (1975)
Halisarca dujardini	Lab	MF from Fucus vesiculosus	I s, a, m	Railkin et al. (in press)
Cnidaria				
Hydractinia echinata	Lab	Bacteria from crab shells	I s, a, m	Müller and Spindler (1972)
H. echinata	Lab	Bacteria Alteromonas espejina	I s, a, m	Leitz and Wagner (1993)
Clava multicornis	Lab	MF from some brown algae	I s	Orlov (1996a)
Laomedec flexuosa	Lab	MF from brown and red algae	F s	Orlov et al. (1994)
Gonothyraea loveni	Lab	MF from Fucus spp.	F s	Orlov et al. (1994)
G. loveni	Lab	MF from artificial substrates	I s, a, m	Chikadze and Railkin (1992)
Aurelia aurita	Lab	Bacteria Micrococcaceae (log-phase of growth)	I o	Schmahl (1985a, b)
Cassiopea andromeda	Lab	Bacteria Vibrio (log-phase of growth)	F s, a, m	Neumann et al. (1980)
Cyanea sp.	Lab	MF	F s	Brewer (1984)
Polychaeta				
Hydroides elegans	Lab	MF, bacteria Roseobacter sp., α-subclass Protobacteria	I s	Lao and Qian (2001)
Ophelia bicornis	Lab	MF	F s	Wilson (1953, 1954, 1955)
Polydora ciliata	Lab	MF	F s	Kisseleva (1967b)
Pomatoceros lamarckii	Lab	Microalgae: Rhinhomonas reticulata, Tetraselmis chui	I s	Chan and Walker (1998)
Spirorbis borealis	Lab	Microalga Chlamydomonas, bacteria Synechococcus	I m	Knight-Jones (1951)
S. borealis	Lab	Diatoms	F s	Meadows and Williams (1963)
S. corallinae	Field	Microalgae	F s	de Silva (1962)
S. tridentatus	Field	Microalgae	F s	de Silva (1962)

Cirripedia

Species	Inducer	Site	Effect	Reference
Balanus amphitrite	MF	Field	F s	Maki et al. (1990)
B. amphitrite	MF	Field	F s	Wieczorek et al. (1995)
Elminius modestus	MF	Field	F s	Neal and Yule (1994)
Semibalanus balanoides	MF, microalgal films	Lab	F s	Crisp and Meadows (1963)
S. balanoides	Diatoms	Field	F s	Oshurkov and Oksov (1983)
S. balanoides	Mature MF	Field	F s	Thompson et al. (1998)

Bivalvia

Species	Inducer	Site	Effect	Reference
Crassostrea gigas	Bacteria	Lab	I s, m	Weiner et al. (1986)
C. gigas	Bacteria	Lab	I s, m	Bonar et al. (1990)
C. virginica	Bacteria	Lab	I s, m	Walch et al. (1987)
Mytilus edulis	MF	Field	F s, a	Dobretsov and Railkin (1994, 1996)
M. galloprovincialis	MF	Field	F s	Braiko (1985)
Placopecten magellanicus	MF	Field	F s	Parsons et al. (1993)
Saccostrea commercialis	MF	Lab, field	F s	Anderson (1996)

Bryozoa

Species	Inducer	Site	Effect	Reference
Bugula neritina	Diatoms	Lab	F s	Kitamura and Hirayama (1987a, b)
Bugula simplex, B. stolonifera, B. turrita	MF	Lab	I s	Brancato and Woollacott (1982)

Echinodermata

Species	Inducer	Site	Effect	Reference
Apostichopus japonicus	Diatoms	Field	F s	Siu (1989)
Arbacia punctulata	Bacteria	Lab	I s, a, m	Cameron and Hinegardner (1974)
Lytechinus pictus	Bacteria	Lab	I s, a, m	Cameron and Hinegardner (1974)

Ascidia

Species	Inducer	Site	Effect	Reference
Ciona intestinalis	Bacteria *Pseudomonas* sp.	Lab	F s	Szewzyk et al. (1991)
Molgula citrina	MF	Lab	F s	Railkin and Dysina (1997)

Notes: Lab — laboratory, MF — microfouling, F — facilitation, I — induction, s — settlement, a — attachment, m — metamorphosis.

Settlement of the serpulid polychaete *Hydroides elegans* is induced by bacteria taken from its habitats (Lau and Qian, 1997). A strong inducing effect under laboratory conditions was shown by *Roseobacter* sp. and the α-subclass proteobacteria. The chemical inductor acted when the nectochaetes came into contact with the bacterial films (Lau and Qian, 2001).

As for barnacles (*Balanus amphitrite* and *Semibalanus balanoides*), data on the influence of biofilms on the settlement of their larvae are contradictory. In particular, the maximum induction of settlement in *B. amphitrite* is caused by young biofilms (Tsurumi and Fusetani, 1998) and in *S. balanoides* by mature ones (Thompson et al., 1998). Young biofilms somewhat inhibited the cyprid settlement in *B. amphitrite amphitrite*, whereas mature ones stimulated it to some extent (Wieczorek et al., 1995). In my opinion, the above contradictions may be caused by unequal sensitivity of the different species studied, and also by the different physical properties, qualitative and quantitative composition, and age of the biofilms (see the details below).

Larvae of the oysters *Crassostrea virginica* and *C. gigas* are rather sensitive to the presence of bacteria on a hard substrate (Bonar et al., 1990). Twice as many pediveligers settle on bacterial films as on a clean polystyrene surface. The study of a great number of bacterial strains has demonstrated that only some of them can stimulate the settlement of these larvae. The active substance proved to be a dioxyphenol that is close in structure to L-dihydroxyphenylalanine (L-DOPA; see Figure 6.11).

Under laboratory conditions, cyphonautes of the bryozoan *Bugula neritina* settle much better on the natural microfouling film than on clean glass (Kitamura and Hirayama, 1987a, 1987b). The difference in the number of larvae that settled on these substrates in parallel experiments increased with the density of diatoms and did not depend much on the abundance of bacteria. Larvae of the closely related species *Bugula simplex* and *B. turrita* did not settle when offered surfaces without a microfouling film (Brancato and Woollacott, 1982; Woollacott, 1984). At the same time, the larvae of *B. stolonifera* settled on surfaces with films and without them in approximately equal amounts. In substrate selection experiments, all three species evidently preferred microfouling films.

Larvae of the ascidian *Ciona intestinalis* can settle on different surfaces, but their settlement is stimulated by the mucous films of the bacterium *Pseudomonas* sp. (Szewzyk et al., 1991). This bacterium produces a polysaccharide that facilitates the attachment and metamorphosis of the larvae.

Very little research has been done on the settlement of motile spores of macroalgae. Nevertheless, natural microfouling has been found to stimulate the settlement and attachment of zoospores of the brown alga *Laminaria saccharina* (Railkin et al., 1985). When different substrates are available in the same vessel, the *Laminaria* spores settle selectively on biological surfaces, including the microfouling film (see Table 5.1). There remain 1.6 times as many spores on it as on smooth Plexiglas and 1.3 times as many as on glass. The stimulating efficiency of the films killed by ultraviolet rays is 1.5 times lower.

The influence of microfouling on macrofouling was studied under marine conditions on surfaces of different types: chemically inert, attractant, repellent, and

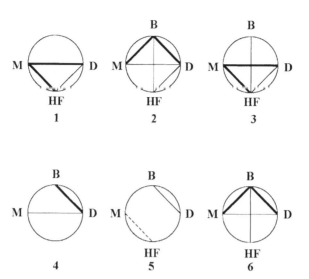

FIGURE 5.5 Correlations between the abundance of micro- and macrofoulers on paint coatings. Coatings: (1–2) with attractants (thiourea – 1, acrylamide – 2); (3) with a repellent (benzoic acid); (4) with an antiadhesive (5,5–diethylbarbituric acid); (5) with biocides (copper and tin); (6) neutral. Foulers: B – bacteria, HF – heterotrophic flagellates, D – diatoms, M – mussels *Mytilus edulis*. Bold solid line designates the correlation coefficient r > 0.8, thin solid line, 0.8 > r > 0.6; broken line, r < –0.8. (After Dobretsov and Railkin, 1994. With permission of the *Russian Journal of Marine Biology*.)

biocidal (Dobretsov and Railkin, 1994). Settlement of the mussels *Mytilus edulis* on all types of surfaces was most strongly affected by diatoms (Figure 5.5). For five out of six surfaces studied, the coefficient of correlation between their abundances (*r*) exceeded +0.6. The next strongest effect may be accorded to bacteria: *r* > +0.8 in two cases out of six. The correlation between the abundance of mollusks and heterotrophic flagellates was found only on a biocidal coating, with the correlation being negative. In another study (Robinson et al., 1985) a positive correlation was shown between the abundance of diatoms and macrofoulers on the surface of antifouling paints. Moreover, a conclusion was made that the presence of a great number of diatoms testified to the inefficiency of such paints.

It should also be noted that larval receptors possess a certain sensitivity threshold to chemical substances (e.g., Burke, 1983; Morse, 1990; Pawlik, 1992; Slattery, 1997; Rittschof et al., 1998). Therefore it is natural to expect the existence of a certain value of microorganism abundance, at which an above-threshold concentration of chemical factors is created, which is sufficient for the induction of settlement. The influence of microorganism abundance in biofilms on the larval settlement is considered in a review by Wieczorek and Todd (1998). Planulae of the hydroid *Hydractinia echinata* start to settle on bacterial film and undergo metamorphosis at the bacterial density exceeding the minimal value, which is 2.5×10^7 cells/cm^2 (Müller and Spindler, 1972). Within the range of $5–30 \times 10^7$ cells/cm^2, the percentage of metamorphosing individuals increases linearly from approximately 5 to 90%. Similar data are also known for diatoms. For cyphonautes of the bryozoan *Bugula neritina*, the threshold density of the alga *Navicula* sp. is 7×10^3 cells/cm^2 (Kitamura and Hirayama, 1987a, 1987b). Above this level, the number of settling larvae increases in semi-logarithmic dependence, i.e., it is directly related to the logarithm of the density of diatoms.

The induction of settlement and metamorphosis of barnacles (e.g., Maki et al., 1988; Olivier et al., 2000) and other invertebrates (Lao and Qian, 1997) is greatly affected by the specific composition of biofilms as well as by their age (e.g., Chan and Walker, 1998; Thompson et al., 1998; Tsurumi and Fusetani, 1998; Olivier et al., 2000). In my opinion, the influence of the age of biofilms may be mediated by the abundance and species composition of inductive microorganisms. For this reason it may be ambiguous; in particular, in some cases (Tsurumi and Fusetani, 1998) the maximum induction of settlement in barnacles is caused by young biofilms and in others (Thompson et al., 1998) by mature ones. Along with abundance, an important factor may be the size of the biofilm. In any case, the percentage of metamorphosed cyprids of *Balanus amphitrite* grows as the volume of the natural biofilm increases up to 0.1 to 1.0 $\mu m^3/\mu m^2$ (Tsurumi and Fusetani, 1998), which corresponds to a biofilm thickness of 0.1 to 1.0 μm, and falls with its further rise.

The action of bacterial cultures may be strain-specific. For example, only some strains induced settlement in the larvae of the polychaete *Hydroides elegans* (Lau and Qian, 1997) and stimulated settlement in pediveligers of the oysters *Crassostrea virginica* and *C. gigas* (Bonar et al., 1990).

Most of the data considered above show that, in many cases, microfouling exerts a positive influence on the initial stages of the colonization process, promoting the formation of communities and concentration of organisms on hard substrates. Yet there are data of another kind that show that biofilms may inhibit larval settlement and attachment — for example, in the barnacles *Balanus amphitrite* and *B. cariosus*, the bryozoan *Bugula neritina* (Maki et al., 1988; Rittschof and Costlow, 1989; Holmström et al., 1992; O'Connor and Richardson, 1998), and the ascidian *Ciona intestinalis* (Holmström et al., 1992) — or, conversely, have no influence on them (see the review in Wieczorek and Todd, 1998).

The reasons for the negative influence of biofilms on larval settlement may be associated with the physical properties of the films (Neal and Yule, 1994; Tsumuri and Fusetani, 1998), their age, the abundance of microorganisms in them, and their species composition (see the reviews in Pawlik, 1992; Wieczorek, 1994; Wieczorek and Todd, 1998). Some strains of the bacterium *Deleya marina* are known to suppress settlement and attachment of the larvae of *Balanus amphitrite* and *Bugula neritina* (Maki et al., 1988; Rittschof and Costlow, 1989).

In principle, it is not surprising that larvae are indifferent to some microorganisms and avoid others. It should appear much more surprising that, in most known cases, microfouling films attract larvae, often stimulating and even inducing not only their settlement but also their attachment and metamorphosis. This may be the result of the prolonged conjugate evolution of communities of micro- and macroorganisms inhabiting the same substrates.

5.6 THE INFLUENCE OF PHYSICAL SURFACE FACTORS ON SETTLEMENT

Submerged objects and industrial constructions have surfaces with different physical properties, which may have different effects on the settlement of dispersal forms of microorganisms. Physical factors, such as material, contour, size, structure and texture, color, wettability, and others, constitute the general characteristics of the surfaces of natural and artificial objects. In principle, they are as individual as the chemical properties of hard substrates.

The physical factors of the surface, unlike the chemical factors, are not specific inductors of settlement for larvae of certain species. They cause transition to a hard surface in any organism, independent of the species. This constitutes their certain universality. The mere mechanical stimulation of larvae during their contact with a hard surface is sufficient for temporary settlement on it. Therefore the generalist species may settle on a great number of substrates. For the larvae of a specialist species to remain on the surface, the surface should possess a certain set of properties, including physical ones. For example, planulae of hydroids settle better on rough surfaces than on smooth (Orlov, 1996b). Sand-dwelling polychaetes are rather particular about the granulometric composition of their substrate. Under experimental conditions, larvae of *Polydora ciliata* selectively settle on sand with grain size 0.25 mm (Kisseleva, 1967b) and have a sufficiently high survival rate on such a substrate. Sand consisting of larger grains (0.5–2.0 mm) is less attractive for *P. ciliata*. Larvae of the polychaete *Eupolymnia nebulosa* attach to grains larger than 0.25 mm but use grains smaller than 0.08 to 0.10 mm to build their tubes (Bhaud, 1990).

It is often difficult to judge which particular stage of fouling (settlement or attachment) is affected by certain physical factors of a surface without special laboratory observations and experiments. It can be suggested that such factors as the material, in the case of a chemically inert surface, and its wettability cannot be recognized by the mechano- and chemoreceptors of a settling larva; most probably they act on its attachment.

The influence of shape, size, texture, and color of the surface on settlement can be demonstrated by a number of instances. The Italian scientists S. Riggio and G. di Pisa (1981) studied the settlement of larvae on asbestos flat plates, cylinders, and corrugated surfaces that were placed in Palermo Harbor in illuminated and shaded places. All of the substrates had approximately the same area — about 600 cm². It was found that convex structures were the first to be colonized in the light, whereas concave structures remained free for a long time. A certain selectivity with regard to the surface shape and its spatial orientation also has been observed. Algae settle better on the upper side of flat plates while barnacles and ascidians prefer the convex areas of corrugated surfaces. Barnacles settle mainly on the slopes and ascidians on the tops. Both groups prefer illuminated sites. Concave surfaces are readily inhabited by calcareous sponges and tube-building polychaetes. These animals are more abundant in shaded places. On flat surfaces, a random distribution of organisms is observed.

TABLE 5.3
Settlement of Bryozoan Larvae (%) on Concave and Convex Areas of Macroalga Surface

	Alga			
	Pelvetia canaliculata		*Gigartina stellata*	
Bryozoan Species	Convex Sites	Concave Sites	Convex Sites	Concave Sites
Alcyonidium hirsutum	56	44	5	95
Alcyonidium polyoum	14	86	6	94
Flustrellidra hispida	0	100	7	93
Celleporella hyalina	14	86	9	91

Calculated from the data of Ryland (1959).

Pediveligers of mussels of the Mytilidae family prefer to settle on threadlike and cylindrical substrates: filamentous macroalgae, byssus of adult mollusks, colonies of hydroids, and arborescent bryozoans (e.g., Chipperfield, 1953; Seed, 1976; Berger et al., 1985). Using three-dimensional plastic models of hydroids and filamentous macroalgae, it was shown that the density of mollusk settlements was higher when the diameter of their branches was lower and the degree of ramification was greater (Harvey et al., 1995). The distribution of settled *Mytilus edulis* on non-branching kapron threads of various diameters followed a similar pattern (Railkin and Zubakha, 2000): the density of recruits increased as the thread diameter decreased from 1.2 to 0.15 mm. This may be one of the reasons why the primary settlement of mollusks of the family *Mytilidae* occurs on filamentous substrates (Bayne, 1964).

J.S. Ryland (1959) presented convincing proof of the fact that bryozoan larvae selectively settle on concave surfaces. He carried out laboratory experiments with four species of bryozoans and two species of algae. The ratio of the number of larvae that settled on the concave and convex areas of the thalli was shown to be 10:1 or even higher in five of the eight cases (Table 5.3).

The size of the substrate also has a certain influence on the quantitative characteristics of fouling. The general rule is more intensive settlement on small surfaces (Jackson, 1977b; Braiko, 1985), provided that the surface area does not limit the attachment and growth of the organisms (Hills and Thomason, 1998). This was established, in particular, for the Black Sea polychaete *Polydora ciliata*, the barnacle *Balanus improvisus*, and the mussel *Mytilus galloprovincialis*, which settled on flat plates made of artificial materials with areas varying from 50 to 1500 cm^2 (Braiko and Kucherova, 1976). The barnacles settled in greater numbers on small plates measuring 5 × 10 cm, on which their density during the autumn settlement period was about 700,000 ind./m^2. Mussels and polychaetes were the most numerous on plates measuring 10 × 15 cm, reaching densities of 24,000 and 8,000 ind./m^2, respectively.

Similar results were obtained from a study of prolonged oceanic fouling of differently sized parts of an experimental buoy (Reznichenko, 1981). In particular, on small surfaces the biomass of hydroids was 20 kg/m^2 and that of the barnacle

Chirona evermanni was 7.5 kg/m². The biomasses of these species on large-sized parts were less than 0.1 and 1 kg/m², respectively.

The influence of the surface's size and contour on settlement intensity may be accounted for by the peculiarities of larval behavior (see Chapter 4) and the distribution of currents around the surface (see Chapter 7). The tendency of larvae to settle on and become attached to concave areas of natural and artificial substrates seems to be associated with the negative phototaxis and possibly the positive geotaxis that are shown by many of them during the settlement period.

The larvae of some species of sponges (Ilan and Loya, 1990), hydroids (Rudyakova, 1981; Braiko, 1985; Orlov and Marfenin, 1993; Orlov, 1996b; Koehler et al., 1999), polychaetes living on sandy substrates (Kisseleva, 1967b; Rudyakova, 1981; Bhaud, 1990; Koehler et al., 1999; Wahl and Hoppe, 2002), many cirripedes (Crisp and Barnes, 1954; Rudyakova, 1981; Oshurkov and Oksov, 1983; Wethey, 1986; le Tourneux and Bourget, 1988; Hills and Thomason, 1998; Lapointe and Bourget, 1999), mollusks (Kisseleva, 1967a; Oshurkov and Oksov, 1983; Rudyakova, 1981; Braiko, 1985; Dobretsov and Railkin, 1996; Koehler et al., 1999; Lapointe and Bourget, 1999), some bryozoans (Koehler et al., 1999; Lapointe and Bourget, 1999), and a number of ascidians (Svane and Young, 1989; Lapointe and Bourget, 1999) prefer rough surfaces. They actively crawl into pits, cracks, and crevices and settle there. This phenomenon is known as *rugophilic* behavior. There are brief reviews on the rugophilic behavior of invertebrates in the books of V.D. Braiko (1985) and G.B. Zevina (1994); the rugophilic behavior of hydroids is described by D.V. Orlov (1996b).

However, the rugophilic mode of settlement is not a general rule. Almost every large taxonomic group includes species that settle better on smooth surfaces and those that are fairly indifferent to the microrelief. This is known of hydroids (Oshurkov and Oksov, 1983), cirripedes (Wahl and Hoppe, 2002), mollusks (Kisseleva, 1967a), bryozoans (Ryland, 1976), and ascidians (Braiko, 1985).

In a number of cases larval settlement in substrate pits appears to be associated with their behavior, that is, their tendency to crawl into narrow crevices, while in other cases it may be conditioned by the fact that the larvae are more easily washed away by the currents from open sites than from pits, where the flow velocity is reduced and, consequently, the lift force is lower (see Section 7.1). Rough surfaces have been found to reduce the mortality of recruits and juveniles, including that caused by predation (Wahl and Hoppe, 2002); thus, rugophily is of adaptive significance. It is widespread among barnacles, bivalves, and bryozoans, whose adults are sessile and may be subject to considerable hydrodynamic loads.

The phenomenon of rugophily is also inherent in macroalgal spores (Linskens, 1966; Harlin and Lindbergh, 1977; Railkin et al., 1985; Figueiredo et al., 1997). If a Plexiglas plate with cells 10 to 30 μm in diameter and 5 μm deep, which are quite comparable with the size of zoospores of the brown alga *Laminaria saccharina* (5–7 μm), is placed in suspension of the zoospores, practically all of them will settle in these pits (Railkin et al., 1985). As many spores are to be found in each pit as can fit in: usually one to two, sometimes three to four. The number of spores that settle on a cellular surface is 1.4 times as many as those that settle on a smooth surface (see Table 5.1). In this case, the cause of rugophily is obviously behavioral, since there is no current around the plates.

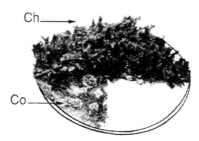

FIGURE 5.6 Development of the macroalgae *Chondrus crispus* (Ch) and *Corallina officinalis* (Co) on an acrylic polymer, covered with particles of silicon dioxide. The surface free from fouling is smooth. (After Harlin and Lindbergh, 1977. With permissions of *Marine Biology* and Prof. M.M. Harlin.)

The microrelief has a positive influence on the settlement of other macroalgal spores: the brown alga *Chondrus crispus*, the green alga *Ulva lactuca*, and the red algae *Corallina officinalis* and *Polysiphonia harveyi* (Harlin and Lindbergh, 1977). Considerably more plants will develop on acrylic disks covered with fine particles of silicon dioxide than on smooth surfaces (without the fine particles of silicon dioxide) of the same material (Figure 5.6). True selectivity is demonstrated by *C. officinalis*. The proportion of this alga is 44.8% on coating with particle sizes of 0.1 to 0.5 mm, 29.1% at particle sizes of 0.5 to 1.0 mm, and 17.6%, at particle sizes of 1.0 to 2.0 mm. The proportion of this alga on a smooth surface is only 8.4%. Conversely, the abundance of two other algae, *C. crispus* and *U. lactuca*, grows as the surface roughness increases, reaching 43.6 and 49.8% at particle sizes of 1 to 2 mm, respectively. The abundance of these forms on a smooth surface is 0.5 and 1.0%, respectively. Unfortunately, these interesting and demonstrative data, obtained from marine experiments, do not explain which processes are stimulated by microrugosity: settlement, attachment, growth, or all of them.

Laboratory observations (Linskens, 1966) showed that spores of brown (*Ectocarpus siliculosus*), green (*Ulva fasciata*, *Enteromorpha* sp.), and red (*Nitophyllum punctatum*, *Polysiphonia deusta*) algae have a different degree of rugophily. In particular, the brown alga *E. siliculosus* and the red alga *P. deusta* have a greater tendency to settle on slightly rough glass, whereas, on the other hand, the rest of the species studied prefer glass plates of a coarser relief.

The color of the substrate may also exert an influence on larval settlement, though it is weak and often insignificant. In a number of cases, a slightly greater number of larvae of sponges, barnacles, ascidians (Dahlem et al., 1984), and mollusks (Dobretsov and Railkin, 1996) will settle on light than on dark surfaces.

5.7 COMBINED INFLUENCE OF SURFACE FACTORS ON SETTLEMENT. THE HIERARCHY OF FACTORS

Natural hard substrates, such as macrophytes, rocks, ship bottoms, etc., on which invertebrate larvae and macroalgal spores settle are usually the source of not only physical but also chemical signals, acting on the larvae both in contact and distantly.

Surface material is well known to influence its fouling. For example, wood is fouled more than concrete and stones (Zevina, 1994). Yet these data usually characterize not only the action of the material itself but also of the different properties of its surface. Unfortunately, there are few studies in which the simultaneous actions of several surface factors on settlement are considered. The data cited do not always make it clear which process, settlement or attachment, is affected. Nevertheless, for many reasons, it is justifiable to consider them here.

Canadian workers (Hudon et al., 1983) studied the distribution of settled larvae of *Balanus crenatus* on natural (shells of *Mytilus edulis*, fronds of *Fucus evanescens*) and artificial (laminated panels) substrates in the St. Lawrence estuary on the Atlantic coast of Canada. Taking into consideration the parameters of substrates on which the larvae did and did not settle, they evaluated the distribution of *B. crenatus* on these substrates using the χ^2 test. The authors assumed that the variables showing lower χ^2 values were more strongly preferred by the larvae. Based on their data, the variables can be arranged in the following series, in decreasing order of significance for larval settlement: absence of detritus and diatoms; texture of the surface; factors associated with the barnacles (cover area, abundance of settled larvae, recruits, and juveniles). It should be noted that the substrates studied were compared not by a single characteristic but by a complex of features, both physical and biological. The high χ^2 values of the factors associated with the barnacles were probably caused by their aggregated distribution over the substrates, whereas the low values may be accounted for by the random distribution of detritus and diatoms.

Settlement of the barnacle *Semibalanus balanoides* was studied in two areas of the Canadian Atlantic coast (Chabot and Bourget, 1988; Le Tourneux and Bourget, 1988). Using a scanning electron microscope, they examined 231 microsites on surfaces that were inhabited and uninhabited by this crustacean. The physical characteristics of the substrates were taken into account, such as the presence of macro- and microcrevices, and detritus, as were the biological characteristics, determined by the presence of micro- and macroalgae and adults of *S. balanoides*. The adult barnacles of the same species were found to exert the greatest influence on settlement. The second important factor was the number of large- and medium-sized (about 1.5–10.0-cm deep) crevices in the rocks, and the third was the absence of a large amount of detritus or diatoms. Finally, the permanent attachment of larvae occurs mainly in pits and crevices that are comparable in size to the larvae. This study suggests the existence in *S. balanoides* of a three-stage choice of habitat: distant search (the choice of the settlement zone positioned below the *Urospora* belt, which is exposed for a long time during low tide), medium-range search (the choice of areas where adults of the same species settle or large crevices are present), and, finally, close search (the choice of microsites with the minimum amount of detritus and diatoms, with preference for microdepressions).

The dependence of larval settlement on the four main characteristics of the surface — material, microrelief, color, and microfouling film — and their combinations was studied in the mussel *Mytilus edulis* (Dobretsov and Railkin, 1996). Experimental plates were given different surface properties by coating them with a layer of hydrophobic paraffin or hydrophilic polyvinyl alcohol. Prior to exposing

TABLE 5.4
Effect of Surface Factors on the Abundance
of Settled *Mytilus edulis*

Surface Factor		Average Abundance, ind./dm²
Material	Hydrophilic	27.8 ± 1.2
	Hydrophobic	86.3 ± 6.2
Microfouling	Present	77.4 ± 7.1
	Absent	36.6 ± 3.8
Microrelief	Present	74.5 ± 7.1
	Absent	39.5 ± 4.8
Color	White	63.7 ± 7.0
	Black	47.6 ± 5.9

After Dobretsov and Railkin, 1996. With permission of the *Zoologicheskii Zhurnal*.

TABLE 5.5
ANOVA Results for the Influence of Surface Factors
on the Settlement of *Mytilus edulis*

Surface Factors	Sum of Squares	Total Influence (%)
Main factors:	19028.958	84.2
Material	10024.594	44.4
Microfouling	4887.760	21.6
Microrelief	3589.260	15.9
Color	527.344	2.3
Two-factor interactions:	1970.729	8.7
Material + microfouling	956.343	4.2
Material + microrelief	625.260	2.8
Microfouling + microrelief	219.010	1.0
Microfouling + color	78.844	0.3
Remaining factors	1593.052	7.1

After Dobretsov and Railkin, 1996. With permission of the *Zoologicheskii Zhurnal*.

the plates in the White Sea, they were covered with microfouling, using the self-assembly process (Railkin, 1998b; see also Section 2.3).

As a result, two important facts were established. First, mussel larvae prefer hydrophobic, hardly wettable paraffin surfaces, having microfouling and microrelief (Table 5.4). Second, the factors in question affect the abundance of settled mollusks significantly ($p < 0.05$) and independently of each other (Table 5.5). In descending order of their influence on the number of settled mussels, the factors studied form the following series: coating (44.4%), microfouling (21.6%), microrelief (15.9%), and color (2.3%).

TABLE 5.6

Influence of Microrelief, Microfouling (MF), and Spatial Orientation on Experimental Plates in a Parallel Flow on the Density of *Mytilus edulis* Postlarvae (ind./m²)

No.	Orientation of Plates	Plate Side	Microrelief	MF	Density of Mollusks
1	Vertical	Side	Fine	+	800 ± 229
2	Vertical	Side	Coarse	+	1657 ± 490
3	Vertical	Side	Fine	–	686 ± 194
4	Vertical	Side	Coarse	–	743 ± 345
5	Horizontal	Upper	Fine	+	24171 ± 3720
6	Horizontal	Lower	Fine	+	1257 ± 390
7	Horizontal	Upper	Coarse	+	18514 ± 2024
8	Horizontal	Lower	Coarse	+	1657 ± 246
9	Horizontal	Upper	Fine	–	667 ± 381
10	Horizontal	Lower	Fine	–	229 ± 107
11	Horizontal	Upper	Coarse	–	12286 ± 436
12	Horizontal	Lower	Coarse	–	1086 ± 522

Note: + and – indicate the presence and absence of microfouling on the plates, respectively.

It should be pointed out that wettability is the integral characteristic of the surface (see Section 6.2), which depends not only on the covering material but also on its rugosity and the microfouling film. Therefore, it should be assumed that substrate wettability must have been the leading factor affecting the colonization of hard substrates by the mussel *M. edulis* in the above experiments (Dobretsov and Railkin, 1996). Wettability affected the settlement of the larvae indirectly, by enhancing or reducing the attachment of their byssus threads.

The results that I obtained on the same object in the White Sea (Kandalaksha Bay, Chupa Inlet) showed that the settlement of *M. edulis* larvae was simultaneously affected by three factors: microrelief, microfouling film, and spatial orientation of the experimental plates (Table 5.6). Plexiglas plates (0.1 × 5 × 7 cm) with a fine microrelief (irregularly spaced mounds and pits 50–70 μm deep) and polystyrene plates (0.2 × 5 × 7 cm) with a coarse microrelief (regular furrows 0.44 mm deep, positioned 1.6 mm apart) were made. Half of the plates were covered with mature microfouling films using the self-assembly process (Railkin, 1998b), while the other half was soaked in sterile seawater before their use. All the plates were placed in the sea on a hydrovane (see Section 7.3, Figure 7.2), a hydrobiological device that allows the plates to be exposed at a constant angle (in this case parallel) to the tidal currents. The duration of exposure was 10 days.

The fouling of the Plexiglas and polystyrene plates of the same types did not differ ($P \geq 0.05$). The upper side of the plate was fouled more than the lower one. The coarse microrelief in combination with microfouling intensified the settlement on poorly fouled vertical plates and the lower sides of the horizontal plates. In the absence of microfouling, the number of mussels on the upper side of a plate was

18 times greater on surfaces with coarse microreliefs than on those with fine microreliefs. In the presence of microfouling, the number of mollusks settling on the upper sides of plates with fine microrelief increased by 36 times (experiments 5 and 9). The results of analysis of the available data show that the main stimulating factor of mussel settlement on the upper sides of horizontal plates is rugosity (the influence of this factor is 52.1%), and that on the lower side of horizontal plates and on the vertical plates is microfouling (about 21% of the total influence).

The above data demonstrate that the problem of simultaneous influence of several characteristics of the hard substrate on larval settlement is still far from being solved. The hierarchy of surface factors can be changed to the opposite one, depending on the ecological situation. This may reflect some sophisticated mechanisms of settlement regulation. The known examples of concurrent actions of several (many) surface factors upon larval settlement are restricted to only two species of barnacles and one species of mussels. Macroalgal spores have been little studied in this respect. The problem of multifactor control of settlement is important for the prediction of colonization of both natural and anthropogenic substrates, which is of great significance for mariculture and the protection of technical objects from biofouling.

5.8 SETTLEMENT ON THE SURFACES OF TECHNICAL OBJECTS

Toxic coatings or toxins released into the space surrounding the protected object are commonly used for protection against biofouling (see Chapter 9). Let us try to understand why larvae and spores settle on toxic surfaces, which are unfavorable for them, and on surfaces situated in highly toxic environments.

When discussing the causes of fouling of technical objects, it should be borne in mind that both surfaces that were initially weakly protected and those with exhausted toxic coverings are subject to biofouling. In both cases, the concentration of toxins around them does not appear to be high enough to deter colonization processes. Among the biological reasons, the following should be mentioned. First, many technical objects (oil and gas platforms, navigation guards, harbor craft, vessels, pipes, etc.) are located in shallow depths, in coastal waters, i.e., exactly in those places where the concentration of dispersal forms is the highest. This certainly facilitates the meeting of larvae and macroalgal spores with them. During the period of mass release of main macrofouling species into the plankton, their abundance is very high. For example, in the coastal waters of the White Sea, the abundance of *Semibalanus balanoides* larvae may be as great as 1000 ind./m³ and that of *Mytilus edulis*, 3000 to 5000 ind./m³, whereas the total abundance of meroplankton in the upper 5-m layer of water reaches several thousand individuals per cubic meter (Shilin et al., 1987), and the larval abundance is even higher for the Black Sea meroplankton (Braiko, 1985).

Second, during the settlement period, the dispersal forms usually have adhesive coverings (e.g., Fletcher et al., 1984; Lindner, 1984) and could, for that simple reason, adhere to the surface of an industrial object.

Third, of course, not all of the larvae, especially the planktotrophic ones, encounter the appropriate surfaces and settle on them in a short period of time. Many of them can stay in the plankton for a long time, with the result that their selectivity of substrates is lowered and part of them may settle on surfaces that are not exactly suitable or possibly even completely unsuitable (see the reviews in Crisp, 1984; Pechenik, 1990). For instance, if the lecithotrophic larvae of the polychaete *Spirorbis borealis* fail to settle 3 h after hatching, their selectivity under laboratory conditions is reduced by almost 42%, according to my estimate based on the data of E. W. Knight-Jones (1953a).

Fourth, the microfouling film that is always present on the external surface of technological objects creates a type of barrier that may lessen the toxic action of the chemical substances of antifouling coatings. This is observed when the organic film forming the basis of the covering from which toxins are released is not used by the attached microorganisms, mainly bacteria, as food. Otherwise, the opposite effect is observed: bacteria consume components of the basic coating and thus promote the leaching of toxins from it, reinforcing its protective action (Gorbenko, 1963, 1981).

Fifth, some macroorganisms that colonize the surfaces of technological objects are resistant to toxins and are therefore especially difficult to control (see Section 9.2 for details). Examples of these macroorganisms include the green alga *Enteromorpha*, the brown alga *Ectocarpus* (Evans, 1981; Callow, 1986), cirripedes, and some bivalves (Dolgopol'skaya et al., 1973; Callow, 1986; Beamont et al., 1987).

N.A. Rudyakova (1981) describes barnacles settling on the underwater part of a ship's hull:

> To all kind of unfavorable actions the cyprid responds with a stereotypic non-specific reaction — settlement on the substrate and an attempt to complete metamorphosis. … Herein the very cue which caused its settlement not infrequently becomes the cause of the subsequent death of the organism. … As observed on ships, mass settlement of cyprids on toxic surfaces almost always results in the formation of a layer of dead juvenile barnacles, which have just completed metamorphosis. … Their shells form a neutral interlayer which creates favorable conditions for the growth and development of other barnacles (p. 50).

Sixth, even small patches of attached macrofoulers are sufficient for other, less resistant organisms to settle on them. Having settled, they will be elevated above the surface and therefore will experience a lesser toxic effect from the coating. The approaching larvae may receive chemical cues to settlement from adults that have settled earlier, either from a distance or during the contact. In addition, as the surface is fouled it becomes rougher and more uneven, which intensifies the turbulization of the water flow carrying the larvae and spores, increases the frequency of their contact with the surface, and reduces the flow velocity. As a result, the range of propagules that are able to get attached is expanded, and the probability of fouling grows.

Thus, the colonization of technical objects, including those protected against biofouling, is the same regular process as the biofouling of natural hard substrates.

However, foulers begin to colonize protected surfaces only after their defenses have been exhausted and become insufficient for colonization suppression. Therefore, to maintain sufficient concentrations, the chemical protection of technical objects requires antifouling substances to be continuously released on their surface or in the surrounding water. In nature, similar protection from epibionts is carried out by the release of toxic secondary metabolites.

6 Attachment, Development, and Growth

6.1 ATTACHMENT OF MICROORGANISMS

The main mechanism of transport of motile foulers, including bacteria, toward hard substrates is the current, since their swimming velocity is low. Yet locomotion also may play a certain role in this process (see Section 3.2). Motile bacteria, as well as other microfoulers, are to some extent selective toward the substrates on which they settle, being attracted to one of them and repelled from others (e.g., Gromov and Pavlenko, 1989).

On the surfaces of any objects submerged in the ocean, be it an experimental plate, a scientific device, or a submerged part of the ship, the adsorption of ions and other dissolved substances, such as sugars, amino acids, proteins, fatty and humic acids, starts immediately (Khailov, 1971; Raimont, 1983). This process is fast, and saturating concentrations of substances on the surface are achieved within tens of minutes (Marshall, 1976; Baier, 1984).

Some sugars in the D configuration and L-amino acids, which are adsorbed on the surface, are known to attract bacteria (Blair, 1995). For instance, attractants for *Escherichia coli* are the sugars galactose, glucose, and ribose and the amino acids serine, aspartate, and glutamate. Unfortunately, fouling bacteria have not been studied in this respect, and the substances attracting them have been studied very little. Yet, following M. Wahl (1989), it is possible to suggest that positive chemotaxis to substances adsorbed on submerged surfaces facilitates the settlement of motile bacteria and other microorganisms.

Most species of marine fouling bacteria are motile (Gorbenko, 1977). They have been found to possess negative chemotaxis to indole, hydroquinone, thiourea, phenylthiourea, tannic and benzoic acids, and other compounds (Chet and Mitchell, 1976). These problems will be considered in greater detail when we discuss repellent protection from marine biofouling in Chapter 10. Immobile suspended microorganisms (spores and aflagellate bacteria, diatoms, and amoebae) settle on any substrates on which they are brought by the current. It is quite another matter that such microorganisms adhere more strongly to some surfaces than to others. Therefore, they may concentrate on certain substrates. Organisms that are immobile at the dispersal stage are supposed to choose their substrate mainly by means of selective adhesion.

Among microorganisms, attachment to a hard surface has been most studied in bacteria, which is reflected in a number of reviews (Zviagintzev, 1973; Marshall,

1976; Fletcher, 1979, 1985; Chuguev, 1985; Harborn and Kent, 1988). Though the majority of the studies were performed in the laboratory, their results and conclusions may be provisionally applied to marine conditions. D.G. Zviagintzev (1973) has shown that the strongest adherence to glass is found in the genera *Micrococcus, Pseudomonas*, and *Bacterium*. It seems to be quite natural that these bacteria are the most frequent marine foulers (Gorbenko, 1977).

C.E. ZoBell (1946) was the first to suggest the existence of two phases of adherence in bacteria: reversible and irreversible. This suggestion was proved by K.C. Marshall and his colleagues (1971). Further investigations showed that the first stage of attachment to a hard surface was mainly controlled by physical mechanisms (see Figure 2.1); therefore, it is quite justly called *adhesion*. In physics, this term means the process of heterogeneous surfaces attaching to each other (Derjaguin et al., 1985; Derjaguin, 1992); in the case under consideration, it would refer to those of a bacterium and a hard body. In the second (irreversible) stage of adhesion, bacteria release extracellular polymers that ensure a stronger attachment. Thus, the leading mechanisms of adhesion are physical in its reversible phase and biological and physical in its irreversible phase.

First let us consider the physical phase of attachment, not infrequently referred to as *sorption* or *adsorption* (Zviagintzev, 1973; Wahl, 1989). The collision of a bacterium with a hard surface is a fairly random event. Therefore, it is quite natural that the probability of such a collision and consequently the successful adhesion should be directly dependent on the abundance of microorganisms in the water surrounding the hard surface. The laboratory experiments of M. Fletcher (1977) with marine *Pseudomonas* sp. support this assumption. At the different stages of culture development, the abundance of attached bacteria grew with the increase in their concentration in the water and the duration of the experiments. The probabilistic nature of the adhesion of marine bacteria is also revealed by analysis of their occurrence on the planktonic diatoms to which they attach (Vagué et al., 1989).

M. Fletcher (1977) developed a simple model, according to which the rate of bacterial adhesion is directly proportional to the concentration of bacteria in the water and the fraction of surface that is free of microorganisms. The experimental data that she obtained are well approximated by this model. The regularities revealed suggest that bacterial adhesion may be described by the same quantitative dependencies as Langmuir adsorption.

There are other facts that point to the prevalence of physical mechanisms in the first phase of bacterial adhesion. For instance, with all other conditions being equal, bacteria killed with ultraviolet attach in the same way as living bacteria (Meadows, 1971); i.e., they behave like inert physical objects. In addition, it should be pointed out that the values of adhesion force in different microorganisms are close to those known for the adhesion of similar-sized inert particles to hard surfaces (Zviagintzev et al., 1971). When related to contact unit area, the force of bacterial adhesion to a hard surface is from 0.8 dyn/cm^2 for *Pseudomonas pyocyanea* to 100 dyn/cm^2 for *Serratia marcescens* (remember that 1 dyn/cm^2 is approximately equal to 0.001 g/cm^2).

It is well known (Derjaguin, 1992) that the main forces determining physical adhesion are electrostatic and dispersive (Van der Waals) interactions, even though

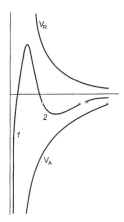

FIGURE 6.1 Total energy of interaction between a bacterium and a hard surface. (1) Primary and (2) secondary energy minimum. (V_A) energy of dispersive attraction; (V_R) energy of electrostatic repulsion. Abscissa – distance from the surface; ordinate – total energy.

there may be a total of more than 10 different forces participating in it (Lips and Jessup, 1979). The forces are considered to be electrostatic because bacterial cells and most hard surfaces in the water medium are negatively charged and therefore should repulse each other. These forces act at a relatively great distance. The main problem that arises when the theory of electrostatic forces is applied to adhesion events is determining the distribution of ions on isolated surfaces and describing their redistribution when the surfaces come close to one another. According to the theory of dispersive forces, the energy of mutual attraction of the bacterium and the surface is very low when they are sufficiently far from each other. However, at a relatively short distance, these forces increase sharply as a result of the unification of the electromagnetic fluctuations of the interacting bodies, which are determined by the corresponding quantum-mechanical effects.

Adhesion on the basis of electrostatic and dispersive forces is described by the DLVO theory (Derjaguin et al., 1985; Derjaguin, 1992), the name being an acronym of its authors' names: Derjaguin, Landau, Vervey, and Overbeek. The theory was initially formulated to explain the behavior of lyophobic colloids. According to this theory, the total energy of a system consisting of two closely positioned surfaces is the sum of energies of their electrostatic and dispersive interactions (Figure 6.1). The resultant curve shows two intervals of minimum energy in which adhesion of the two bodies is observed: primary and secondary. For adhesion to occur, the bacterium must be positioned at a distance corresponding to the secondary (10–15 nm) or primary (0.5–1 nm) energy minimum.

The surface of bacteria is hydrophobic and carries electrostatic charges. The size of bacteria is about 1 µm, with their lower size limit overlapping the upper size limit for colloid particles (Marshall, 1976). The superficial similarity of bacterial cells and colloid particles gave reason to apply the theory of lyophobic colloids to bacterial adhesion. At present, the DLVO theory explains the main experimental facts quite satisfactorily (Zviagintzev, 1973; Marshall, 1976, 1980; Fletcher, 1985; van Loosdrecht et al., 1990). On this basis, it is possible to discuss many biological mechanisms of adhesion that are associated, for instance, with the presence of macromolecules (such as polysaccharides and glycoproteins) on the surface of bacterial cells, with positive and negative polyvalent charges, and with other features (Lips and Jessup, 1979).

One of the reasons in favor of the theory of bacterial adhesion is the experimentally observed action of cations on the adhesion. As the concentration of cations in the series NaCl, CaCl₂, AlCl₃ decreases or increases, the adsorption of bacteria of the genera *Sarcina* and *Micrococcus*, found in fouling (ZoBell, 1946), decreases or increases, respectively (Zviagintzev, 1973). For example, when trivalent cations are added in the medium, bacterial adhesion increases more profoundly than when bivalent and especially univalent cations are introduced; in other words, adhesion is influenced not only by the sign of the charge but also by its magnitude.

These effects are explained by the DLVO theory (Derjaguin et al., 1992). As noted above, many surfaces in the water medium are negatively charged, and so are bacterial cells. Therefore they are mutually repulsive, and a layer of counterions is formed around them. Thus, interacting charged surfaces are surrounded by a double diffusive layer. According to the DLVO theory, an increase in the electrolyte concentration or the cation charge results in either a reduction of the electrostatic potential on the surface, owing to the counterion adsorption; or in a compression of the double diffusive ion layer; or in both phenomena simultaneously. In any case, the threshold of repulsion is reduced.

An important role of calcium ions in bacterial adhesion has been shown, which is conditioned by the non-specific neutralization of the negative charge of the double electric layer, on the one hand, and by the specific interaction of calcium with protein and polysaccharide adhesive molecules, on the other (Geesey et al., 2000).

The opposite action of cations has been reported in a number of cases. For example, lanthanum (Fletcher, 1979), cobalt, and nickel (Railkin et al., 1993b) cations may not intensify but, on the contrary, may suppress the adhesion of marine bacteria.

The presence of bacteria within the range corresponding to the secondary energy minimum usually does not ensure its adhesion to the surface, since, in this case, van der Waals attraction only slightly exceeds the electrostatic repulsion. The bacterium may be detached owing to external perturbations or its own locomotion. Conversely, in the primary minimum area, when the bacterium approaches the surface, at a distance of less than 1 nm, adhesion is faster. These energy minima correspond to the temporary (reversible) and irreversible forms of adhesion.

The latter term should not be taken literally. Indeed, when adhesion is irreversible the attachment of bacteria is faster. Yet they may be detached from it mechanically without any visible damage (Neu, 1992). This is due to the fact that the cell is detached from the polymer, rather than from the surface proper. Consequently, it is only the adhesive material that is disrupted, whereas the cell itself remains intact. The "footprints" of the detached bacteria are visible on electron micrographs (Neu, 1992).

The existence of two forms of attachment (reversible and irreversible) had already been suggested by ZoBell (1946). Yet they were experimentally demonstrated on marine bacteria much later by K.C. Marshall and his colleagues (1971).

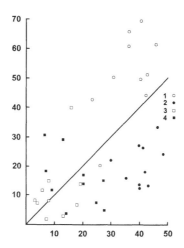

FIGURE 6.2 Selective adhesion of bacteria to glass (%) under laboratory conditions. (1) Short rods, (2) large rods, (3) curved rods, (4) cocco-bacilli. Abscissa: reversible; ordinate: irreversible adhesion of bacteria. (After Marshall et al., 1971. With permission of the *Canadian Journal of Microbiology* and NRC Research Press.)

These workers observed in the laboratory and in the ocean that part of the microorganisms adhered to the hard surface temporarily, detached from it, and could reattach again later. Such temporary adhesion happened fast, usually within 15 to 30 min. On the contrary, irreversible adhesion required much more time. Yet, in a day, attachment was fairly secure. Bacteria sampled directly from the ocean showed a varied adhesion capacity. Some morphological types revealed a greater and some a smaller ability for reversible and irreversible adhesion (Figure 6.2). The greatest selectivity, i.e., the earlier attachment, was characteristic of small rod-shaped bacteria that, together with large rod-shaped ones, dominated on the substrates during the first day of observation. They were followed by cocco-bacilli and curved rods and, finally, by stalked bacteria.

My observations and laboratory experiments (Railkin, 1998b) on the colonization of hard surfaces by natural microfoulers from cell suspensions support and supplement the data of Marshall and his colleagues (1971). Indeed, rod-shaped bacteria reveal quite distinctly a selective attachment to hard surfaces. As a result, they can adhere to the bottom of a Petri dish in as little as 15 min, though many cells soon detach themselves. The processes of attachment and detachment of bacteria during the first hours are rather dynamic. In 3 h, the mass detachment of rods can be observed and the adherence of cocci and spirilli starts. Nevertheless, within the first day of observations, rod-shaped forms dominate in the fouling over other morphotypes. Occasional stalked forms appear in just 24 h. During the first 3 to 6 h, bacteria of different morphological groups are not yet strongly attached. According to my data, irreversible adhesion of rods and cocci occurs in 9 to 12 h, and this time does not noticeably depend on the surface material (glass, polystyrene, polyvinylchloride). According to M. Fletcher (1979), the bacteria *Pseudomonas* sp. attach irreversibly to both hydrophobic and hydrophilic surfaces in 5 h.

Experiments performed in marine conditions (Marshall et al., 1971; Laius and Kulakowski, 1988; Railkin, 1998b) have shown that rod-shaped bacteria are the first to colonize on hard surfaces (first small, then large rods). Following them, cocci settle and become attached, and then vibrios and spirilli. The last to colonize the substrates are stalked bacteria of the genera *Caulobacter* and *Hyphomicrobium*. As a result, bacterial succession in temperate waters is completed in several days. Thus, the above data suggest that the succession sequence of morphological groups of bacteria under laboratory and probably marine conditions is determined by selective adhesion of bacteria.

The final (irreversible) attachment of bacteria to the surface involves biological mechanisms. In order to overcome electrostatic repulsion from a negatively charged surface and approach it from a distance corresponding to the primary energy minimum, where adhesion is facilitated, the motile bacterium can use its own kinetic energy. The approach to a hard body surface by immotile and motile bacteria or their spores is facilitated by Brownian motion, turbulent pulsations in the viscous sublayer (see Section 7.1), and the presence of cell outgrowths and polymer threads (Abelson and Denny, 1997). In M. Fletcher's estimation (1979), the kinetic energy of a moving bacterium is sufficient for overcoming the repulsion forces. According to her data, in *Pseudomonas* sp., which are devoid of flagella, the number of attached cells is reduced threefold and more.

The surface of bacteria is to some extent hydrophobic. Therefore, they reveal particular adherence capacities toward hydrophobic materials, such as teflon, paraffin, etc., and usually stick to them strongly (Marshall, 1976). Adhesion to hydrophilic surfaces (glass, metals) is reduced. Hydrophobic interactions between surfaces may be carried out by means of hydrophobic bridges, as a result of the polar group and functional group interaction (Fletcher, 1979), and also by means of polymers (Marshall, 1976). According a hypothesis of J. Maki and his colleagues (1990), polymer molecules used by bacteria for attachment are heterogeneous by their composition and local adhesive properties. Some domains of these molecules take part in attachment to hydrophobic materials or their hydrophobic sites, and others, to hydrophilic sites. Therefore, the abundance of microorganisms adhering to surfaces with different properties would be different.

In the common fouling bacteria *Pseudomonas* (marine) and *Caulobacter* (freshwater), filiform structures known as *fimbria* or *pili* have been described (Corpe, 1970). These proteinaceous outgrowths act as a kind of probe and may provide contact with the hard surface and irreversible adherence of bacteria. Another structure serving the same purpose is the base of the stalk in *Hyphomicrobium* and *Caulobacter*, which contains sticky material and represents an analog of the rhizoid of macroalgae.

Yet the general mechanism of irreversible adhesion (biological attachment) is the release of extracellular polymers, which strengthen the adhesion achieved at the first stage (physical attachment). Such adhesive materials may be acid polysaccharides and glycoproteins (see the review in Lock et al., 1984). The synthesis of these polymers does not depend on the taxonomic position or morphotype of the bacteria. Numerous filaments of polymers on the surface of bacteria ensure their fast attachment (Figure 6.3a).

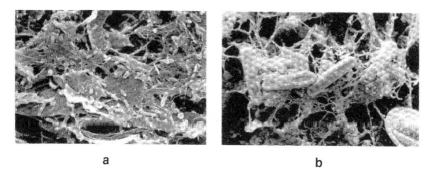

FIGURE 6.3 Attachment of microorganisms by means of polymers. (a) Bacteria (after Boyle and Mitchell, 1984; with permission of the United States Naval Institute); (b) diatoms (after Underwood et al., 1995; with permission of *Limnology and Oceanography* and the American Limnological Society).

It is interesting to note that the production of exopolymers in bacteria depends on the type of surface to which they attach. It was found (Maki et al., 2000) that *Halomonas marina* on polystyrene revealed increased binding with the lectin concanavalin A as compared to the same bacteria attached to the tissue culture polystyrene.

The stage of final (irreversible) attachment of bacteria is biological by its nature and mechanisms. The above facts testify in favor of this opinion. Nevertheless, in the literature, it is regarded as a purely physical phenomenon of adhesion, together with the reversible adhesion stage. Without rejecting the physical nature of the adhesion of heterogeneous surfaces (that of a bacterium and some hard substrate), I will try to give additional arguments to support my point of view.

First, the irreversible attachment of bacteria is a selective process (Zviagintzev, 1973), and different morphotypes are capable of it to different degrees (Marshall et al., 1971; Railkin, 1998b). Second, it involves the metabolic activity of cells, manifested by the secretion of exopolymers, which provide attachment. These macromolecules may be synthesized both before and after contact with the hard surface (Corpe, 1970). The bacterium–surface connection becomes stronger in time, owing to the continuing synthesis of the exopolymers. Third, the attachment of bacteria depends on their physiological state (Fletcher, 1977). Fourth, interaction with the hard surface may deform the bacterial cell wall, changing its permeability and adhesive properties (Lips and Jessup, 1979). On attachment to surfaces with different surface energies, the production of adhesive polymers in the bacterium *Halomonas marina* was changed (Maki et al., 2000). Fifth, bacterial adhesion and detachment are active biological processes, which are controlled at the genetic level (O'Toole et al., 2000). The above peculiarities of bacterial adhesion show that, together with purely physical mechanisms, biological mechanisms also play an important role. Thus, bacterial adhesion must be different from that of non-living colloid particles (Visser, 1988a, 1988b).

Unfortunately, the mechanisms of adhesion and attachment in diatoms, which together with bacteria constitute the major component of microfouling film, are much less studied. They can be discussed only on the basis of a small number of investigations and also by comparing them to what is known about bacterial adhesion. Diatoms are approximately 10 to 100 times, and maybe even more, larger than bacteria, i.e., their size considerably exceeds that of colloid particles. Therefore, it would be extremely incorrect to speak of their attachment in terms of the DLVO theory, which is applicable to colloids and comparable systems. Yet it is impossible not to admit that the process of diatoms sticking to a hard surface represents adhesion in the physical sense. Biological mechanisms appear to play an even more important part in the adhesion of microalgae than in bacteria (see Figure 2.1), but unfortunately, they are still little studied.

All solitary raphid diatoms are motile when they come in contact with a hard surface. In accordance with the capillary model (Gordon and Drum, 1970; Gordon, 1987), the gliding movement of diatoms is caused by the secretion of the mucopolysaccharide, which is synthesized by the Golgi apparatus and released through the anterior or posterior pore of the raphe. The viscous polymer is ejected at a high velocity from the cell and adheres to the surface with which the diatom comes in contact. As a result, the cell slides in the opposite direction. Thus, the mucopolysaccharide is used simultaneously both for movement and for temporary attachment (Avelin, 1997). The direction of sliding is determined by which pore the polymer is ejected from. The force necessary for movement is provided by two mechanisms. First, the mucopolysaccharide flows out of a very fine capillary and, consequently, has a great extrusion rate. Second, the polymer is hydrated before extrusion, which increases its volume and the pressure developed as it leaves the cell.

To support the sliding of diatoms, a constant inflow of calcium ions from the outside is necessary (Cooksey, 1981); this also holds true for other forms of cell movement — amoeboid, ciliary, and flagellar (Seravin, 1971). Therefore, if the calcium transport is somehow interrupted, movement will stop as soon as the internal calcium pool is exhausted.

In motile diatoms, movement and adhesion to the substrate appear to be closely connected, since they are mediated by the polymers released on the surface of the substrate. Therefore, the agents influencing the motility of the diatoms may be expected to affect their adhesion in a similar way. Indeed, the presence of calcium ions in the medium was shown to intensify the adhesion of diatoms (Cooksey et al., 1984; Geesey et al., 2000).

Adhesion was studied in greater detail on the diatom *Amphora coffeaeformis* (Cooksey, 1981; Cooksey et al., 1984; Cooksey and Cooksey, 1986). In calcium-free sea water there is no adhesion at all. The agent blocking calcium transport into the eukaryotic cell, known as D-600, also suppresses adhesion. When the calcium ion concentration in water is 0.25 mM, adhesion is weak, and few cells are able to attach to glass. As the calcium concentration is raised to 2.5 mM, adhesion increases fivefold and does not significantly change any further, even when the Ca^{2+} concentration is as high as 10.0 mM. Different agents blocking protein synthesis in eukaryotes (i.e., cycloheximide), respiration, and photosynthesis (carbonylcyanid 3-chlorophenylhydrazon) also suppress adhesion. Tunicamycin, an inhibitor of glycoproteid synthesis, is known to inhibit adhesion as well. Analysis of available data suggests that the adhesion of *A. coffeaeformis* depends on cell metabolism and, consequently, is an active biological process.

Adhesion of solitary diatoms may be carried out differently (Chamberlain, 1976): by means of a sticky mucous case and stalk and, additionally, mucopolysaccharide polymers (Figure 6.3b). Of some importance for the attachment of diatoms is the structure of their theca (Stevenson and Peterson, 1989). Among pennate diatoms, araphid forms have a certain advantage over monoraphids in this respect, judging by their relative abundance on hard surfaces and in plankton. In some species of biraphid diatoms this ratio is greater, and in others smaller, than in the araphids and monoraphids. The reasons for this are not clear.

The above peculiarities of attachment of diatoms show that biological factors play the leading role in irreversible adhesion in them as well as in bacteria.

Bacteria, preceding diatoms in the fouling succession owing to their hydrophobic properties, on the one hand, and the release of extracellular polymers, on the other, evidently change the adhesion properties of the surface and probably make it more favorable for the adhesion of diatoms. Thus it is highly probable that, in the succession of non-swimming, passively settling microorganisms, an important role is played by the adhesion processes.

In the ocean, one of the most important factors preventing temporary adhesion of protists, as well as other microorganisms, is the current. The cells coming into contact with a hard surface are acted upon mainly by shearing stress, which is directed parallel to the surface (Schlichting, 1979; see Figure 7.1). This stress arises from the inertia properties of the liquid, which is slowed down while it flows over the surface, forming the so-called boundary layer. Calculations show that the current velocity that is usually observed in natural reservoirs is sufficient for the detachment of bacteria adhered to aquatic vegetation (Silvester and Sleigh, 1985). Larger cells of diatoms and protists are affected by a greater shearing stress; therefore, in order to stay at the surface, they should have special adaptations.

The adhesion mechanisms in protists are still less studied than in diatoms and especially bacteria. According to the reviews (Dovgal and Kochin, 1995, 1997; Dovgal, 1998b), the first group of adaptations for attaching in current comprise settlement and attachment in places sheltered from the current, the secretion of sticky substances, the development of special structures and organelles, and the formation of structures that protect the junction of the body and the stalk (papillae, loricae, endostyles, etc.). Mucous polymers play the main role in the attachment of vagile as well as sessile forms of protists. Choanoflagellates and some other hetero- and autotrophic flagellates possess adhesive stalks. Ciliates are remarkable for the variety of ways in which they attach to the surface: by thigmotaxis of cilia, secretion of exopolymers, scopula (in Peritricha), fixation rings (in Peritricha and Suctoria), tentacles (in Rhinchodida), stalks, suckers, hooks, and other structures (Faure-Fremiet, 1952; Dovgal, 1998b).

The second group of adaptations allows the protists to not only keep to the surface but also to experience less hydrodynamic action from the current. These adaptations include a flattened body shape and spreading over the surface, as, for instance, in many motile amoeboid organisms and heterotrophic flagellates; the ability to bend under great hydrodynamic stress, which is observed in, e.g., vorticellid ciliates with a flexible stalk; elongation of the flexible stalk, which makes it possible to occupy an optimal position in the current and change it according to the parameters of the flow, thereby reducing the overall resistance.

Various adaptations of protists to life under the conditions of the boundary layer may considerably reduce the topical and trophic competition between the different species and facilitate the formation of a multilayered spatial structure of the microfouling communities (Dovgal, 1998a, 2000; Railkin, 1998b).

6.2 MECHANISMS OF ATTACHMENT OF LARVAE AND SPORES OF MACROORGANISMS

Attachment is an elementary process of biofouling, following settlement and preceding growth (see Section 2.1 and Figures 8.1 to 8.4 later). It determines the maintenance of the settled larvae of invertebrates and spores of macroalgae on the surface. Adhesion and temporary attachment are the crucial processes that, as it were, fix the choice of habitat and the conditions of further development of dispersal forms of macroorganisms. Permanent attachment makes irreversible the choice of hard substrates by sessile species, which usually dominate in fouling communities (see Chapter 1).

The distinct association of settlement and metamorphosis on a hard surface with attachment, a frequent coincidence of these processes in time, and their high rate may have been the reason for considering attachments a stage of settlement, on the one hand, (Crisp, 1984; Lindner, 1984; Davis, 1987; Pawlik, 1992; Zimmer-Faust and Tamburri, 1994, etc.) or as a stage of metamorphosis, on the other hand (Burke, 1983; Orlov, 1996a, b, etc.). There are objective reasons for such grouping. Indeed, in many cases, metamorphosis takes place in attached or motionless individuals, whereas settlement and moving on the surface inevitably involve temporary attachment, without which the very movement along the substrate would be impossible.

Yet, on the grounds of such arguments, it would be incorrect to put attachment together with settlement and metamorphosis. It should be emphasized that attachment and settlement (as defined in Section 2.1) characterize different aspects of the activity of larvae and spores settled on the surface: their physical connection (adhesion) to the substrate and their movement across it (until they become permanently fixed, in the case of sessile species). Attachment undoubtedly accompanies metamorphosis when the latter takes place on a hard surface and is one of its conditions, but it is not a process of transformation from a larva into a juvenile, which is what is referred to as metamorphosis. Therefore, uniting attachment and metamorphosis would not be correct. The adhesive properties of the surface are already manifested in a larva and, in the case of sessile species, is only intensified with its development into an adult (Young and Crisp, 1982). Similarly, the attachment of macroalgal spores does not represent a stage of their germination. With the growth of algae, their attachment to the hard surface becomes more durable. This is an additional argument in favor of treating settlement, attachment, and metamorphosis as independent processes of colonization (see Section 2.1).

Together with the common term "attachment," the term "adhesion" is also used in the literature. Strictly speaking, *adhesion* refers to a purely physical process of two heterogeneous bodies sticking together (Derjaguin, 1992). As early as at the stage of temporary attachment of propagules of macroorganisms, biological processes begin to prevail over physical ones (see Figure 2.1). Therefore, I will use the term "attachment" where possible to emphasize this fact. The term "adhesion" should refer only to the first stage of contact between the settling macroorganisms and a hard surface and sticking to it owing to the adhesive properties of their external structures. Starting with the period of induction and stimulation of secretion of additional adhesives after contact with the surface, it seems to be more correct to speak of attachment.

Thus, considering the interaction mechanisms of propagules of invertebrates, ascidians, and macroalgae with hard substrates, one can distinguish between adhesion, temporary (reversible) attachment, and permanent (irreversible) attachment. The first mechanism is constantly present, since it is the beginning of the physical interaction with the surface and it is the adhesion force that determines the durability and reliability of adherence. However, after coming into contact with the hard surface and adhering to it, the biological mechanisms are put in action (we will discuss this in more detail later in this section). They significantly change the nature of the interaction between the foulers and the hard substrates and, as a rule, increase the adhesion force, in particular by the secretion of adhesive polymers. Therefore, it seems quite reasonable to distinguish as independent adhesion, temporary (reversible as to its mechanism) attachment, and permanent (irreversible as to its mechanism) attachment. It should be noted that such views on the problems of terminology are also shared by other writers (e.g., Abelson and Denny, 1997).

The above does not mean, of course, that adhesion should be rejected as a physical mechanism of interaction between the external surface of the larva (or spore) and the surface of a hard body. It only emphasizes the fact that biological mechanisms included in the processes of attachment start to play a major role as the larva (or spore) starts to interact with the hard substrate, and become more important than the physical processes of adhesion, from a biologist's point of view.

Yet it should be remembered that the proposed distinction between adhesion and temporary and permanent attachment, though more or less evident in theory, may evoke certain difficulties when applied in practice. For instance, it may be difficult to distinguish between temporary and permanent attachment: an attached and motionless larva may suddenly become detached and move to another place or even swim away.

To avoid any misunderstanding, it should be emphasized that temporary attachment as it relates to its phenomenology does not necessarily correspond to temporary attachment as it relates to its mechanism. For example, the permanent attachment of adult bivalves is considered to be temporary by its phenomenology because these mollusks may become detached from the substrate and move to another place when the conditions change. However, their attachment is permanent with regard to its mechanism: it is carried out by means of secretions of definitive (adult) glands and is in fact irreversible. The detachment in this case is associated not with breaking of the attachment, but with the rupture of the byssus threads, which usually occurs close to the attachment disc (Young and Crisp, 1982).

It should be noted that all vagile forms possess only temporary (reversible) attachment, whereas, in the postlarval stages of sessile species, temporary attachment during their movement over the surface is finally replaced by permanent attachment.

When related to its mechanism, temporary attachment may be defined as the process of reversible adherence to the hard surface, allowing the dispersal (juvenile and adult) forms to remain and move on it by means of sticky adhesives produced by special larval, juvenile, or definitive glands. The sequence of the stages of reversible (temporary) attachment may be expressed as follows: adherence → detachment → adherence, or reversible attachment of larvae → reversible attachment of juveniles and adults. It should be noted that calling an attachment temporary does not imply that it is short-term, only that it is reversible.

Permanent attachment is the process of irreversible adherence of larvae and algal spores to the hard surface, which is usually intensified as they develop into juveniles and adults. It is carried out by means of secretions (adhesives) produced by special glands and may be expressed as follows: irreversible attachment of larvae → irreversible attachment of juveniles and adults.

Permanent attachment is observed in the postlarval stages of many echinoderms, while juveniles and adults reveal temporary attachment. This may be represented schematically as follows: irreversible attachment of larvae → reversible attachment of juveniles and adults.

Taking all of the above into consideration, we can understand adhesion as the interaction of propagules, juveniles, and adults with the hard surfaces to which they stick owing to mere physical mechanisms. Distinguishing between adhesion and temporary and permanent attachment makes it possible to consider physical and biological mechanisms separately and concentrate our attention on the latter.

The simplest adaptation to attachment is the stickiness of covers, described in all the spores (Fletcher et al., 1984) and larvae (Lindner, 1984) studied in this respect. The adhesive polymers that they secrete are usually complexes of polysaccharides with proteins and in many cases belong to the group of mucopolysaccharides or glycoproteids; sometimes they are simple polysaccharides (Baker and Evans, 1973). It should be noted that mucopolysaccharides consist mainly of carbohydrates (70–80%) and proteins, while glycoproteids are complex proteins in which the carbohydrate content is considerably lower. Mucopolysaccharides and glycoproteids also differ in other properties, such as localization, function, etc. They facilitate contact and keep the propagules on the hard surface during settlement. The adhesives of spores of green, brown, and red algae contain sulphated polysaccharides, which distinguishes them from terrestrial and freshwater plants as well as from animals (Kloareg and Quatrano, 1988). The protein–carbohydrate complexes on the surface of spores of brown, green, and red algae are not infrequently aggregated into scales or plaques, which, in some authors' opinions (Oliveira et al., 1980), may be considered as a kind of specialized structure analogous to the attachment discs in bivalves.

Some larvae possess temporary appendages in the form of long sticky threads or "tails" (Crisp, 1984; Rittschof and Bonaventura, 1986) that are similar in function to the cell outgrowths of microorganisms. They serve to increase the probability of contact with a hard surface and facilitate adherence to it. Such mucous structures, which are usually several millimeters in length, are described in the larvae of the hydroid polyp *Clava squamata* (Williams, 1965), the soft corals *Xenia macrospiculata* and *Parerythropodium fulvum fulvum* (Benayahu and Loya, 1984), the polychaete *Spirorbis borealis* (Knight-Jones, 1951), and the bryozoan *Bugula neritina*

(Lynch, 1947). I observed sticky transparent threads in pediveligers of the bivalve *Mytilus edulis* at their swimming-crawling stage, and also in planulae of the hydroids *Dynamena pumila* and *Gonothyraea loveni*. Drawing a preparation needle close to the larvae, it is not difficult to catch them by those "tails" and pull them in any direction. The function of mechanical location of the surface and anchoring to it must be also performed by the flagella of macroalgal spores, which are small but still extend beyond the cell outline. When the zoospores of the brown alga *Laminaria saccharina* are settling, the flagellum is the first to touch the surface and adhere to it. A brief review of the attachment mechanisms of algae and invertebrates by means of temporary polymer appendages was presented by A. Abelson and M. Denny (1997).

Let us assume, for the sake of simplicity of calculation, that a larva's contact with some surface depends only on its linear dimensions (y). According to this assumption, if the larva has appendages of length β, the probability of its contact with the surface will increase by $(1 + \beta/y)$ times. Thus, if at the settlement stage the larvae is 1 mm (*D. pumila*) or 0.25 mm (*M. edulis*) long and the threads are 2 mm long, the probability of contact with a hard surface will rise by a factor of 3 and 9, respectively. The above estimations are mostly illustrative. Yet the presence of appendages in propagules obviously increases their chances of finding a favorable habitat for settlement.

In flowing water, the initial stage of adhesion to the surface after coming into contact with it is the crucial event of the larvae and spores passing over to periphytonic existence (see Section 4.1), i.e., to life on a hard surface. The above properties of propagules of foulers (stickiness of covers, small size, adhesive appendages, etc.) have an adaptive significance when under the influence of currents (see Section 7.1): they reduce the action of the hydrodynamic forces that impede adhesion.

After the initial adhesion by means of sticky polymers, connection with the surface is intensified by the secretion of additional portions of adhesives, which is considered in detail in reviews devoted to macroalgal spores (Fletcher et al., 1984) and larvae of invertebrates (Lindner, 1984). Thus, mechanical contact with the surface induces and stimulates not only adhesion but also the subsequent, more durable attachment. In the simplest case, this is associated with an increased production of the adhesive material. The synthesis and secretion of adhesives proceed comparatively fast. In the zoospores of the green alga *Enteromorpha intestinalis*, a fairly dangerous ship fouler, a new portion of sticky polymer is secreted within several minutes after their settlement on the surface (Christie et al., 1970). The additional secretion of adhesive material by spores of brown and red algae also starts quickly, within minutes or tens of minutes after settlement (Oliveira et al., 1980). Secretion in larvae occurs as early as in the stage of exploration of the substrate, which usually lasts from several minutes to 1 to 2 h in different species (Foster, 1971; Lindner, 1984); this fully corresponds in duration to the secretory period in algal spores. The above values agree with the data obtained by direct videotaping in an experiment conducted by J.M. Hills et al. (1998). Cyprids of *Semibalanus balanoides* were observed to occupy nearly half of the pits containing the settlement factor in as little as 10 min, whereas the mean time of their settlement was about 30 s.

Of great importance for attachment is such an integral characteristic of the surface as wettability, whose value depends not only on the material of the substrate but also on its roughness and the properties of the microfouling film covering it. For example, the barnacles *Balanus perforatus* and *Elminius modestus* attach more strongly to dense multispecific microfouling films formed in the fast current than to loose films that develop in the slow current (Neal and Yule, 1994). Increasing roughness causes greater wettability of the material, i.e., greater hydrophily.

Foulers show real selectivity with regard to surfaces with different wettability (Crisp et al., 1985). If adult mussels *Mytilus edulis* in aquaria are offered different materials in pairs (for instance, slate–paraffin or glass–paraffin), the mollusks form twice as many attachment discs on the more wettable glass and slate (Young and Crisp, 1982; Young, 1983). Juvenile and adult barnacles *Semibalanus balanoides* also adhere more firmly to hydrophilic surfaces (Crisp et al., 1985).

Though the connection between the wettability of a surface and attachment to it has been studied less in larvae than in adult organisms, the available data suggest that planulae of the jellyfish *Cyanea* (Brewer, 1984) and pediveligers of the mussel *M. edulis* (Dobretsov and Railkin, 1996), on the contrary, adhere better to hydro-phobic surfaces. A similar trend also has been shown by cyprids of *Semibalanus balanoides*. They attach weakly only to the poorly wettable beeswax (Crisp et al., 1985). It is interesting to note that zoospores of the green alga *Enteromorpha* also prefer to settle on hydrophobic substrates (Callow et al., 2000). They settle in groups, with these groups being larger on those low-energy (hydrophobic) substrates than on hydrophilic surfaces.

Most larvae possess specialized structures for temporary attachment to the sur-face, which are also used for the final (permanent) attachment. These structures are usually connected with larval glands producing adhesive secretions, not infrequently called "cements."

Yet neither sponges nor hydroid polyps have larval glands whose secretions would provide their attachment; instead, this function is performed by secretory ectodermal cells, and also by nematocysts in hydroids (Chia and Bickell, 1978; Yamashita et al., 1993). In stagnant water in the laboratory, larvae usually attach with their anterior ends or, rarer, by their sides (Ivanova-Kazas, 1975). They change their shape, spread, and achieve close contact with the substrate. The planulae of hydroids flatten, assuming the shape of a disc, from which a stolon with the primary polyp grows later (Figure 6.4). According to my observations, the larva at the disc stage is rather difficult to detach from the substrate. The above peculiarities of the attachment of sponges and hydroids certainly have adaptive significance, since they increase the area of contact with the surface and provide firm attachment. In an adult solitary polyp or a colony, the hydrorhiza becomes attached to the substrate by means of an adhesive polymer secreted onto the fouled surface (Figure 6.5).

The larvae of polychaetes of the family Sabellariidae, in the process of crawling, adhere to the substrate with their ventral side, so that it may be difficult to tear them away from it (Eckelbarger, 1978). When the larvae find metamorphosing young or adult individuals of their own species, they stop and firmly attach to the substrate. The metamorphosing larvae secrete a semi-transparent mucous cocoon around them-selves. It serves as a base during the building of the tube, to which small sand grains easily adhere and finally form the tube of the adult worm. Permanent attachment to the hard surface (Figure 6.5) is carried out by means of definitive gland secretion.

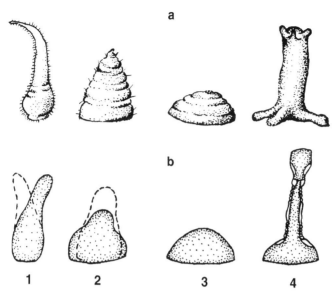

FIGURE 6.4 Attachment and metamorphosis of planulae. (a) Solitary polyp *Hydractinia echinata* (after Burke, 1983; with permission of the *Canadian Journal of Zoology* and NRC Research Press); (b) colonial hydroid *Gonothyraea loveni* (after Marfenin and Kosevich, 1984; with permission of the Publishing House of Moscow State University). Stages of attachment and development: (1) adhesion, (2) temporary attachment, (3) disc stage, (4) stolon growth and development of the hydranth.

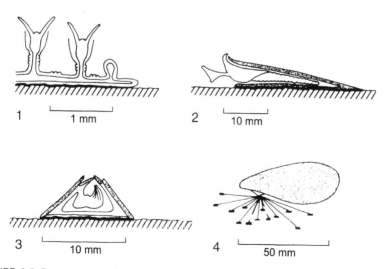

FIGURE 6.5 Permanent attachment of adult invertebrates. (1) Colonial hydroid, (2) polychaete in a tube, (3) barnacle in its shell, (4) bivalve attached by byssus threads. The layer of definitive adhesive is shown as a bold line between the animal and the substrate; in the bivalve, on terminal attachment discs of the byssus threads. (Modified from Young and Crisp, 1982. With permission of Prof. G. A. Young.)

Cyprids of barnacles remain on the substrate by means of the attachment disc (Figure 4.3) located on the third antennular segment, which seems to act like a sucker (Saroyan et al., 1968) and at the same time represents an adhesive pad, since sticky secretions of the larval glands are released onto its surface (Nott and Foster, 1969). After completion of metamorphosis, secretion of juvenile and later adult glands starts, which results in firmer attachment of the barnacles to the substrate (Figure 6.5).

Settled pediveligers of bivalves at first crawl on their foot, the ventral surface of which is continuously supplied with mucus secreted by its glands. At this stage of temporary attachment, the mollusks can be comparatively easily torn off from the substrate. The permanent firm attachment of settled pediveligers in all bivalves is ensured by byssus threads. Such a means of attachment is preserved in adult mollusks of the families Mytilidae (Figure 6.5), Pectinidae, Heteranomiidae, Hiatellidae, Nuculanidae, and Arcidae. All adult mollusks of these and other families hold reliably onto the substrate while crawling or staying motionless on it. This is possible because of the fact that the ventral surface of the foot is covered with a sticky mucous secretion, and the foot itself acts as a suction cap (Lindner, 1984). Chitons (class Loricata) and some motile gastropods of the order Patelliformes, for example, the widely spread limpets *Patella pontica* (family Patellidae) and *Testudinalia tessellata* (family Tecturidae), can attach to a hard surface especially fast.

The glandular apparatus and the processes of byssus formation have been most studied in the mussel *Mytilus edulis* (Waite and Tanzer, 1981; Lindner, 1984; Crisp et al., 1985; Berger et al., 1985; Waite, 1991). Other species (*M. galloprovincialis, M. californianus, M. trossulus, Modiolus modiolus, Pinna nobilis, Geukensia demissa*) have been studied in less detail (Cook, 1970; Waite et al., 1989; Pardo et al., 1990; Bell and Gosline, 1996). Yet the structure and formation of their byssus are known to have much in common with those of *M. edulis*. Therefore, they will be discussed by the example of the latter species.

The byssus apparatus consists of a stem with cuffs, byssus threads, and internal glands participating in their synthesis (Figure 6.6). The byssus threads branch from a common stem. They include an expandable part and a terminal adhesive disc of a constant size. In the mollusk's foot are located five glands, which are arranged from its base to its distal end: byssus, collagen (white), auxiliary, polyphenol (purple), and mucoid. All of these glands open near the distal pit (Figure 6.6), from which a groove passes toward the foot base, and it is there that the byssus thread is formed. The adhesive disc is formed in the distal pit. The core of the byssus thread is produced by the white and byssal glands and consists of collagen, which to a great extent determines its elastic properties and high mechanical strength. The thread is more extendable in its proximal part, located closer to the foot base and, on the contrary, is more rigid in its distal part, near the adhesive disc. Secretions of other glands form the thin outer layer (cortex) of the byssus thread. The polyphenol and auxiliary glands secrete polyphenolic proteins that are rich in aromatic amino acids (phenylalanine derivatives), and also the enzyme polyphenoloxidase. The mucoid gland enriches the byssus material with mucopolysaccharides, which seem to take part in temporary attachment as well.

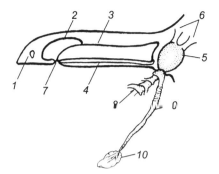

FIGURE 6.6 Schematic view of the byssus apparatus of the mussel *Mytilus edulis* (after different authors). (1) Mucoid gland, (2) polyphenol gland, (3) collagen gland, (4) auxiliary gland, (5) byssus gland, (6) retractors, (7) distal pit, (8) stem, (9) byssus thread, (10) adhesive disc.

According to modern conceptions that we owe mostly to J.H. Waite (Waite and Tanzer, 1981; Waite et al., 1985; Waite, 1991), the polyphenolic protein, which determines the adhesive properties of the byssus in *Mytilus edulis*, has a molecular mass of 125,000 D. It consists of 75 to 85 short peptide chains, which represent a combination of two peptides, one containing ten amino acids and the other six. The decapeptides are repeated throughout the length of this protein approximately 70 times and around hexapeptides about 13 times (Figure 6.7). The protein is rich in lysine, threonine, proline, L-DOPA, contains tyrosine, and is rather specific in composition (Amato, 1991).

Polyphenolic proteins of 15 studied species of various marine mollusks have a similar structure (Waite et al., 1989; Pardo et al., 1990). They are formed by two to three or more short repeating peptide fragments and have a molecular mass close to that of the corresponding protein in *M. edulis*. In *M. californianus*, the attachment protein consists of a multiply repeated decapeptide. It should be emphasized that all polyphenolic proteins are rich in 3,4-dihydroxyphenylalanin (L-DOPA; see Figure 6.11 later). It is the presence of the L-DOPA that largely determines the sticky properties and mechanical strength of the byssus and allows the bivalves to hold fast to hard surfaces (Waite et al., 1989).

In the process of forming the attachment thread (about 150 μm in diameter), the outer layer of byssus (10–20-μm thick) is hardened due to the binding of protein molecules with the phenol derivative o-quinone in the presence of oxygen (Lindner, 1984; Waite, 1991; Fant et al., 2000). The process is as follows. The enzyme catechol oxidase catalyzes the oxidation of peptidyl-DOPA into peptidyl-DOPA-quinone (Figure 6.8). Concurrently, the protein is bound with hydroxyl groups that are always present on a hard surface in an aquatic medium. Then, the peptidyl-DOPA-quinone undergoes condensation, interacting with peptidyl-lysine or some other peptide with a nucleophilic terminal amino group. As a result, the two protein molecules become bound, and the resulting protein complex is capable of repeating the binding cycle. Thus, the above cyclic process embraces more and more protein molecules.

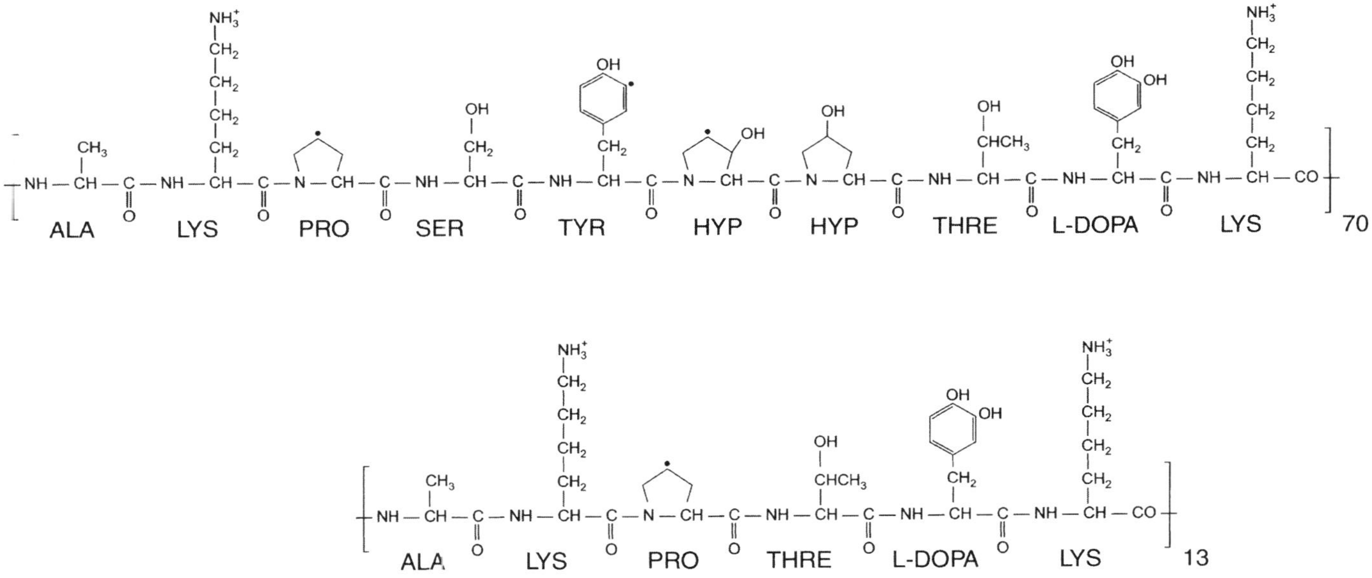

FIGURE 6.7 Repeated peptide fragments in the polyphenolic protein of *Mytilus edulis* byssus. Above: decapeptide; below: hexapeptide. Figures designate the number of repetitions, dots, hydroxylation sites. ALA – alanine, L-DOPA – 3,4-dihydroxyphenylalanine, LYS – lysine, HYP – hydroxiproline, PRO – proline, SER – serine, TYR – tyrosine, THRE – threonine. (After Waite, 1991. With permission of *Chemistry and Industry*.)

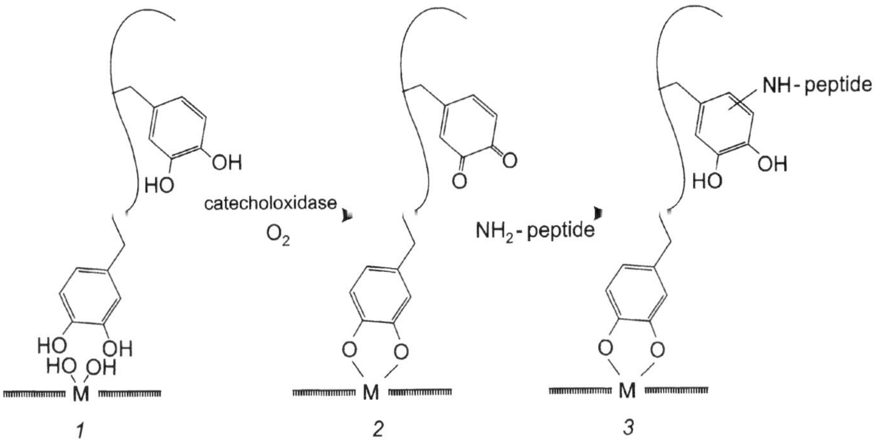

FIGURE 6.8 Linking of polyphenolic protein molecules. (1) Adsorption of polyphenolic protein (DOPA-protein) on the surface, (2) oxidation of DOPA-protein and its binding with the surface, (3) binding of DOPA-protein and lysine-containing protein. For explanations, see text. (After Waite, 1991. With permission of *Chemistry and Industry*.)

In adults of *M. edulis*, each attachment disc is about 1 mm in diameter, and the area of its contact with the hard substrate is about 3 mm². The settled pediveliger produces not one but many byssus threads and attaches strongly to the chosen surface. According to different estimations (Price, 1981; Young and Crisp, 1982), the stress necessary to detach a single thread of juvenile and adult mollusks is 10^4 to 10^6 N/m², i.e., approximately 1 to 100 g/mm², depending on its thickness and environmental conditions. As many as 10 to 20 threads may be formed daily. The attachment of juvenile and adult mussels is stronger on polar, well-wettable materials (glass, slate) and weaker on non-polar materials (paraffin, polytetrafluorethylene) (Young, 1983).

Its great force of adhesion, which does not yield much to that of synthetic glues, in addition to its unique property of adhering well under water and on wet surfaces may form the basis for the use of byssus in industry (Waite, 1991). The problem of tight coupling of construction elements in water and moist atmosphere is known to be rather urgent. Methods of gene engineering have made it possible to introduce into yeast cells a gene of the polyphenolic protein (Amato, 1991). In the future, this may make it possible to obtain this protein in great quantities and use it as an underwater glue. A commercial preparation of polyphenolic protein is being produced that is used for enzyme immobilization and adhesion of cells and tissues in experimental medicine and biology.

A unique feature of cyprid larvae of cirripedes is a specialized organ, an attachment disc (see Figure 4.3), that is located on the lower side of the third antennular segment (Lewis, 1978; Young and Crisp, 1982; Elfimov et al., 1995). It functions simultaneously as a sucker and as an adhesive pad. The disc is about 50 µm in diameter and bears chemoreceptors. In addition, numerous ducts of larval antennular glands open onto its surface; they produce a proteinaceous secretion for temporary attachment (Walker and Yule, 1984; Clare et al., 1994). As it moves along the

FIGURE 6.9 The sequence of morphological changes during metamorphosis in the barnacle *Semibalanus balanoides*. (a) Antennula. (After Burke, 1983. With permission of the *Canadian Journal of Zoology* and NRC Research Press.)

substrate, the cypris presses its antennulae close to it, assessing its suitability for settlement, and may attach to it temporarily. The second and the third segments of antennulae possess communicating cavities. When the volume of the second segment increases due to muscle action, negative pressure is developed in the third segment and the attachment disc located on it sticks to the substrate. The numerous setae on the attachment disc and the secretion released between them are conducive to strong attachment, since they increase the area of contact between the larva and the surface. A considerable force, which may reach 3 kg/cm², is required to pull the cypris away (Crisp, 1984). The permanent attachment of cirripedes is carried out by a proteinaceous secretion that is produced first by larval and later by definitive glands.

According to their composition, the adhesive proteins of the cyprids of *Semibalanus balanoides* (Walker and Yule, 1984), *Balanus eburneus* (Hillman and Nace, 1970), and *B. crenatus* (Cook, 1970) are rather similar. The antennal (larval) glands consist of two types of cells: one produces protein and the other, together with protein, also produces phenol compounds (for instance, the amino acid tyrosine) and the enzyme polyphenoloxidase. Similar structures and production of adhesives are described for *Megabalanus rosa* (Okano et al., 1998). A series of biochemical reactions result in quinone tanning of protein, similar to that just described for bivalves. Thus, the adhesive complexes of cirripedes and mollusks seem to be fairly similar (Cook, 1970); they both consist of proteins that are rich in L-DOPA and have a low content of lipids and carbohydrates.

During metamorphosis in cirripedes, the larval gland cells dedifferentiate and then become specialized into definitive cement gland cells (Lindner, 1984; Elfimov et al., 1995). The body turns in such a way that its axis becomes vertical (Figure 6.9), yet the site of antennulae attachment to the substrate is preserved, whereas the main gland duct elongates toward the periphery of the shell base. Up to seven stages of attachment have been distinguished in *Semibalanus balanoides* (Rzepishevsky et al., 1967). At the sixth stage, the young barnacle with well-distinguishable shell plates is still attached by means of the cement produced by its larval glands.

In the process of the subsequent growth of the barnacle, its definitive glands enlarge and the produce of their secretion increases. The cement fills all of the space between the growing base and the hard surface to which the barnacle has attached (see Figure 6.5). And, although the thickness of the adhesive layer under the animal is not great (only about 5 μm), the force of adhesion of *S. balanoides* may reach as much as 10 kg per 1 cm² (Crisp et al., 1985). Similar values have also been found

TABLE 6.1
Attachment of the Barnacle *Semibalanus balanoides* to Slate

Colonization Stage	Means of Attachment	Force Needed for Detachment	Adhesion Force (kg/cm²)
Exploration of the surface by the cypris	Temporary adhesion by antennulae	20–50 mg	1.5–3.0
Stopping and attachment of the cypris	Permanent attachment with antennulae	1.5 g	9.7
Development after attachment			
7-day-old barnacle	Attachment with the shell base (secretion of juvenile glands)	6 g	0.64
Juvenile barnacle (2 months)	Attachment with the shell base (secretion of definitive glands)	600 g	1.2
Adult barnacle (5–10 months)	The same	5–20 kg	4.9

After Crisp et al., 1985, with modifications. With permission of the *Journal of Colloid & Interface Science.*

in other studies (Yule and Crisp, 1983; Yule and Walker, 1984). Adhesion reaches its maximum before the start of metamorphosis. At the same time, the force necessary for detaching the barnacle steadily increases as it develops and grows (Table 6.1). Adult barnacles attach to hard surfaces very quickly and belong to the few invertebrates, together with bryozoans, hydroids, and polychaetes, that are able to grow in large quantities even on high-speed ocean-going ships.

DOPA-containing proteins participating in attachment were also revealed in other foulers: in the polychaete *Phragmatopoma californica* during the building of its tube (Jensen and Morse, 1988) and in the ascidian *Pyura stolonifera* (Dorsett et al., 1987). Quinone tanning, the chemical mechanism of adhesion best studied in mollusks and barnacles, is much more widespread among animals. It was described in some radiolarians, sponges, gorgonarians, flatworms, polychaetes, and ascidians (see Lindner, 1984). The DOPA-containing proteins and the mechanism of quinone tanning appear to be quite universal. They are used by animals whenever it is necessary to reinforce the mechanical connection between the elements of biological constructions or to ensure fast attachment of sessile forms to the hard surface.

Bryozoan cyphonautes have the pyriform organ (see Figure 4.5), which feels the substrate and assesses its suitability for settlement (Ivanova-Kazas, 1977). It practically performs the additional function of temporary attachment (Reed, 1978). When the site is chosen, the suckerlike organ of the larva, releasing abundant mucous, is turned inside out and pressed close to the substrate, thus accomplishing the final attachment of the cyphonautes. The secretion released by the larva of *Bowerbankia gracilis* is of a mucopolysaccharide nature (Woollacott, 1984). After the completion of metamorphosis, the young bryozoan remains strongly attached to the substrate, but due to the definitive cement.

Echinoderm larvae (brachiolaria, pluteus, ophiopluteus) crawl over the substrate while exploring it. Their connection to the substrate is temporary, provided by the papillae that are located at the anterior end of the larva and are covered with a sticky secretion. The permanent attachment during metamorphosis is accomplished by specialized organs. The attachment apparatus of the starfish brachiolaria (see Figure 4.6) consists of the attachment disc, three brachiolae, and lateral papillae (Kasyanov, 1984b). Having chosen a site on the surface, the larva moves its brachiolae apart and presses its attachment disc to the substrate. Close contact is achieved by the action of the attachment papillae located laterally on the disc and covered with the sticky secretion. The final attachment of the brachiolaria is provided by a proteinaceous secretion released by the cells of the attachment disc. As a result, the larva attaches to the surface with a stalk. After the completion of metamorphosis, the juvenile starfish breaks away from the stalk by means of its primary ambulacral podia and starts a free moving life. Larvae of the starfish of the families Luidiidae and Astropectinidae have no specialized attachment organs (Strathmann, 1978) and keep themselves on the surface by the almost completely developed juvenile ambulacral podia.

Attachment of pluteus larvae of sea urchins is carried out by means of five sticky podia (Strathmann, 1978). Their total surface is relatively large, which provides a sufficiently strong adherence to the substrate during metamorphosis.

The tadpole larvae of ascidians have different ways of remaining attached to the hard surface (Cloney, 1978; Svane and Young, 1989). Primitive species of the family Molgulidae may not have any specialized structures. In this case, a secretion of subepidermal ampoules scattered all over the larva's body is released onto the surface, providing the stickiness of their tegument. Many ascidian larvae possess adhesive papillae, which may be classified into three groups, according to their structure and type of secretion (Cloney, 1978). They may be glandular, eversible, or non-eversible, or they may have no glandular structures at all. In more complex larvae, a special suckerlike fixation organ is developed (see Figure 4.7), which is controlled by the nervous system (Svane and Young, 1989). This allows the larva to become attached and detached repeatedly in search of the appropriate site for settlement, after which it finally fixes on the substrate and starts its metamorphosis.

Macroalgae also develop specialized structures for permanent attachment — rhizoids. They provide a firm connection between the alga and the hard surface. For instance, spores of red algae are kept on the surface by mucous adhesives. They are washed away by a flow of about 20 cm/s (Polishchuk, 1973). The situation changes drastically when the germinating spores form rhizoid cells. In this case, they cannot be washed away by a flow of the same intensity.

Studies of 20 species of green, brown, and red algae have shown that their cells have already differentiated after the first division: the lower (closer to the substrate) cell secretes mucus and later develops into the rhizoid; the upper cell gives rise to the thallus (Fletcher, 1976). The rhizoid first adheres to the substrate by means of mucus but subsequently attains firm attachment. It was shown experimentally that, if the upper cell (the precursor of thallus) were destroyed, the lower one would divide and differentiate again. Even if the whole thallus were eliminated, the cell may give rise to a complete plant that would be capable not only of vegetation but also of reproduction. Another type of attachment also has been observed. Fragments

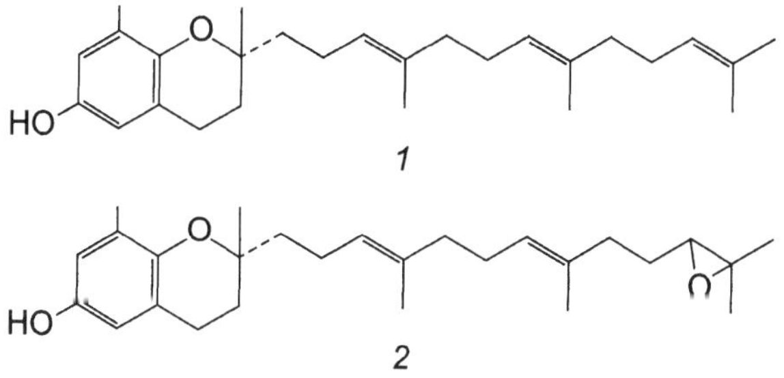

FIGURE 6.10 Natural inductors of settlement, attachment, and metamorphosis of the hydroid polyp *Coryne uchidai*. (1) δ-Tocotrienol, (2) epoxy-δ-tocotrienol.

of the thallus of the red alga *Callithamnion corymbosum*, consisting of five to nine cells, may form rhizoid cells within several hours and thus achieve firm attachment to the substrate (Polishchuk, 1973).

Subsequent studies (Brownlee et al., 1994) supported and extended these views. In *Fucus serratus*, the existence of sulphated polysaccharides responsible for the attachment of the rhizoid cell to the substrate was shown. The factor of differentiation was found to already exist at the two-cell stage. Under its influence, at the early stage of development, thallus cells may be re-differentiated into rhizoid cells. This factor acts under certain external and internal conditions. In particular, blue light locally activates the electron transport chain of the plasmalemma and consequently causes the local elevation of the calcium ion concentration at the site of rhizoid development. At the same time, the apex of the rhizoid possesses mechanosensitive Ca channels, which are activated by its stretching. The local elevation of the cytoplasmic calcium concentration results in the fast growth of the rhizoid and accelerates attachment.

6.3 NATURAL INDUCTORS OF SETTLEMENT, ATTACHMENT, AND METAMORPHOSIS

Substances that induce larval settlement on natural substrates in many cases also will cause their attachment and metamorphosis. Thus, they trigger the most important elementary processes of colonization and thus determine the biofouling of hard surfaces. Let us consider the chemical nature of the best-studied natural inductors.

Planulae of the hydroid polyp *Coryne uchidai* develop mainly on the gulfweed *Sargassum tortile* (Nishihira, 1967, 1968, cit. after: Orlov, 1996a). Under experimental conditions, these algae were preferred to 20 other species of algae offered to the planulae. Extracts of this plant rather effectively induced not only settlement, but also attachment and metamorphosis. The distantly acting, biologically active substances were low-molecular terpenoids: δ-tocotrienol $C_{27}H_{40}O_2$ and epoxy-δ-tocotrienol $C_{27}H_{40}O_3$, that are released by this alga in water (Kato et al., 1975). Their structure is shown in Figure 6.10. This is one of the few cases when the chemical structure of a natural substance causing settlement and metamorphosis in an invertebrate has been determined precisely.

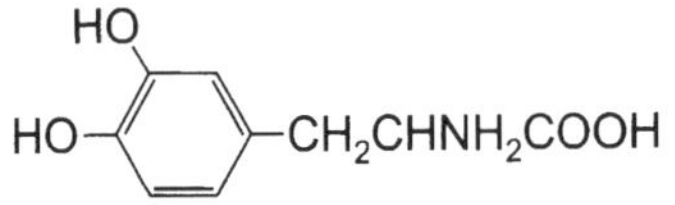

FIGURE 6.11 Inductors of settlement, attachment, and metamorphosis of *Phragmatopoma californica*. (1) 3,4-Dihydroxyphenilalanine (L-DOPA), (2) 2,6-ditretbutyl-4-methylphenol.

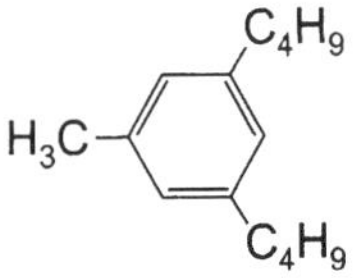

The settlement and attachment of the planuloid bud of the upside-down jelly *Cassiopea andromeda* and its metamorphosis into the scyphistoma can be induced experimentally by a factor that is released into the medium by the marine bacterium *Vibrio* sp., cultivated in suspension (Neumann et al., 1980). The molecular weight of the inductor, determined by ultrafiltration, is from 1,000 to 10,000 D. The inductor is precipitated by acetone and can be destroyed by hydrochloric acid hydrolysis and by treatment with pronase, papain, hyaluronidase, and lipase. The summarized results of the biochemical analyses of the inductor made it possible to conclude that it is either a protein or it contains a peptide.

It was found (Morse and Morse, 1991; Morse et al., 1994) that the molecules inducing settlement and the associated sequence of irreversible events in planulae of the corals *Agaricia humilis* and *A. tenuifolia* are located on the surface of the red coralline alga *Hydrolithon boergesenii*. The compound in question is an insoluble sulphated polysaccharide glucosaminoglycan, which is bound with the cell wall. Gentle hydrolysis of this polysaccharide with enzymes or alkali yields a low-molecular compound that induces settlement, attachment, and metamorphosis. The substance is soluble in water, has a molecular weight of approximately 2000 to 5000 D, and contains 8 to 14% sulphates. Adding it to the vessel with the swimming larvae of *A. humilis* caused their fast settlement and attachment, followed by the normal metamorphosis and subsequent growth, even in the absence of the red alga *Hydrolithon boergesenii*. The low-molecular inductor acted on *A. tenuifolia* in a similar way.

The reef-building activity of polychaetes *Phragmatopoma* is related to the selective settlement and attachment of their larvae to the tubes of adult individuals (Eckelbarger, 1978). The settlement inductors were originally thought (Pawlik, 1986) to be some water-insoluble fatty acids, mainly arachidon and palmitic, released by these polychaetes. Further, more thorough studies showed that the compounds mentioned above were most probably the result of the contamination of the biochemical material, possibly by polychaete pellets and microorganisms (Morse, 1990). If these worms are provided with fine glass particles with which to build their tubes, an absolutely different substance is observed to induce settlement and metamorphosis in *Ph. californica* (Jensen and Morse, 1984). This is a unique protein that is similar to silk in composition and contains many residues of 3,4-dihydroxyphenilalanine (L-DOPA; Figure 6.11). The latter forms numerous links within the molecule, considerably increasing the mechanical strength of the protein. The process of binding (tanning) of proteins was considered earlier in this chapter (see Figure 6.8). L-DOPA was found to induce settlement, attachment, and metamorphosis in *Ph. californica*

(Morse, 1990). A similar effect is shown by another aromatic compound, 2,6-ditretbutyl-4-methylphenol (Figure 6.11).

In the polychaete *Hydroides elegans*, the settlement and metamorphosis of larvae are induced by adult worms, which also live in tubes (Bryan et al., 1997). The inductor is a low-molecular substance with a molecular weight (MW) less than 10,000 D.

The monospecific aggregate settlement of cirripedes is based on the contact chemoreception by their cyprids. A special high-molecular substance, present on the surface of the epicuticle of adult animals, is recognized by the larvae. It belongs to the class of arthropodin proteins, which acquired their name from the name of the phylum Arthropoda. These very specific proteins were first isolated from the cuticles of insects. When they come into contact with the arthropodin, the cirripede larvae obtain information about conspecific individuals and, consequently, about whether the environment and the surface itself are favorable for attachment and metamorphosis.

The best-studied representative of cirripedes is the barnacle *Semibalanus balanoides*. In the presence of the settlement factor, its larvae attach more strongly to the hard surface (Yule and Crisp, 1983; Yule and Walker, 1984). Presently the properties of the arthropodin are fairly well known (Crisp and Meadows, 1962; Larman et al., 1982). This is a glycoprotein resembling the cuticular protein of the crab *Carcinus maenas* in its amino acid composition. Nevertheless, extracts from the crab cuticle cause settlement only in 44.4% of the larvae (Crisp and Meadows, 1962). The cuticular arthropodins seem to be recognized by the "self/non-self" scheme, based on the structure of not only the protein but also the carbohydrate part of the molecule. The arthropodin is thermostable, and its active fraction is resistant to boiling. The isoelectric point of this substance is about 4.5. Detailed biochemical investigations performed by V. Larman and colleagues (Larman et al., 1982) showed that the protein component is a polymorphic system of proteins consisting of subunits with MW 5,000 to 6,000 and 18,000 D.

The red abalone *Haliotis rufescens* inhabits calcareous red algae of the genera *Lithothamnium, Lithophyllum*, and *Hildenbrandia* (Morse et al., 1979). According to investigations carried out by D. E. Morse and his colleagues (Morse and Morse, 1984; Morse, 1992), induction of settlement, attachment, and metamorphosis in the abalone larvae takes place during its contact with the surface of the algae. A similar action is shown by the aqueous extract of their tissues. The natural inductor proved to be a low-molecular substance, γ-aminobutyric acid (GABA), associated with a small protein that contains several unusual amino acid residues. GABA induces not only settlement and temporary attachment, but also the normal metamorphosis of the mollusk. Some structural analogs of GABA (Figure 6.12), particularly δ-aminovaleric acid and ε-aminocaproic acid, possess a comparable inducing activity. Increasing or reducing the length of the hydrocarbon chain lowers the activity. This demonstrates the high specificity of GABA as an inductor of settlement, attachment, and metamorphosis.

The coral polyp *Porites compressa* is a prey of the gastropod *Phestilla sibogae* (Hadfield, 1978). The polyp releases a substance that induces not only settlement but also metamorphosis in the predator (Hadfield and Scheuer, 1985). The natural

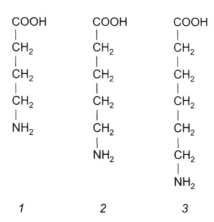

FIGURE 6.12 Organic acids inducing settlement and metamorphosis in larvae of the red abalone *Haliotis rufescens*. (1) γ-Aminobutyric acid, (2) δ-aminovaleric acid, (3) ε-aminocapronic acid.

inductor is a water-soluble compound with a small polar molecule with MW of 200 to 500 D (Hadfield and Pennington, 1990). This compound of a still-unknown nature has a high biological activity; the dilution of 10^{-7} g of purified preparation per 1 l of sea water is quite sufficient to induce settlement and metamorphosis in the mollusk. The active substance is contained in the coral in minute quantities: 65 l of the extract yield only about 2 μg of the inductor. These and some other circumstances have not yet made it possible to obtain enough of the inductor for its more exact identification.

The Japanese scallop *Patinopecten yessoensis* is an object of mariculture in the Far East. In this respect, the study of the factors conducive to the settlement of its larvae under mariculture conditions is of considerable interest. It was established (Zhuk, 1983) that homogenates of the mantle and the adductor muscle attract the larvae of the scallop. If collectors (substrates for collecting larvae and rearing mollusks) are impregnated with these substances, more larvae settle on them. Biochemical studies using different methods made it possible not only to purify the inducing substances but to characterize their properties with some certainty. As a result, up to 15 proteins were isolated, with their isoelectric points varying from 4.0 to 6.0 (Zhuk, 1983).

Studies on the commercial oyster *Crassostrea virginica* have shown (Zimmer-Faust and Tamburri, 1994) that its aggregated settlement on hard substrates is caused by a low-molecular inductor released by the adults. This substance appears to be a peptide with a molecular weight of 500 to 1000 D. Tests involving 21 amino acids have shown that only arginine and lysine at the C end of the natural peptide induce settlement. Its synthetic analog is the tripeptide glycil–glycil–L-arginine, efficient in a low concentration equal to 10^{-10} M.

Development of the king scallop *Pecten maximus* is usually retarded in the absence of inductors (Nicolas et al., 1998). However, the rate of metamorphosis increases noticeably when extracts of some red algae are added to the water (Cochard et al., 1989; Chevelot et al., 1991). For example, of the several inductors isolated

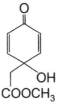

FIGURE 6.13 Jacaranone, an inductor of settlement and metamorphosis in the bivalve *Pecten maximus*.

from *Delesseria sanguinea*, the most active one was jacaranone (Figure 6.13), causing maximum settlement of *P. maximus* at a concentration of 0.5 mg/l.

Adults of the sea urchin *Dendraster excentricus* attract larvae of their own
species from a distance (Highsmith, 1982). The attractant induces both settlement
and metamorphosis. In the laboratory, under the influence of purified preparations,
these processes are initiated within several minutes, and metamorphosis into the
juvenile is completed in less than 1 h (Burke, 1984). The inductor is a small protein
with an MW of about 1000 D.

These examples show that chemical factors of a biological nature play an important role in the formation of communities as well as monospecific aggregations. In
many cases, they act while in contact; in some cases they act from a distance. Many
known inductors are low-molecular compounds. Further studies and applications of
their natural or synthetic analogs will make it possible not only to clarify the
mechanisms of establishing biological relations on hard surfaces, but also to use
inductors for attracting invertebrates and regulating their populations. This may be
of special importance for breeding marine commercial organisms, for instance,
mollusks.

6.4 UNIVERSAL MECHANISMS OF ATTACHMENT

The study of adhesion and attachment is important for understanding the way in
which the colonization of hard natural and artificial surfaces, in particular technical
objects, proceeds in an aquatic (marine) environment. In spite of the peculiarities
inherent in various invertebrate larvae and macroalgal spores, they possess some
common mechanisms of adhesion and attachment. Among those already mentioned
(see Sections 6.1 to 6.3), universal mechanisms can be distinguished that are peculiar
to any propagule: physical adhesion within the range of the secondary and primary
energy minimums and the secretion of adhesives. The stickiness of the surface that
can be observed in microorganisms, larvae of invertebrates, ascidians, and spores of
macroalgae is prerequisitive to their high adhesion ability. All propagules seem to
be characterized by the secretion of adhesive polymers onto their own surface prior
to their contact with the hard surface on which they settle. Another universal mechanism is associated with intensifying the secretion of the attachment polymers after
temporary adhesion. In addition to this, the replacement of larval cements by definitive ones can be observed in macroorganisms. Yet these mechanisms are too general;

TABLE 6.2
The Effect of Chemical Treatment on *Pseudomonas marina*
Bacterial Films and Larvae of the Polychaete *Neodexiospira (Janua)
brasiliensis* on Attachment and Metamorphosis of the Larvae

Substances Used for Treatment of Film	Type of Effect	Substances Used for Treatment of Larvae	Type of Effect
Concanavalin A	Inhibition	D-Glucose	Inhibition
Peanut lectin	—	2-Deoxy-D-glucose	—
		D-Ribose	—
Trypsin	Inhibition	D-Mannose	—
Sodium periodate	Inhibition	D-Galactose	—
		α-Methyl-D-glycoside	—
		α-Methyl-D-mannoside	—
		N-Acetyl-D-glucosamine	—

Note: Dash indicates the absence of effect.

After Maki and Mitchell, 1985. With permission of the *Bulletin of Marine Science.*

they do not reveal the exact ways in which adhesion and attachment in larvae and spores are carried out. However, knowledge of more specific and at the same time sufficiently universal mechanisms may be used both in devising protection from biofouling and in the mariculture industry.

One of the possible universal mechanisms is the interaction between lectins on the surface of the larvae of invertebrates (and maybe also macroalgal spores) and the carbohydrates of the bacterial–algal film (Michael and Smith, 1995). This mechanism was demonstrated for the first time in the larvae of polychaetes of the family Spirorbidae, selectively settling on bacterial films from the surface of the green alga *Ulva lobata*, on which adult worms live under natural conditions (Kirchman et al., 1982; Mitchell and Kirchman, 1984; Maki and Mitchell, 1985). According to the taxonomic revision by A.V. Rzhavsky (1991), these polychaetes should be placed in the genus *Neodexiospira*, preserving the former name *N. brasiliensis*, while the previously used name *Janua (Dexiospira) brasiliensis* should be considered invalid. This species, found in the Far East, is new to the Russian fauna.

Larvae of *N. brasiliensis* develop well on laboratory films of the bacterium *Pseudomonas marina*, with which the main experiments were carried out. Placing the larvae in solutions of D-glucose, D. Kirchman and R. Mitchell (1981) found that they stopped adhering to the bacterial film altogether. The process of fouling was also suppressed if the film was pre-treated with a solution of the lectin concanavalin A.

Studies carried out by these and other scientists (Kirchman and Mitchell, 1981; Kirchman et al., 1982; Mitchell and Kirchman, 1984; Maki and Mitchell, 1985; Mitchell and Maki, 1988) established the following facts (Table 6.2). The incubation of larvae in solutions of simple sugars other than D-glucose had almost no effect on the colonization processes. Only in the case of 2-deoxy-D-glucose was a 20% inhibition of metamorphosis observed. The peanut lectin, showing an affinity to galactose, did not inhibit either attachment or metamorphosis. Effective blockers of these processes were the protein-splitting enzyme trypsin and also periodate, oxidizing 1,2-dihydroxyl groups of carbohydrates.

It should be pointed out that lectins are complex glycoproteids that contain such non-protein components as carbohydrates and metal ions (Lutsik et al., 1981). Some of them, for instance concanavalin A (Con A), do not contain a carbohydrate part. The biological activity of lectins is determined by the presence of two binding sites of metal ions (usually calcium and manganese) and one or several sites of carbohydrate binding in their molecules. For example, the plant lectin Con A, which is widely used in experimental biology, may reversibly bind glucose and mannose. Preparations of plant lectins are routinely used in determining the human blood groups, since they interact with very specific sugar residues on the surface of erythrocytes, causing their agglutination into large clumps, which precipitate easily. Plant lectins are usually dimers or tetrameres, i.e., they consist of two or four identical subunits, whereas many lectins of animal origin are polymers (Lutsik et al., 1981). The structure of lectins in a number of invertebrates has been studied in great detail (see the review in Laenko et al., 1992).

The above data allowed D. Kirchman and R. Mitchell (1981) to formulate the following hypothesis. Lectins similar to concanavalin A are located on the surface of the larva of *Neodexiospira brasiliensis*. They interact by the "lock-and-key" scheme with D-glucose residues of polysaccharides or glycoproteids of the bacterial film, which leads to the attachment of the larva (incorrectly referred to as settlement by these authors) and the subsequent metamorphosis.

At the same time, the results of D. Kirchman, R. Mitchell, and other workers, considered above, leave a number of questions unanswered. They are connected with that part of the hypothesis in question that concerns lectins on the surface of the polychaete *N. brasiliensis*. The fact is they were not isolated and characterized. Still, the lectin–carbohydrate hypothesis seems quite plausible to me. First, lectins have been found and studied in many invertebrates (Laenko et al., 1992). Second, the bacterial–algal films on which larvae settle contain various carbohydrates and have free sugar residues on their surface (Sutherland, 1977; White, 1984; Bhosle et al., 1990; Michael and Smith, 1995). Third, D-glucose was reported to be the main sugar in the exopolysaccharides of the bacterium *Ps. marina*, which induces metamorphosis in *N. brasiliensis* (see Kirchman and Mitchell, 1981). Fourth, carbohydrate specificity in inhibiting metamorphosis was well proved (see Table 6.2). Thus, lectin–carbohydrate interactions between larvae of this polychaete and the bacterium *Ps. marina*, postulated in the hypothesis, seem to be sufficiently credible.

D. Kirchman and R. Mitchell (1981) consider that the mechanism of adhesion and induction of metamorphosis that they suggested is universal for all invertebrates. The general considerations presented above do not contradict such an idea. In support of their opinion, these authors present brief data (unfortunately not specific) on the fact that some simple sugars inhibit attachment and metamorphosis in three species of bryozoans of the genus *Bugula* (Maki and Mitchell, 1985). Settlement of the unicellular green alga *Dunaliella* sp. is stimulated by bacterial films and inhibited by concanavalin A, which suggests that this alga may also possess a lectin–carbohydrate mechanism of attachment (Mitchell and Kirchman, 1984).

TABLE 6.3
Effects of Chemical Treatment of Microfouling Films and Larvae of
Gonothyraea loveni **on Attachment and Metamorphosis of the Larvae**

Substances Used for Treatment of Film	Type of Effect	Substances Used for Treatment of Larvae	Type of Effect
Concanavalin A	Enhancement	Concanavalin A	Enhancement
Trypsin	Inhibition	Trypsin	—
Sodium periodate	Inhibition	Sodium periodate	
D-Glucose	Enhancement	D-Glucose	Inhibition
D-Xylose	—	D-Xylose	Inhibition
D-Galactose	—	D-Galactose	Enhancement
D-Fructose	Enhancement	D-Fructose	—
D-Mannose	—	D-Mannose	—
N-Acetyl-D-glucosamine	—	N-Acetyl-D-glucosamine	—
L-Fucose	—	L-Fucose	—
Blue dextran	Enhancement	Blue dextran	Inhibition

Note: Dash indicates the absence of effect.

After Chikadze and Railkin; compiled from data of 1992–1999.

One more argument in favor of the universality of the lectin–carbohydrate mechanism may be the data of K. Matsumura and colleagues (1998). To study the adhesion of cyprid larvae of *Balanus amphitrite* they used a new assay, associated with the surface sorption of the settlement factor (more accurately, the attachment factor) of adult barnacles. They found that when the attachment factor had been treated previously with lentil lectin, the adhesion of cyprids was inhibited. On the contrary, α-D-mannopyradosine removed this effect. These data suggest that the adhesion of *B. amphitrite* cyprids is also mediated by the lectin–carbohydrate mechanism.

The studies carried out by S.Z. Chikadze and myself (Chikadze and Railkin, 1992; Railkin and Chikadze, 1995, 1999) suggest that the lectin–carbohydrate mechanism is also highly probable for hydroid polyps. It was established on the *Gonothyraea loveni* planulae that the climax microfouling, consisting mainly of bacteria (99.2%) and diatoms (0.16%), induces the attachment of hydroids on the first day. Within 4 days, 100% of the larvae become attached to the microfouling film, whereas only about 40% attach to the initially sterile glass. By this time the attached bacteria appear on the glass, though in small numbers. The preliminary incubation of planulae in D-glucose solutions (from 0.1 to 2.5 mM) suppresses their attachment. The same, though a somewhat weaker, effect is revealed by xylose and blue dextran. Additional experiments with larvae and microfouling films that were treated with different sugars and the lectin concanavalin A show that the lectin–carbohydrate relations (Table 6.3) are more complicated than those following from the hypothesis of Kirchman and Mitchell.

To explain our data, one should suggest that there are bilateral lectin–carbohydrate interactions between the microbial film and the larvae attached to it, i.e., that the lectins and carbohydrates both may be present on the surface of larvae and on the biofilm. Indeed, the enhanced attachment and metamorphosis on biofilms treated with lectin may be accounted for in the following way. The lectin binds with one of its centers to a terminal sugar residue on the film, and with the other center it binds to another sugar on the surface of the larva. The enhancement of attachment reactions and metamorphosis on films treated with some sugars (Table 6.3) may be explained by the presence of lectins on the biofilm and sugars on the surface of larvae. Other analogous data may be interpreted in the same way.

The presence of lectins on the surface of bacteria and other microorganisms has been firmly proved (see the review in Uhlenbruck, 1987). The presence of carbohydrates on the surface of larvae leaves no doubt, since the surface of the plasmalemma of animal cells is known to bear an external glycoproteid layer referred to as the glycocalyx. Moreover, adhesion is one of the functions of the glycocalyx.

The hypothesis of bilateral lectin–carbohydrate interactions (Chikadze and Railkin, 1996) does not contradict that of one-sided interactions (Kirchman and Mitchell, 1981); on the contrary, it supplements it. Assuming the presence of this bilateral mechanism, it would be possible to explain the widespread occurrence of stimulation and induction of larval adhesion and metamorphosis by the microfouling films. The mechanism suggested reveals the structural basis of the biological adhesion of larvae and also shows that firm attachment to microfouling films is achieved as a result of a great number of molecular interactions between the microorganisms and larvae. The lectin–carbohydrate hypothesis supplemented by us makes it possible to explain the known cases of inhibition of larval attachment by bacterial films (see Section 5.5) by the impossibility of forming lectin–carbohydrate bonds between them.

Studies on binding of various plant (concanavalin A) and animal lectins (limulin, *Helix pomatia* lectin) involving marine algae have shown that their surfaces carry carbohydrates with different terminal sugar groups (Michael and Smith, 1995). This accounts for not only the specificity but also the diversity of the lectin–carbohydrate interactions. Data of the cited authors may serve as an additional argument in favor of the idea of the universal lectin–carbohydrate mechanism of adhesion of invertebrate larvae and algal spores in the formation of epibiotic relations.

6.5 GROWTH AND COLONIZATION OF THE HARD SURFACE

Growth is the final stage in the colonization of hard surfaces. It is a process that results in the increase of the biomass of settled and attached macroorganisms, and also the abundance and biomass of microorganisms. It is because of the growth processes that the colonized surface becomes overgrown in the literal sense of the word. Due to fast growth, especially during the initial period after settlement on a hard surface, macrofoulers reach maturity early. Reproduction and the appearance of new propagules lead to the extension of the colonization process to new territories and to the repetition of the whole colonization cycle (Figures 8.1 to 8.4), from ephemeral planktonic dispersal forms to long-living periphytonic organisms.

Due to growth on a hard surface, concentration of biomass takes place, considerably exceeding the biomass of the same groups on soft grounds in the benthos (see Section 1.2). One of the most important mechanisms of such concentrations is associated with the development of a multilayered structure of communities (Partaly, 2003). The growth of macrofoulers results in an increase in their surface, which can then be colonized by foulers of the second, third, and higher orders. This allows a still greater biomass to be concentrated on the initially limited area.

Let us briefly consider how the nutrition and growth of micro- and macrofoulers is carried out. Within the matrix of bacterial–algal film there is a system of microscopic channels (Costerton et al., 1995; O'Toole, 2000). Dissolved nutrients can be carried along these channels by water currents directly to the microorganisms. Another means of transport is diffusion through the matrix. It should be noted that microfouling film adsorbs ions, while enzymes are bound on its matrix and on the hard surface. Therefore, the film should be regarded not as a mere physical frame, but as a biochemically active system (Lock et al., 1984; Costerton et al., 1995; O'Toole, 2000). Substances diffusing through the film may react with its components in various ways. These features of the microfouling film evidently exert the greatest influence on the microorganisms inhabiting it, whereas its action on macroorganisms is less tangible.

The peculiarity of the nutrition of bacteria, and probably also unicellular fungi and diatoms, is the fact that they utilize organic molecules and ions adsorbed on the hard surface. Thus, they use nutrients and growth factors concentrated on the surface, even though their concentration in the surrounding water may be very low. Due to this, and also to the close contact with the food substrate, the metabolism and growth rates of bacteria on hard surfaces are many times higher than the growth rate of the same species feeding in a suspended state (Kjelleberg, 1984). This phenomenon was observed and partly accounted for by C.E. ZoBell (1946). The increase in the number of bacteria on a hard substrate is determined by growth (cell division) to a much greater extent than by accumulation (Punčochař, 1983).

Unicellular animals (protists) inhabiting hard substrates have various ways of feeding, the most common of which are sedimentation (also referred to as filtration) and predation. They are characteristic of both attached and motile forms (see, e.g., Burkovsky, 1984; Fenchel, 1987).

Many multicellular animals belonging to macrofouling communities are suspension feeders (Jørgensen, 1966). This group includes the animals that filter water with small food organisms (bacteria, unicellular algae, etc.) and detritus particles: sponges, polychaetes of the families Serpulidae and Sabellidae, cirripedes, bivalves, bryozoans, and ascidians. The propulsion of water through their filtering apparatus and the delivery of food into the digestive system may be carried out by water currents created by numerous cilia, flagella, or extremities. This mode of suspension feeding is often regarded as "active" filtration (see reviews in Jørgensen, 1966; Shunatova and Ostrovsky, 2001). "Passive" filtration is characteristic of deposit feeders (Jumars et al., 1982). In this case, food particles are captured from the flow by perioral structures. An important role in deposit feeding is played by the mucus, which facilitates the adhesion of particles to the perioral structures. At relatively high flow rates, a number of species may switch from active to passive filtration, as is the case with barnacles (Trager et al., 1990).

Many filter feeders are capable of sedimenting the suspended particles that have passed through them but were not assimilated, or, on the contrary, were rejected. The undigested particles are glued together by the mucus into large lumps, which then sink to the bottom. Sedimentation and filtration are of great importance in the process of removing suspended particles and pathogenic microorganisms from water, a process that usually is performed by invertebrates and ascidians (see the reviews in Jørgensen, 1966; Jumars et al., 1982; Riisgård and Larsen, 1995; Shunatova and Ostrovsky, 2001). Sessile animals inhabiting hard substrates, such as ciliates, heterotrophic flagellates (Fenchel, 1987), sponges (Reiswig, 1971; Thomassen and Riisgård, 1995, Witte et al., 1997), polychaetes (Christensen et al., 2000), cirripedes (Trager et al., 1990, 1994), mollusks (Okamura, 1990; Kulakowski, 2000), and bryozoans (Winston, 1977; Best and Thorpe, 1994; Riisgård and Goldson, 1997; Riisgård and Manríquez, 1997; Shunatova and Ostrovsky, 2001) are capable of fast and effective mechanical cleaning of vast quantities of water. Mass species of filter feeders form real "biofilter stations," which may extend for many kilometers along the shore (Gili and Coma, 1998), such as, for example, blue mussel beds (see Figure 5.2). Macroalgae can chemically purify water by releasing strong oxidants which destroy toxic compounds that are not easily decomposed by microorganisms (Skurlatov et al., 1994; Abrahamsson et al., 1995; Steinberg et al., 1998).

Another form of feeding that is widespread in macrofouling communities is *predation*, i.e., the capture and ingestion of prey that are larger than in the case of suspension feeding. Various forms of predation are observed in hydroids, a number of coral polyps, polychaetes (except the families Serpulidae and Sabellidae), and starfish.

Feeding and growth rates are affected by different environmental factors, the most important of which are temperature, concentration of food, and the current (Odum, 1983; Green et al., 1988; Khailov et al., 1992; Begon et al., 1996). This was demonstrated on both attached microorganisms (Zhdan-Pushkina, 1983) and macroorganisms (Turpaeva, 1987a). The influence of temperature on feeding and growth is not direct but is mediated by the rate of food transport into the organism and its metabolism. As a rule, growth accelerates as the concentration of food in the medium, including dissolved nutrients, increases. In macroorganisms, this is manifested by the growth of body size and mass; in microorganisms it is manifested by the elevated rate of cell division and consequently in the growth of biomass and the number of cells on the surface. At artificially increased concentrations of food the feeding rate usually decreases.

The influence of the current on feeding and growth is more complicated and diverse. In still water, a zone of low concentration of nutrients is formed around the attached bacteria and algae as a result of the food being absorbed by the organisms from the adjacent layer. Replenishment of nutrients, growth factors, and biogens in this impoverished layer occurs owing to diffusion. If there is flow of water around the organisms, the diffusion-related restrictions on the feeding of bacteria and algae are eliminated, since the thickness of the depleted layer diminishes progressively as the flow rate increases. In attached organisms, the mass exchange due to flow is several orders of magnitude higher than that due to diffusion (Schlichting, 1979). The rate of feeding and, consequently, growth increases, since the actual concentration of food around the bacteria and algae increases. These ideas were suggested for the first time and experimentally substantiated by L. Whitford (1960), and later developed by other workers (e.g., Pertsov and Vilenkin, 1977; Fréchétte et al., 1989; Khailov et al., 1992).

The diffusion constraints discussed above can also be applied to the feeding and growth of large filter feeders. In this case, though, they are manifested on a macro scale instead of on a micro scale. In particular, the feeding rate in the bivalves *Mytilus edulis* is limited by the abundance of phytoplankton, on which they mainly feed, in the 1-m near-bottom water layer (Fréchétte et al., 1989). As the current velocity increases, the vertical mixing of the water becomes stronger. This reduces the thickness of the depleted layer and partly removes the diffusion-related restrictions, thus increasing the feeding rate.

The structure of the near-bottom boundary layer remains little studied. Studies performed in the White Sea (Babkov and Golikov, 1984) revealed a three-layer vertical structure of the water: the layer of wind mixing, the layer of tidal mixing, and finally the near-bottom layer. The turbulent oscillations are predominantly determined by the bottom communities and non-living objects (Kazar'yan et al., 2003). Studies of the near-bottom layer have shown that the most significant turbulent oscillations are observed within a 0.4-m-thick layer adjacent to the bottom. This suggests that food depletion in the near-bottom layer may not be as serious a problem for benthic organisms as it is commonly assumed.

The positive influence of the current on growth (and feeding) has been demonstrated in bacteria (Zhdan-Pushkina, 1983), unicellular algae (Ivlev, 1933; Whitford, 1960; McIntire, 1968; Stevenson, 1983, 1984; Railkin, 1991, etc.), and multicellular algae (Whitford, 1960; Khailov et al., 1992, 1995). It has been documented for various species of protists (Ivlev, 1933; Dovgal, 1990) as well as for multicellular invertebrates inhabiting hard surfaces, such as hydroids (Marfenin, 1984, 1993b), barnacles (Vilenkin et al., 1984), bivalves (Wildish and Kristmanson, 1985), and bryozoans (Okamura, 1990).

It is well known that the growth rate is higher on artificial substrates in the water column than in the same species inhabiting natural substrates on the bottom (Zevina, 1994). Therefore, under mariculture conditions, mussels (see, e.g., Sukhotin and Maximovich, 1994; Kulakowski, 2000) and macroalgae (South and Whittick, 1987) grow faster and reach biomasses that are several times greater than in the bottom communities, where they live in large aggregations. How is it possible to account for these well-known facts, which are of such practical importance?

Based on the above data, let us try to understand the main reasons for the higher productivity of attached macroorganisms on artificial substrates. First, marifarms are situated in coastal areas. The cultured species are grown on special collectors, usually placed within the well-lit and well-heated surface layer. The water temperature in this layer is higher and the illumination is greater than at the bottom. Therefore, marifarms provide more favorable conditions for the feeding and growth of both algae and invertebrates. Second, the concentration of phyto- and bacterioplankton, on which the mussels feed, as well as biogens, which are needed for algal growth, is higher near the surface than it is close to the bottom. This also gives the cultured organisms an advantage in feeding and growth. Third, the current velocity is higher in the surface layer. This provides better conditions for the supply of food, oxygen for respiration, and carbon dioxide for photosynthesis. In addition, fast-flowing water effectively removes the excreta. Fourth, in accordance with the above, the communities of hard substrates are dominated by suspension feeders, whereas predatory forms are not abundant. The ratio of predators to suspension feeders appears to be greater on the bottom (on soft and hard grounds) than on hard substrates in the water column. Summing up, it can be said that the reasons for the elevated productivity of species under mariculture conditions are not associated with the artificial substrate *per se*. Instead, they are due to the positive influence exerted on the feeding and growth of organisms by certain environmental conditions, such as a more optimal temperature, light, hydrodynamic, and trophic regimes, as well as reduced predation.

Of course, it would be incorrect to think that increasing the current speed would inevitably result in better feeding and growth. In reality, this is true only within a certain range, with its upper limit determined by some critical value of current speed, which varies depending on the species. Therefore, where the critical current velocity is exceeded, the feeding and growth rates decrease. This may be accounted for by both physical and biological reasons, or a combination of the two. For instance, a reduction and even a cessation of feeding and growth in colonial hydroids at current rates above 0.3 to 0.5 m/s can be largely accounted for by mechanical reasons: stretching of the whole colony along the flow and compression of its branches, and also a drastic reduction of the probability of food sticking to the tentacles as a result of the contraction of hydranths (Marfenin, 1984, 1993b). The giant scallops *Placopecten magellanicus* close their shells and stop feeding and growing when the flow speed is too high for them (0.14 m/s) (Wildish and Saulnier, 1992). Studies performed in different areas of the world (Yakubenko and Shcherbakova, 1981) have shown that the biomass of fouling on the inner walls of ship pipelines increases with the flow rate, reaching a maximum value at 0.1 to 0.5 m/s (Figure 6.14). An analysis that will be performed in Chapter 7 makes it possible to assume that the increase in biomass is caused by an intensification of the accumulation (settlement and attachment) and growth of organisms on hard substrates when the current speed increases. On the contrary, a reduction in the fouling biomass at current speeds that exceed the above values, which can be considered as critical values for different species, is probably caused by a violation of adhesion and feeding conditions (see Sections 7.1 and 7.2). Fouling usually is not possible at current velocities greater than 1.5 m/s. Similar values of the critical and limiting current rates are known for accumulation, feeding, and growth on various types of hard substrates for both microfoulers (Ivlev, 1933; Patrick, 1948; McIntire, 1968; Stevenson, 1983, 1984) and macrofoulers (Crisp and Barnes, 1954; Crisp, 1984; Trager et al., 1990, 1994; Pawlik and Butman, 1993; Thomason et al., 1998). However, the maximum current rates at which settlement and attachment are still possible for motile spores of the green alga *Enteromorpha* sp. (Houghton et al., 1972) and cyprid larvae of the goose barnacle *Conchoderma* (Dalley and Crisp, 1981) are considerably higher than those mentioned above and amount to 5.5 and 7.0 m/s, respectively. The maximum current

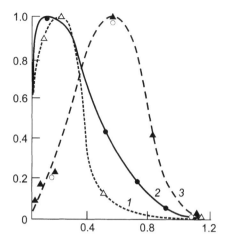

FIGURE 6.14 Relation between the biomass of fouling of inner walls of ship piping (2 cm in diameter) and the flow rate. Abscissa – flow rate in the pipes, m/s; ordinate – fouling biomass in relation to the maximum value. (1) the Black Sea, 25 days; (2) the Gulf of Guinea, the Atlantic Ocean, 25 days; (3) open region of the Atlantic Ocean, 43 days. (After Yakubenko and Scherbakova, 1981. With permission of Sudostroenie, St. Petersburg.)

rates for different species may certainly differ, and sometimes considerably; nevertheless, the above analysis of the literature suggests that the current rate range of 0.1 to 0.5 m/s is indeed optimal for the development of hard-substrate communities of most of the species studied, and for many of them the range can be narrowed down to 0.2 to 0.5 m/s.

The population growth of microorganisms (Bryers and Characklis, 1982; Odum, 1983) and the individual growth of macroorganisms (Green et al., 1988) after their settlement on a surface are subject to similar regularities and are usually described by an S-shaped curve (Figure 6.15) or, more rarely, by a J-shaped curve, when no deceleration of growth occurs. The S-like growth is known to include a number of stages that differ in the rate of growth: (1) the lag phase or slow growth, (2) the log phase or exponential phase of the fastest growth, (3) the phase of decreasing growth rate, and (4) the plateau phase (Figure 6.15).

Settlement on a surface is usually accompanied by rapid growth. For example, the shell length of a young mussel *Mytilus edulis* increases 100-fold during 80 days of growth on a suspended collector under the mariculture conditions (Buyanovsky and Kulikova, 1984)! The same high growth rate during the initial period of growth on a hard surface is demonstrated by the barnacle *Balanus improvisus* (Kashin and Kuznetsova, 1985). During the first 8 days after metamorphosis, its linear size, which is estimated by the rostrocarinal diameter, increases 14-fold (Figure 6.15b). The higher rate of growth on hard substrates in the initial period is also observed in other invertebrates: protists and hydroids (Turpaeva, 1987a), sponges (Osinga et al., 1998), corals (Lam, 2000), cirripedes (Lignau, 1924–1925; Barnes and Powell, 1953; Hines, 1979; Turpaeva, 1987a), mollusks (Kulakowski, 1987, 2000; Sukhotin and Maximovich, 1994), and bryozoans (Lignau, 1924–1925; Winston, 1977; Denisenko, 1988). A review of this question was presented by V. E. Zaika (1985).

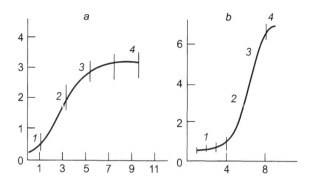

FIGURE 6.15 S-like growth in invertebrates. (a) Population growth in vorticellid ciliates (after Turpaeva, 1987a; with permission of the Oceanological Institute of P.P. Shirshov, Moscow); (b) individual growth in the barnacle *Balanus improvisus* (after Kashin and Kuznetsova, 1985; with permission of the Russian journal *Okeanologiya*). (1) lag-phase, (2) logarithmic phase, (3) decreasing rate phase, (4) plateau phase. Abscissa – time, days; ordinate – a: abundance of ciliates, $\times 10^3$ cells/cm^2, b: shell size, mm.

In prothalli and young individuals of *Laminaria*, fast growth is observed both under natural conditions and in mariculture (Kain, 1979). The growth rate is about 1 cm/day in *Laminaria hyperborea* and more than 10 cm/day in *L. angustata var. longissima*. The higher rate of growth during the initial period of substrate colonization is also known for other macroalgae (Khailov et al., 1992).

The main ship foulers also show the highest growth rate after settlement and attachment to the ship's hull within the first 1.5 months (Rudyakova, 1981). It is during this period that a sharp rise in the water resistance is observed.

In solitary species, the growth rate decreases with size and age. In colonial species, growth is not necessarily restricted by age and usually does not slow down (Jackson, 1977a; Marfenin, 1993a). This is evidently associated with a modular organization of a colony, in which the growth is manifested not in the enlargement of some body parts (with their proportions changing or being preserved), as in the case of an individual organism, but in the augmentation of the number of similar modules (parts of a colony). Of course, the almost unlimited growth of a colony depends on environmental conditions. Under unfavorable conditions, the greater part of a colony of hydroid polyps can be resorbed and decrease drastically both in linear size and in biomass. However, rapid growth is observed when the conditions improve.

One of the consequences of macroorganism growth is the increase of the hard surface area created by the foulers themselves. This results in the subsequent settlers being able to settle and become attached to those that settled earlier. Thus, settlers of the second, third, and higher orders appear on the surface. As a result of the progressive colonization of the surface, communities with a complex spatial structure, sometimes comprising as many as nine layers, are formed (Partaly, 2003). The appearance of a multilayered structure probably results in more complicated quantitative regularities of the community growth. The biomass of a new layer may be growing exponentially (log phase of the S-like growth curve; see Figure 6.15) when the preceding layer has already entered the plateau phase or even the subsequent suppression phase. Therefore, it is quite probable that the growth of multilayered communities cannot be described in general by an S-like curve. It also should be suggested that certain models of individual growth (see, e.g., Zaika, 1985; Urban, 2002) are not applicable to the generalized description of unlimited colonial growth.

Growth is a result of the rapid colonization of the surface by foulers, especially those growing along it (the so-called lateral growth), such as sponges, hydroids, corals, bryozoans, and also colonial ascidians (e.g., Sebens, 1985b; Paine, 1994). In bryozoans, thin sheet colonies are more susceptible to overgrowth than are arborescent ones. Therefore, with all other conditions being equal, the result of competition between them depends on the texture of the algal thallus on which they live. Arborescent forms have an advantage on flat surfaces, whereas sheet colonies win on elevated parts (Walters and Wethey, 1986, 1991). The compound ascidian *Aplidium* grows over other attached organisms (except the sea anemone *Metridium*), winning the overall competition (Sebens, 1985b).

As a rule, colonial species are more resistant to competition and readily overgrow the solitary forms. This is connected with a number of biological features of the colonial forms, primarily with their high fecundity and rapid growth, the rate of which often shows no decrease with age (Jackson, 1977a). Therefore, the area of a colony tends to increase exponentially with time. Non-colonial species that have a high biotic potential or settle in aggregations, such as serpulid polychaetes, barnacles, oysters, mussels, and sea urchins, may completely cover with their bodies the surface and the other organisms inhabiting it, and not infrequently cause their elimination (e.g., Jackson, 1977a; Sebens, 1985a, 1985b; Paine, 1994).

A comparative study carried out by G. R. Russ (1982) in Port Phillip Bay (Australia) allowed him to characterize the relative ability of animals to overgrow other species. In descending order of this ability, referred to as the *epibiotic potential*, animals form the following series: ascidians, sponges, bryozoans, cirripedes, polychaetes, and hydroids. Thus, according to this author's data, which are frequently cited, ascidians can overgrow organisms of all other groups whereas hydroids are, on the contrary, more subject to being overgrown. Basibionts may be resistant to the species growing on them. Consequently, the competition between species does not follow such a simple linear scheme as that observed for the epibiotic hierarchy.

The result of intra- and interspecific competition for space often depends on the relative growth rates of the overgrowing and overgrown organisms (e.g., Paine, 1994; Barnes and Dick, 2000). Of great importance is also the presence of chemical and physical protection against epibionts (see Section 10.1). Among the attached foulers from various groups characterized by a hard exoskeleton (polychaetes, mollusks, bryozoans, and ascidians), there is a clear tendency toward an upward elongation of body, which allows them to reduce competition for space and probably also trophic competition (Sebens, 1985b). This feature was observed earlier by the Russian scientist S. A. Zernov (1949) and called "the principle of upward ascendance of attached forms." Some other authors (e.g., Jackson, 1977a), who observed this phenomenon independently, referred to it as "escape to size."

The growth mechanisms of foulers considered in this chapter facilitate the concentration of vast biomasses on hard substrates. This is an important contribution to maintaining high species diversity and also the production of such communities, especially in the coastal area and in mariculture. On the other hand, since the concentration of vast biomasses of foulers is accomplished by growth processes, it may be assumed that protection of man-made structures from biofouling is best achieved at the stages of settlement and attachment, preceding growth (see Chapter 10).

7 Fundamentals of the Quantitative Theory of Colonization

7.1 MATHEMATICAL MODELS OF ACCUMULATION

The quantitative aspects of colonization of hard surfaces and the concentration of organisms on them are difficult to explain without employing mathematical models. Such models may be useful for predicting the development of biofouling and for analyzing the measures of antifouling protection (see Chapter 11).

The accumulation and growth of foulers can be described in terms of statistical (Gorbenko and Knyshev, 1985), dynamic (Characklis, 1984; Kent, 1988; Eckman, 1990), allometric (Zaika, 1985), and other (Caldwell, 1984; Stevenson, 1986; Zainullin, 1992) models. The common drawback of all these models is the great number of unknown parameters, the changing of which may yield virtually any result. According to V. D. Braiko (1985), the models should include the following easily measurable parameters: the abundance of larvae in the plankton, the current velocity, the volume of water flowing past the surface during a certain time interval, the size (area) of the surface, the duration of the fouling process, and others.

Since colonization usually takes place in moving water, it is essential to consider some concepts of hydrodynamics before proceeding to colonization models. According to the boundary layer theory (Schlichting, 1979), the velocity of the water flow near the surface decreases as a result of the friction between the flowing water and the surface. Therefore, a layer of slowed-down fluid is formed around any object that is placed in a flow. For example, near a flat plate this layer increases in thickness from the leading edge (i.e., the edge exposed to the flow, also termed the *upstream edge*) to the trailing (*downstream*) edge, whereas the flow along the surface decreases gradually owing to the slowing down of the fluid (Figure 7.1).

The thickness of the slowed-down layer, termed the *boundary layer* in hydrodynamics, depends on the flow velocity, its characteristics (laminar or turbulent), the shape of the body, the texture (smoothness) of its surface, the direction of flow, and the viscosity of the water (which, in its turn, depends on the temperature and salinity) (Vogel, 1981). Calculations based on the approximation formula (Equation 7.8 below) (Schlichting, 1979) show that, in a 0.2-m/s laminar flow of sea water along a smooth, thin (no more than 0.2 mm thick), and infinitely long plate, the boundary layer thickness is 0.43 mm at a distance of 1 mm from the leading edge, 1.4 mm at a distance of 1 cm, 14 mm at a distance of 1 m, and 140 mm at 100 m from the leading edge; in these calculations, the kinematic viscosity was assumed

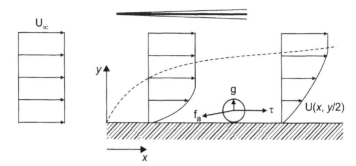

FIGURE 7.1 Distribution of flow velocity in the boundary layer of a smooth flat plate positioned parallel to the flow (above).

to be 1.5×10^{-6} m^2/s. As the flow velocity decreases by n times, the thickness of the boundary layer increases by a factor of $\sqrt{n}$. The turbulent stagnated layer is thicker than the laminar layer.

These estimations of the boundary layer thickness are in good agreement with the known size range of propagules, the largest of which being the larvae of invertebrates. The size of settling larvae usually does not exceed 1 to 2 mm and often is below 0.5 mm (Fraser, 1962). It should also be borne in mind that, for hydrodynamic reasons, the most probable position of larvae (and other propagules) during attachment in the water flow is parallel to the surface or at some angle to it. Therefore, it may be quite reasonably assumed that the settlement and attachment of propagules (microorganisms, larvae, and macroalgal spores) proceed within the stagnated boundary layer that always exists near the hard surface. The shape of the propagules also plays an important role. In particular, spherical and elongate symmetrical models imitating the larvae were found to rotate in the flow, whereas asymmetrical elongate models assumed a stable position parallel to the flow vector (Abelson and Denny, 1997).

The currents and other water flows in seas and oceans are nearly always characterized by some degree of turbulence owing to the wind, the non-uniform heating of water, tides, bottom relief, and other factors (see, e.g., Abelson and Denny, 1997). Contrary to what is observed in the laminar flow, the velocity and direction of the current in the turbulent flow change in time. The turbulence in seas and oceans is isotropic within a certain range of dimensions (Ozmidov, 1968). One of these zones of isotropy is 0.01 to 1 m, which corresponds to the size of the experimental plates used in the fouling studies.

It should be mentioned that the turbulent boundary layer is always underlain by a thin viscous sublayer, which is characterized by strongly smoothed turbulent pulsations and has virtually the properties as a laminar-flow layer (Schlichting, 1979). Even though propagules settle in a turbulent flow, their very contact with a surface almost always takes place within this viscous sublayer. Therefore, for the sake of simplicity, we shall consider biofouling in a laminar flow around a surface. The variant of a turbulent flow would be more realistic, but it would make the description much more sophisticated. As will be shown in this chapter, the laminar models of colonization agree with many experimental facts and may be considered satisfactory within the framework of this study.

The swimming velocity of propagules (microorganisms, larvae of invertebrates, and spores of macroalgae) is usually one to three orders of magnitude lower than that of the current in the seas and oceans (see Chapter 3). Therefore, with a certain approximation, it may be assumed that the horizontal transport of propagules, including their movement along a hard surface, is accomplished only by the current.

Let y be the linear dimension of the propagule, with its shape neglected for simplicity. The x-axis in this model is directed along the flow. The current velocity at an arbitrary point of a boundary layer with the coordinates $(x, y/2)$ is designated by $U(x, y/2)$, or simply U. With these assumptions, any propagule transported by the current along the surface will be moving at a velocity U along the x-axis, at the point $(x, y/2)$. Let us designate the concentration and biomass of the propagules of the i-th species per unit volume, referred to in hydrobiology as the abundance and biomass of that species in the plankton, as c and b, respectively. We shall also assume that the spatial distribution of locomotion activity of the propagules does not change when they get into the boundary layer. Let α be the proportion of organisms moving toward the surface, with $0 \leq \alpha \leq 1$. The coefficient α characterizes the selectivity of settlement in motile propagules: it can be equal to 1 in the case of an attractant surface and 0 in the case of a repellent surface. For passively settling (sedimentating) forms, $\alpha = 1$. It is assumed that ascending flows are absent. In this model, the abundance and biomass of the i-th species that will be transported toward the surface through the unit area per unit time in the direction x are given by the following equations:

Abundance:

$$Q_{c,x} = \alpha \, c \, U \, ; \tag{7.1}$$

Biomass:

$$Q_{b,x} = \alpha \, b \, U \, . \tag{7.2}$$

Let us assume that propagules only move in the vertical direction under their own locomotion. The transversal flow velocity in the boundary layer can be neglected as compared to the velocity of propagules. Let V be the velocity of swimming or sedimentation of the propagules of the i-th species. Then the quantity of that species settling onto the surface through the unit area per unit time along the y-axis will be given by the following equations.

Abundance:

$$Q_{c,y} = \alpha \, c \, V \, ; \tag{7.3}$$

Biomass:

$$Q_{b,y} = \alpha \, b \, V \, . \tag{7.4}$$

It is necessary to emphasize that fouling is possible not at any velocity of the current. The reason for this is shearing stress, τ, which results from the gradient of flow velocity in the boundary layer, as a function of distance from the surface. It affects the propagules by preventing their attachment. The flow velocity is 0 at the surface and virtually equal to the free-stream velocity at the outer boundary of the slowed-down layer. Thus, the shearing stress increases normally to the surface and decreases tangentially to it in the direction of the flow. Because of this, small-sized propagules, for example microorganisms and minute macroalgal spores, are more likely to adhere to the surface in a flow than large invertebrate larvae.

In the previous chapter (see Section 6.5), we considered the problem of the critical value of current velocity as it concerns colonization. Although the critical flow velocity is species-specific, in many cases it is about 0.1 to 0.5 m/s. At a flow velocity that exceeds the critical value, the biomass of fouling diminishes drastically because of the augmentation of the shearing stress within the boundary layer and the violation of the conditions necessary for adhesion. Finally, at a flow velocity greater than 1.5 m/s, fouling usually does not occur at all (see Figure 6.14).

Of course, the ability of propagules to remain on the surface in the water flow does not depend only on their dimensions and the current velocity; to a considerable extent it is determined by the adhesion of their integuments to the surface. Let the buoyancy of propagules be near-neutral. The possibility of adhesion and remaining on the hard surface during the critical period of initial contact with the surface is determined mostly by the resultant of three forces acting within the boundary layer: the adhesion force f_a, the shearing force τ, and the lift g (Figure 7.1). The last force is determined by the gradient of pressure directed normally to the surface. Thus, for the fouling of any object, the following inequality is true:

$$f_a \geq \tau + g. \tag{7.5}$$

Since the shearing stress is usually much greater than the lift force ($\tau \gg g$), the latter can be neglected in many cases, which results in the inequality $f_a \geq \tau$. Formula 7.5 defines one of the main conditions of fouling of any hard substrate in the water flow.

It is evident that only a very small portion of the propagules that come into contact with a fouling-free surface adheres to it. The proportion of foulers that, upon coming into contact with a surface, have succeeded to retain a hold on it is designated as the coefficient of adhesion K_a. Theoretically, $0 \leq K_a \leq 1$, but in fast-flowing water the value of this coefficient is close to 0. According to the results of field experiments that I performed in the White Sea coastal area (Railkin, unpublished data), in which diatom fouling was observed on the upper side of microscope slides placed in the water flow ($K_a \ll 1$) or protected from direct flow in a transparent box open from above and from below ($K_a = 1$), the estimated coefficient of adhesion of various diatoms in the flow (average velocity 4 cm/s, maximum velocity 20 cm/s) varied but did not exceed 0.0002 to 0.01.

Considering K_a, the number of propagules of the i-th species that are being transported by the current, settling, and attaching at an arbitrary point of the surface owing to adhesion forces during the period t will be determined by the following equations,

Abundance:

$$Q_c = Q_{c,x} + Q_{c,y} = \alpha K_a\, c\, (U + V)\, t. \tag{7.6}$$

Biomass:

$$Q_b = Q_{b,x} + Q_{b,y} = \alpha K_a\, b\, (U + V)\, t. \tag{7.7}$$

These are the general local equations that describe the accumulation of the i-th species on a surface placed in a laminar flow. They hold for any fouled surface and characterize an arbitrary point within the boundary layer.

In order to determine the mean values of fouling abundance and biomass from these equations, one must know how the current velocity U at an arbitrary point within the boundary layer depends on its coordinates and on the free-stream velocity U_∞. If the abundance or biomass values for all points of the surface are calculated, their sum will characterize the abundance of the i-th species on the whole surface. Furthermore, the abundance (biomass) of the i-th species per unit area will be determined as this sum related to the total surface area whereas summation by i species will produce the total abundance (biomass) of the community.

For example, let us consider a surface in a tangential flow. Within the flow velocity range observed in the seas and oceans, the inertia forces in a flow prevail over friction forces; in other words, fouling proceeds at large Reynolds number values. In the case of a laminar flow parallel to a flat, thin (no thicker than 0.2 mm), and smooth plate, the boundary layer thickness δ will be approximately equal to

$$\delta = 5\sqrt{\gamma\, x\, /\, U_\infty} \tag{7.8}$$

(Schlichting, 1979), with γ being the kinematic viscosity of water, x the distance from the leading edge of the plate (i.e., that directed to the flow), and U_∞ the velocity of the approach flow outside the boundary layer (the so-called potential flow velocity, hereinafter referred to as the current velocity).

Using the known Blasius formula for flow velocity distribution within a boundary layer over a plate in a parallel flow (Schlichting, 1979), let us select a rectilinear area for which

$$U / U_\infty = 5y/3\delta, \tag{7.9}$$

where y is the distance from the plate, $y \leq \delta/3$. Substituting the expression for δ from Equation 7.8 in Equation 7.9, we obtain

$$U = (y/3)\sqrt{U_\infty^3/\gamma x} . \tag{7.10}$$

A similar dependence can be derived from formula (4) in D. J. Crisp's work (1955).

Finally, a propagule of size y at the point $(x, y/2)$ of the boundary layer will be transported by the flow at the velocity

$$U = (y/6)\sqrt{U_\infty^3/\gamma x} . \tag{7.11}$$

Substituting the expression in Equation 7.11 into Equations 7.6 and 7.7, we obtain the equation of fouling by propagules of the i-th species for an arbitrary point of a flat plate placed in a longitudinal flow:

by abundance

$$Q_c = \alpha K_a c[(y/6)\sqrt{U_\infty^3/\gamma x} + V]t, \tag{7.12}$$

by biomass

$$Q_b = \alpha K_a b[(y/6)\sqrt{U_\infty^3/\gamma x} + V]t \tag{7.13}$$

Then, for a plate with width a, length l, and area $S = a \cdot l$, propagules of the i-th species will have the average abundance

$$Q_c / S = 1/S \int_0^{z=a} \int_0^{x=l} Q_c(x)\,dz\,dx = a/S \int_0^{x=l} Q_c(x)\,dx = \alpha K_a c[(y/3)\sqrt{U_\infty^3/\gamma l} + V]t \tag{7.14}$$

and the biomass

$$Q_b / S = K_a \alpha b[(y/3)\sqrt{U_\infty^3/\gamma l} + V]t . \tag{7.15}$$

If the plate is fouled by n species, the average biomass can be determined as the sum of the individual mean biomass values.

It should be noted that the model of accumulation describes the processes of biofouling preceding growth, and, therefore, it holds true only at small values of t, before growth starts. Additionally, this simplified model does not take into account the spatial orientation of a fouled surface.

Let us determine the adhesion force needed to keep a propagule attached to a plate in a parallel laminar flow, provided that this propagule is located entirely within the boundary layer. It is known (Schlichting, 1979) that the shear stress at an arbitrary point $(x, y/2)$ is

$$\tau = \mu \, \partial U(x, y/2) / \partial y \,, \tag{7.16}$$

with μ being the dynamic viscosity of water, or simply viscosity.

Let us substitute the expression for U from Equation 7.11 into Equation 7.16, putting $\gamma = \mu / \rho$, where ρ is the density of water. It should be noted that the density of the propagule is close to ρ. Taking into consideration the fact that the lift force g in Equation 7.5 can be neglected (see comments above), we obtain from Equation 7.16 the condition of physical adhesion of a propagule to the plate by taking the partial derivative by y:

$$f_a \geq (\rho / 6) \sqrt{\gamma U_\infty^3 / x} \,. \tag{7.17}$$

An equivalent expression can be obtained by substituting the expression $\sqrt{U_\infty^3 / x}$ from Equation 7.11 into the inequality 7.17:

$$f_a \geq \mu U / y \,. \tag{7.18}$$

The equality in Equation 7.17 and Equation 7.18 is reached at some critical value of $U_{\infty cr}$ (U_{cr}), which must, however, comply with the condition of fouling in Equation 7.5, namely, $f_a = \tau$. If this condition is met, then

$$f_a = (\rho / 6) \sqrt{\gamma U_{\infty cr}^3 / x} \,, \tag{7.19}$$

or

$$f_a = \mu U_{cr} / y \,. \tag{7.20}$$

The shearing force may be estimated using Equations 7.19 and 7.20, as in the following example. For sea water, putting $\mu = 1.5 \cdot 10^{-2}$ g/cm·s and $y = 0.1$ cm, the value $U_{cr} = 19$ cm/s at a distance of 1 cm from the leading edge ($U_{\infty cr} - 20$ cm/s) will result in $f_a = 3$ g, which corresponds well to the force needed for the detachment of a cyprid larva (see Table 6.1).

Taking into account the fact that the maximum flow velocity at which biofouling is still possible is $U_{\infty cr}$ and using Equations 7.14 and 7.15, it is easy to obtain mean estimates of the maximum abundance and biomass for the i-th species:

$$Q_c / S_{max} = \alpha K_a c [(y/3)\sqrt{U_{\infty cr}^3 / \gamma l} + V] t \, , \qquad (7.21)$$

$$Q_b / S_{max} = \alpha K_a b [(y/3)\sqrt{U_{\infty cr}^3 / \gamma l} + V] t \, . \qquad (7.22)$$

Let us determine the necessary and sufficient conditions for fouling of an unprotected hard surface. The biological features of colonization (considered in Chapters 2 to 6) suggest that all of the elementary processes dealt with above, such as transport by flow, settlement, adhesion, attachment, development, and growth, are necessary for biofouling to occur. It is their regular sequence that determines biofouling and the concentration of organisms on hard substrates.

However, are all the processes mentioned really mandatory for fouling to occur? An analysis of the mathematical models gives a negative answer. The current brings the propagules to hard surfaces. Yet contact with a surface may also be achieved with no current, since the propagules settle from the plankton on their own. The crucial moment of transition to periphytonic life is adhesion, which marks the transition from the plankton to the hard surface. Contact with the surface and adherence to it induce a chain of consequent events: induction of the release of adhesive secretions onto the surface, stronger attachment, development (metamorphosis), and the subsequent growth of organisms.

Thus, firm adhesion to a hard surface is a sufficient condition for colonization. Of course, this assertion can be true only if the surface is not toxic and the environment is favorable for the development and growth of a particular fouler. The extremely important role of adhesion naturally follows from an analysis of Equations 7.6, 7.7, 7.14, and 7.15: the arrival of propagules with the flow and their settlement on a surface result in its colonization only if the propagules attach to it due to adhesion forces, i.e., if $K_a > 0$. The significant role of adhesion and temporal attachment in the transition from the plankton to the hard surface was considered theoretically in the literature (see the reviews in Young and Crisp, 1982; Abelson and Denny, 1997).

In my judgment, for biofouling to take place, the adhesion force in all points of a surface must be significantly greater than the shearing stress and the lift force combined:

$$f_a \gg \tau + g \, . \qquad (7.23)$$

This is the main condition of biofouling expressed in mathematical form. It holds true with respect to the colonization of any natural (both living and non-living) surface as well as surfaces of technical objects.

The combined analysis of Equations 7.6, 7.7, 7.14, and 7.15 and the biological data presented in Chapters 3 to 6 allow one to make the following conclusions. At the accumulation stage in flowing water, the colonization of hard surfaces by foulers occurs predominantly due to their transport by marine currents within the boundary layer, because $U \gg V$. In the general case, the impact of the horizontal and vertical flows of foulers on their accumulation on an arbitrary surface is determined by the ratio of the rates of their transport on the horizontal and vertical planes, i.e., by the ratio U/V. On the contrary, in the absence of flow, for example, during the short period of slack water, accumulation results from the active migration of propagules, as was established by observations and experiments (Orlov and Marfenin, 1993; Orlov, 1996b; Belorustseva and Marfenin, 2002). The key condition of biofouling is not transport by the current or settlement of propagules by themselves, but the combination of transport by the current, settlement, and adhesion. In short, biofouling is impossible without adhesion. Thus, the most important conditions of accumulation of propagules are their contact with the hard surface and their attachment to it. Adhesion should be regarded as a sufficient condition of accumulation, provided that it meets the conditions in Equations 7.5, 7.19, and 7.20.

Let us consider some other consequences of the accumulation models. It is well known (see, e.g., Gaines et al., 1985; Dobretsov and Miron, 2001) that the abundance of postlarval and juvenile forms on the substrates is proportional to that of larvae in the plankton during the presettlement period, which agrees with the models.

With all other conditions being equal, as reflected in Equations 7.21 and 7.22, smaller forms have an advantage at the accumulation stage because they settle at lower flow velocities near the surface.

The nonlinear dependence of fouling abundance on the current velocity (according to the model, in the degree 3/2), considered in Section 6.5, can be demonstrated by the left parts of the curves shown in Figure 6.14. The abundance of fouling in a moderate current (at near–critical-velocity values) must be much higher than in a slow current, and especially in stagnant water; this agrees with observations of microfouling (Stevenson, 1983, 1984; Dovgal, 1990; Railkin, 1991) and macrofouling (Whitford, 1960; Okamura, 1990; Pawlik and Butman, 1993; Khailov et al., 1995; Thompson et al., 1998; Qian et al., 1999, 2000).

It is well known (see, e.g., Jackson, 1977b; Braiko, 1985) that small-sized objects, for example, experimental plates, are more strongly fouled compared to larger ones, which was demonstrated in our model using the example of a plate in a parallel flow (Q_c/S or $Q_b/S \sim 1/\sqrt{x}$).

Thus, the models of accumulation developed in this chapter agree with the known facts and can be used to describe the process of biofouling. Moreover, they predict the non-uniform (gradient) distribution of fouling on surfaces streamlined by longitudinal flow. This follows from Equations 7.21 and 7.22, which describe the abundance of fouling as a function of the plate length (in fact, of the distance from its leading edge). Such a pattern of the biomass and density distribution of biofouling can indeed be observed experimentally (see Section 7.3), which confirms the validity of our models and the concepts on which they are based.

Some practical conclusions may be derived from the accumulation models. On the one hand, processes of colonization in aquaculture may be regulated to a certain extent by increasing or suppressing the larval flow past the substrate. On the other hand, the same method may be used to suppress the biofouling of man-made structures.

7.2 MATHEMATICAL MODELS OF FEEDING AND GROWTH

In order to develop the general models of continuous feeding and growth, we shall consider a feeding organism. Let us assume that dissolved or particulate food is transported at a rate w to the mouth (or to the surface absorbing the nutrients) with an area s. In the case of bacteria, micro-, and macroalgae, w designates the rate of diffusion and active transport whereas, in the case of invertebrates and ascidians, w is either the rate of filtration (sedimentation) or the rate of catching the prey and transporting it to the mouth. Let us suppose that the feeding of algae is limited only by the availability of biogens, but not the quantity of light. If p is the concentration of food in the organism's environment, then the total amount of food brought per unit time to the mouth opening or to another absorbing surface is swp. Let us assume that the amount of food being ingested depends on the degree of "satiety" h, i.e., on the fraction of food that the organism could still take up to reach complete satiation:

$$h = (q - q_i)/q,   \tag{7.24}$$

where q is the greatest possible amount of ingested food and q_i is the current amount of ingested food. Thus, the quantity of food consumed by one organism per unit time (the individual feeding rate) I will be determined as

$$I = dq_i / dt = (q - q_i)\, s\, w\, p / q.   \tag{7.25}$$

Separation of variables gives

$$d(q - q_i)/ q - q_i = -(s\, w\, p / q)\, dt.   \tag{7.26}$$

Solving the differential equation (7.26), we obtain

$$q \ln(q - q_i) = -s\, w\, p\, t + C,   \tag{7.27}$$

where C is an integration constant. Obviously, $C = q \cdot \ln q$ with the starting conditions $t = 0$, $q_i = 0$. In view of this, the quantity of food q_i consumed by one organism during the period t is equal to

$$q_i = q\,(1 - e^{-swpt/q}).   \tag{7.28}$$

This model has the same principal structure as the well-known one proposed by N. Rashevsky (1959) to describe feeding by predatory fishes, but it differs in a number of features. In particular, our model applies not only to predators swallowing their prey but also to filter and suspension feeders, and even to microorganisms and algae. This has been made possible by the introduction of two general parameters that are applicable to organisms with various ways of feeding: the area of the structure responsible for the intake of food (body surface, mouth opening, etc.) and the rate of food transport to that structure. The assimilation capacity of algae, which is determined by the efficiency of photosynthesis and membrane transport, is indeed directly related to their surface area (South and Whittick, 1987).

Let us consider two basic cases: an organism that feeds in stagnant water and one that feeds in flowing water. In the first case, the velocity w of movement of food toward the structure through which it is consumed is determined by the rate v of its transport by the organism itself; in the second case, w also depends on the velocity of the surrounding water flow. We shall assume that the flow is oriented parallel to the structure that serves for feeding. This is observed in many real cases, because the flow is directed along a surface, and the structures taking in food are usually oriented parallel to the surface on which the organism lives. Applying the parallel-ogram rule to the composition of the velocity of food transport by the organisms v and the flow velocity (which is equal to U for the organisms feeding inside the boundary layer and U_∞ for the organisms feeding outside it), we obtain the following formulas for flowing water:

$$w = \sqrt{U_\infty^2 + v^2} \ \text{ or } \ w = \sqrt{U^2 + v^2} \ , \tag{7.29}$$

and, correspondingly, for stagnant water:

$$w = v. \tag{7.30}$$

Substituting the expressions for w from Equations 7.29 and 7.30 into Equation (7.25), we find that the rate of feeding must be greater in flowing than in stagnant water. In addition, Equations 7.29 and 7.30 indicate that the feeding rate increases with the current speed, which is mainly confirmed by experimental data for invertebrates (e.g., Marfenin, 1984, 1993b; Vilenkin et al., 1984; Wildish and Kristmanson, 1985; Okamura, 1990) and algae (e.g., Whitford, 1960; Pertsov and Vilenkin, 1977; Khailov et al., 1992, 1995). However, it should be borne in mind that feeding is slowed at current velocities that are too high and may stop completely in a very rapid flow, primarily for biological reasons. This is usually not observed at a free-stream velocity $U_\infty \leq 0.1$–0.5 m/s (see Section 6.5). Thus, we shall assume that the model of feeding, as well as the model of accumulation, is valid for current velocities that do not exceed these critical values.

The growth rate in animals is determined by the difference between the consumed A and non-assimilated K masses of food and depends linearly on the rate of feeding (Zaika, 1985). Non-assimilated, or catabolized, food is in fact that fraction which is used for respiration and excreted as unmetabolized wastes. Consequently, according to Equation 7.25 for the feeding rate I, the growth rate will be determined as

$$dW / dt = A - K = \beta I = \beta(1 - q_i / q) s w p , \tag{7.31}$$

where W is the mass of an individual and β is a coefficient describing the dependence of growth on feeding and metabolism.

Equation 7.31 shows that the necessary conditions of individual growth include, in particular, a certain rate of food transport into the organism and its assimilation. These conditions are always met in nature, because the propagules settle in biotopes that are favorable for their development, feeding, and growth.

It should be noted that, if the specific growth rate (i.e., rate per unit mass or volume of an organism) is determined by dividing both parts of Equation 7.31 by W, we find that the specific growth rate of algae, dW/Wdt, is a function of the specific surface area s/W. This conclusion agrees well with the theory of functional morphology of algae (Khailov et al., 1992, 1994, 1995, 1999).

The authors of this theory proposed a novel approach to studying and characterizing feeding and growth. They considered the exchange of matter with the environment within the inhabited space around the organism (coinciding with its outlines) in which the exchange is most intensive. According to their theory and the experimental data (Khailov et al., 1992, 1995), the growth parameters depend on the amount of materials coming in through the surface of an alga in relation to its unit volume. Thus, the specific surface area determines the growth rate. K. M. Khailov and his colleagues (1995) found that the phytomass concentration in the inhabited space shows a regular dependence on its volume: namely, it decreases as the volume is reduced. The general approach proposed by these researchers to describe the relationship of algae with their environment can be applied to animal foulers as well.

In order to determine the immediate relation between the individual's mass (biomass) W and the variables included in Equation 7.31, we shall substitute the expression in Equation 7.28 for q_i in Equation 7.31 and then separate the variables W and t:

$$dW = \beta s w p \cdot e^{-swpt/q} \, dt \, . \tag{7.32}$$

Solving Equation 7.32 for W, we find that at the initial conditions $t = 0$, $W_0 = 0$,

$$W_t = W_\infty (1 - e^{-swp(t-t_0)/q}), \tag{7.33}$$

where t_0 is the initial moment of the time.

The developed model describes the growth asymptotically approaching the definitive biomass of an adult individual. During an initial period (at small values of t), the growth is rather rapid and then slows down. Equation 7.33 structurally corresponds to a linear modification of the von Bertalanffy equation (Bertalanffy, 1938), which describes the growth of animals, for example, mollusks (e.g., Weinberg and Helser, 1996; Urban, 2002), in terms of their linear size as a function of time. Our model, and the resulting formula, which is analogous to the von Bertalanffy equation, describe growth not only in animals (e.g., Rumrill, 1989; Sukhotin and Maximovich,

1994; Roegner and Mann, 1995; Lam, 2000) but also in plants (e.g., Hurd et al., 1996), being directly derived from the generalized model of feeding.

By analyzing Equation 7.31, one may note that a faster growth is observed in those species that use their food more efficiently, i.e., have a high coefficient of assimilation β and a low proportion of non-assimilated food K. Furthermore, the growth rate depends directly on the intensity of feeding and is determined by the physiological state of an organism, as shown by the coefficient $(1 - q_r/q)$. The amount of food assimilated per unit time is proportional to the product of three independent variables, swp, which means that efficient feeding can be accomplished in different ways in different species. For example, in algae, the augmentation and subdivision of their surface s increase their assimilative capability and growth rate, owing both to the photosynthetic processes and the membranous transport of biogens, of which nitrogen is the general limiting element in the marine environment (South and Whittick, 1987). For sponges, bivalves, ascidians, bryozoans, and cirripedes, which filter off food particles through small openings (mouth, incurrent siphon, etc.), the main control parameter is the rate w of the food transport into the alimentary system. Colonial animals (sponges, hydroids, bryozoans, and ascidians) have an immense total intake surface s. Of course, one must remember that it is the relative value of swp, rather than the absolute value, that is important, since the assimilated food is distributed over the whole organism or colony. A microscopically sized organism or colony needs a small amount of food, whereas a large one needs a large amount of food.

The available data concerning the dependence of feeding and growth of micro- and macrofouling on the flow velocity or, more broadly, on hydrodynamic activity (see Section 6.5; McIntire, 1968; Yakubenko and Shcherbakova, 1981; Thirb and Benson-Evans, 1982; Stevenson and Peterson, 1989; Dovgal, 1990; Railkin, 1991; Ereskowsky, 1994; Khailov et al., 1995; Thomassen and Riisgård, 1995; Hurd et al., 1996, etc.) generally confirm the above mathematical models of colonization processes, based on the consideration of laminar flows. According to the models of colonization, the contribution of growth to abundance for those microfoulers that, due to their small size, feed within the boundary layer depends on the flow velocity within that layer. Other conditions being equal, the rate of their feeding and growth will be determined by the rate of food transport in the boundary layer, i.e., by the current velocity gradient within it. For macrofoulers and those microfoulers that feed outside the boundary layer, the growth rate does not depend on their position on a surface. It is evident that, in both cases, growth cannot distort the quantitative interrelations that are reflected in the accumulation model.

Indeed, food particles are transported to the accepting structure by the two flows (horizontal and vertical), similar to those considered in the settlement model with respect to propagules. It is important to note that these two food flows exist both within the boundary layer and beyond it. Therefore, the rate of transport of food particles can be described by the models that are structurally similar to the models of settlement, while the processes of capturing and holding food can be described by the models that are similar to those of adhesion. With certain reservations, a similar reasoning may be applied to diffusion feeding as well. Therefore, one may assume that some general corollaries of the settlement and adhesion models will also be valid for the models of feeding and growth.

7.3 GRADIENT DISTRIBUTION OF FOULERS OVER SURFACES IN A FLOW

The development and analysis of colonization models (see Sections 7.1 and 7.2) allow one to make some conclusions concerning the gradient distribution of the fouling abundance and biomass along the surface. This hypothesis and its experimental proof for flat surfaces positioned parallel to the flow, and the bodies that can be equaled to such surfaces, are the subject of this chapter.

The boundary layer formed near the surface in a flow is not uniform from a hydrodynamic viewpoint, and therefore the conditions of transport, settlement, attachment, and growth inside this layer are also not uniform. In particular, in the case of a surface placed in a longitudinal flow, such as an experimental plate, the inner wall of a pipe, or the submerged part of a ship hull, the flow velocity within the boundary layer is maximum at the leading edge of the surface (i.e., the one facing the flow) and diminishes regularly toward the trailing edge (Schlichting, 1979). Conversely, the thickness of the boundary layer is minimum at the leading edge and gradually increases down the flow (Figure 7.1). Therefore, according to the equations of the colonization models, the conditions for accumulation and growth will be unequal in different parts of the boundary layer, resulting in non-uniform, or gradientlike, fouling.

It is well known that the margins of experimental plates are more strongly fouled than their middle areas. This phenomenon even has a special name: the edge effect (see, e.g., Zevina and Lebedev, 1971; Munteanu and Maly, 1981). There are quantitative data concerning the non-uniform distribution of the density of freshwater diatoms over the circumference of cylindrical objects in a transverse flow (Gessner, 1953). A number of publications report a non-uniform but at the same time regular pattern of fouling distribution. In many cases, the abundance of fouling was observed to decrease gradually along the experimental plates (Mullineaux and Garland, 1993), even along such large objects as ship hulls (Rudyakova, 1967; Zvyagintsev and Mikhailov, 1980; Zvyagintsev, 1981; Alibekova et al., 1985; Zvyagintsev et al., 1990) and pipes (Yakubenko and Shcherbakova, 1981; Yakubenko et al., 1984).

Still, these data cannot be regarded as strict proof of the hypothesis in question. First, the non-uniform distribution of fouling may be caused not only by hydrodynamic factors, but also by other reasons, for example, the impact of competitors or predators; second, in most cases described in the literature, the direction of the flow was not constant.

In view of this, I conducted special observations and experiments in which the unidirectional flow over experimental plates was maintained. This was achieved by means of an original device called the hydrovane (Railkin and Fateev, 1990), which operates on the following principle (Figure 7.2). A horizontal frame with holders for experimental plates is equipped with a rudder and two rotation points. Depending on the positive or negative buoyancy of the device, two symmetrical guy-ropes are

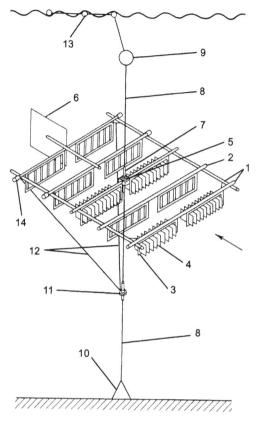

FIGURE 7.2 Total view of the hydrovane, a device used for exposing experimental plates at a fixed angle to the flow. (1) Rectangular frame, (2) holder, (3) cassette, (4) experimental plate, (5) main (central) rotation point, (6) rudder, (7) main (central) axis, (8) cable, (9) float, (10) anchor, (11) additional rotation point, (12) guy-rope, (13) marking drogue, (14) balance float. (After Railkin and Fateev, 1990, with modifications. With permission of St. Petersburg University's Publishing House.)

attached above or below the frame. All of the parts are made of chemically inert and corrosion-resisting materials. If the direction of flow changes, the rudder is affected by the moment of force, which returns the whole frame to the initial position relative to the flow. Thus, the principle of a vane is realized in the aquatic medium, allowing the plates to be exposed at a constant pre-defined angle to the flow. The hydrovane frame may be horizontal (as shown in Figure 7.2) or vertical, and its size may vary from several tens of centimeters to 1 m or more. The frame itself can be replaced with a set of holders with experimental plates. The axis of rotation can be fixed on an underwater base, eliminating the need for a second point of rotation, guy-ropes, and a suspender hawser. The use of this device over a number of years by the author and his colleagues has shown the hydrovane to be very reliable and easy to operate in coastal marine conditions. A similar device was used for the study of settlement of invertebrate larvae by L.S. Mullineaux and E.D. Garland (1993).

An experimental study of the distribution of foulers on experimental plates in a longitudinal flow was carried out in the White Sea between 1987 and 1990, near the Biological Station of St. Petersburg State University (Chupa Inlet, the Kandalaksha Bay). Experiments with microfouling were conducted for two weeks in the summer and autumn and, with macrofouling, during the periods of mass larval settlement in summer. The average current velocity in the study area was 5 to 10 cm/s.

Microfouling was studied using microscope slides ($1 \times 26 \times 76$ mm) and plates ($0.4 \times 26 \times 76$ mm) made of mica and transparent polyvinylchloride. Experiments with macrofouling were performed using plates ($1 \times 50 \times 70$ mm) made of polymeric materials such as Plexiglas or polystyrene. The plate texture was represented by parallel notches 0.45 mm deep and positioned 1.6 mm apart. The experimental substrates were fixed in the hydrovane on the vertical and horizontal planes and submerged at a depth of 1 to 3 m. The margins of the polymeric plates were rounded to ensure a more streamlined flow. The distribution of diatoms was studied in 30 to 60 microscope fields-of-view (about 0.04 mm^2 each) arranged in transversal transects, and that of multicellular animals was studied over the entire plate surface.

The experiments showed that the density of diatoms on differently oriented surfaces increased in the following series: vertical, lower, and upper surfaces. Both horizontal and vertical plates in a parallel flow were non-uniformly fouled by diatoms (Figure 7.3). Observing the plates under the microscope, one could easily see that the maximum density of diatoms was reached as early as the second day within the first transect (0 to 5 mm wide), counting from the leading edge. The density gradually decreased with the distance from the leading edge and rose again slightly near the trailing edge. Despite the continuing settlement, growth, and movement of some vagile species over the surface, this pattern of distribution was preserved for 16 days and longer on differently oriented plates. This was possible due to an almost complete absence of phytophagous protists, which could have distorted the pattern observed.

Essentially the same pattern of distribution over vertical and horizontal plates in a parallel flow is shown by the mussels *Mytilus edulis*, whose mass settlement in the White Sea often takes place in the second half of July and early August (Figure 7.4). Their density is maximal near the leading edge of the plates and diminishes gradually toward the trailing edge. The density of mussels on differently oriented plates increases in the following series: side surfaces, lower, and upper surfaces. Despite this, the overall distribution of mollusks on differently oriented plates in a parallel flow is the same.

Settlement of *Aurelia aurita* planulae was studied experimentally by the author together with A. E. Fateev in 1984. Our observations showed that the planulae settled predominantly on the lower side of experimental plates. There they developed into scyphistomae, which later released the ephyrae into the plankton. The distribution of these cnidarians, only temporarily included in the biofouling community, was also non-uniform along the flow. In this case, however, the density of individuals was low at the leading edge and increased progressively toward the trailing edge (Figure 7.5). Thus, the distribution of scyphistomae along a surface with respect to the flow vector is the opposite of that described for diatoms and mollusks.

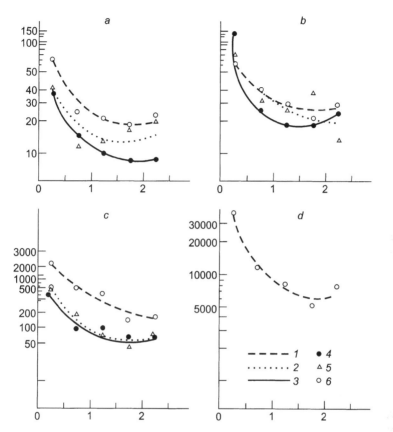

FIGURE 7.3 Density gradient of diatoms on smooth plates in a parallel flow. (a) 2, (b) 4, (c) 9, (d) 16 days. *1, 4* – upper and *2, 5* – lower side of horizontal plates; *3, 6* – vertical surface. *1–3* – empirical regression lines; *4–6* – experimental data. Abscissa – distance from the leading edge, cm; ordinate – density of diatoms, cells/cm^2. (After Railkin, 1991. With permissions of the *Botanicheskii Zhurnal* and the Publishing Firm Nauka, Moscow and St. Petersburg.)

These experimental data show that biofouling develops non-uniformly over the surface of small objects placed in a longitudinal flow. There is an expressed gradient of density and probably biomass. In some cases, it is directed *toward* the flow (direct gradient) while in others it is directed *along* the flow (inverse gradient). The direct density gradient is typical of diatoms, heterotrophic flagellates, and the bivalves *Mytilus edulis* on vertical and horizontal surfaces. The inverse density gradient is observed in scyphistomae of *Aurelia aurita* on the lower surface of experimental plates.

It was proved experimentally that the non-uniform pattern of the fouling distribution is related to the water flow over the surface. On plates protected from the flow by the side walls of a transparent box that is open above and below, the fouling was distributed uniformly in all cases.

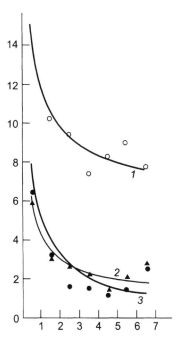

FIGURE 7.4 Density gradient of the mussels *Mytilus edulis* on rough plates in a parallel flow. (1) Upper and (2) lower side of horizontal plates; (3) vertical surface. Abscissa – distance from the leading edge, cm; ordinate – density of mollusks, ind./cm².

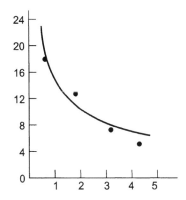

FIGURE 7.5 Density gradient of *Aurelia aurita* scyphistomae on rough plates in a parallel flow. Abscissa – distance from the trailing edge, cm; ordinate – density of scyphistomae, ind./cm².

Observations show that one of the plate margins gets more strongly fouled than the middle area (see Figures 7.3 to 7.5). In marine experiments, the conditions of unidirectional flow are not usually observed, and in the tidal zone the plates are subject to an alternating flow. It is obvious that, in this case, the "edge effect"

(Munteanu and Maly, 1981) will be observed on the entire periphery of a plate. This evidently results from the gradient of fouling on the surfaces that are alternately affected by differently directed flows.

The hydroids *Tubularia crocea* and bryozoans *Schizoporella unicornis* on plates positioned parallel to the flow during the entire tidal cycle develop in great numbers near the leading edge (Mullineaux and Garland, 1993). Their abundance decreases abruptly within the first several centimeters from the edge and then diminishes more smoothly. The distribution histograms of these species correspond to the data reported above for the mussel *M. edulis*. The authors noted that the pattern of distribution of hydroids and bryozoans is related to the distribution of current velocity within the boundary layer. At the same time, the distribution of the bryozoan *Bugula turrita* and the polychaete *Hydroides dianthus* did not show an expressed gradient (Mullineaux and Garland, 1993).

In the above experiments, conducted by me and other authors, the distribution of only several mass species of foulers was observed over relatively short periods. It was shown that the gradient of biofouling abundance can be both direct and inverse. A question arises about whether the gradient distribution represents a feature of the entire fouling assemblage or whether it occurs in individual species and may be leveled at the community level owing to the opposing directions of gradients in different populations.

Let us consider the available literature. According to the data of a number of authors (for example, Rudyakova, 1967; Zvyagintsev and Mikhailov, 1980; Zvyagintsev, 1981; Yakubenko and Shcherbakova, 1981; Yakubenko et al., 1984; Alibekova et al., 1985), the biomass and density of fouling communities may be distributed over the parts of vessels and pipes subject to water flow in the same manner as observed on experimental plates.

A.R. Yakubenko and his co-authors (Yakubenko and Shcherbakova, 1981; Yakubenko et al., 1984) studied the dynamics of fouling on ship piping over a one-month period. Using a diagrammatic representation of their data, the gradient character of biofouling distribution on the inner walls of large-diameter pipes can be clearly observed at a flow rate of 0.15 m/s (Figure 7.6). The area near the inlet is fouled most intensively, with the fouling biomass decreasing noticeably over the first several meters. Thus, the fouling of ship piping is a good example of the direct biomass gradient. At a flow velocity of 0.5 m/s, which apparently exceeds the critical value, a trend toward an inverse gradient of biomass is observed.

A.Ju. Zvyagintsev and S.R. Mikhailov (1980) used SCUBA gear to collect undisturbed fouling, and examined in this way over 100 vessels. A gradient biomass distribution of the fouling, dominated by the cirripede *Chthamalus dalli*, was observed on the submerged part of a BK-153 towboat after about a year of operation (Figure 7.7). The gradient was directed along the boards and was inversed on the ship's bottom. Further examination of ocean-going vessels (Moshchenko and Zvyagintsev, 2001) revealed a more complicated pattern of fouling related to the water flow, which, however, was gradientlike.

Thus, both the author's and the literary data leave no doubt that gradient distribution may be observed not only for individual species, but also for entire communities. The

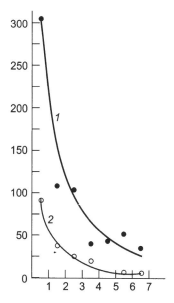

FIGURE 7.6 Gradient of fouling biomass on an inner wall of a pipe. Pipe diameter: (1) 8.3 cm, (2) 4.6 cm. Abscissa – distance from intake, m; ordinate – biomass of fouling, g/m². (Based on the data of Yakubenko et al., 1984.)

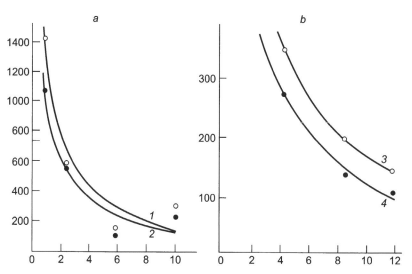

FIGURE 7.7 The gradient of fouling biomass on a ship's hull. (a) Direct and (b) inverse gradient of biomass. Board of the vessel: (1) at sea, (2) docked; bottom of the vessel: (3) at sea, (4) docked. Abscissa – distance, m (*a* – from the bow, *b* – from the stern); ordinate – biomass of fouling, g/m². (Based on the data of Zvyagintsev and Mikhailov, 1980.)

direct gradient is characteristic of the upper side of a horizontal surface and of vertical surfaces (experimental plates, board of a vessel), and the inverse gradient is characteristic of the lower side of a horizontal surface (plates, the ship bottom). This rule probably cannot be applied to the inner walls of pipes. In pipes, a direct gradient of fouling is observed at flow rates below the critical value, and an inverse gradient is observed at higher velocities.

Summing up, I shall note the following. On bodies of different sizes, materials, and surface textures that have been fouled in the ocean from several days to 1 year or more under the conditions of longitudinal flow, the distribution of biomass and density of the fouling is non-uniform. In the cases when this phenomenon was investigated in more detail using quantitative methods, it was possible to observe a direct (toward the flow) or inverse (along the flow) gradient of abundance. They are determined by the pattern of distribution of mass species. The gradient distribution is revealed both by individual species and by communities that are dominated by invertebrates or macroalgae. This is generally true for microfoulers as well. Thus, the gradient distribution of fouling along surfaces in a longitudinal flow is an overall regular phenomenon.

The principal cause of the formation and maintenance of the gradient distribution of biofouling appears to be the non-uniform (gradientlike) distribution of the flow velocity within the boundary layer. The direct gradient of abundance is established if the main condition of biofouling (Equation 7.23) is observed for the entire surface: namely, if the force of adhesion between the fouler and the surface is much greater than the shearing stress and lift stress combined (see Section 7.1). If this condition is violated, the abundance of more sensitive species, possessing a smaller adhesive capacity, is more strongly reduced. In this case, an inverse gradient of fouling is formed.

Such a formation mechanism of the direct and inverse gradients is confirmed by a similar functional dependence of the abundance (biomass) and adhesion force on the flow velocity. In particular, according to Equations 7.14, 7.15, and 7.19, both values are directly related to the velocity to the 3/2 power.

It is also quite probable that the gradient distribution of fouling biomass cannot be distorted by growth processes. Indeed, with all other conditions being equal, the feeding and growth rates of macrofoulers (and large microfoulers), which obtain their food from outside the boundary layer, will be on average the same, regardless of whether they are located in the front, middle, or back part of a surface with respect to the flow. As for microorganisms (and small macroorganisms), which consume dissolved or particulate food within the boundary layer, the rate of water exchange around them is determined by the flow velocity in the boundary layer. Therefore, the gradient of their abundance, which results from accumulation processes, may be only enhanced due to feeding and growth within the boundary layer.

The data obtained in my experiments and provided by other authors were additionally analyzed in order to find to what degree of precision our models of colonization describe the process of biofouling. It was necessary to test the hypothesis according to which, as shown by Equations 7.14 and 7.15, the abundance and

TABLE 7.1

Empirical Equations of Regression between the Density of Diatom Fouling (ind./cm²) on Horizontal and Vertical Plates and the Distance (cm) from the Leading Edge of a Plate

Side of the Plate	Duration of Experiment, days	Equation of Regression	Correlation Coefficient, r
Upper	2	$Q_c / S = -16.55 + 44.69 / \sqrt{x}$	−0.951
	4	$Q_c / S = -56.45 + 91.95 / \sqrt{x}$	−0.947
	9	$Q_c / S = -646.89 + 1140.92 / \sqrt{x}$	−0.992
	16	$Q_c / S = -13201.17 + 24989.52 / \sqrt{x}$	−0.978
Lower	2	$Q_c / S = 0.99 + 18.11\sqrt{x}$	−0.800
	4	$Q_c / S = -14.50 + 50.78 / \sqrt{x}$	−0.920
	9	$Q_c / S = -260.97 + 412.68 / \sqrt{x}$	−0.987
Side	2	$Q_c / S = -7.80 + 21.29 / \sqrt{x}$	−0.987
	4	$Q_c / S = -3.56 + 38.41 / \sqrt{x}$	−0.978
	9	$Q_c / S = -188.50 + 320.71 / \sqrt{x}$	−0.961

After Railkin, 1991, with modifications. With permission of *Botanicheskii Zhurnal.*

biomass per unit area, Q/S, is an inverse function of the square root of the distance x from the leading (trailing) edge of the surface placed in a longitudinal flow:

$$Q / S = K_1 + K_2 / \sqrt{x} , \qquad (7.34)$$

where K_1 and K_2 are some coefficients describing the conditions of biofouling. The regression analysis revealed a high and significant correlation between the predicted and observed values of fouling abundance at different distances from the leading (trailing) edge (Tables 7.1 to 7.4). The correlation coefficient (r) varies from −0.769 to −1.000, in most cases exceeding 0.900 in absolute value. Therefore it is possible to say that the relation described by Equation 7.34 is essentially a quantitative law.

It is clear from the above that, in the marine environment, the horizontal component of propagule accumulation is determined entirely by the current, so that propagules may be regarded as passively transported particles. On the contrary, the vertical component of accumulation does not depend on the current and characterizes the ability of the dispersal forms to settle actively onto a surface. Certainly, settlement may be followed by the redistribution of foulers over the surface (e.g., Mullineaux and Butman, 1990; Walters, 1992). However, it may be assumed that this process is not directional if the surface is homogeneous and the settlers do not possess a

TABLE 7.2

Empirical Equations of Regression between the Density of Fouling (ind./cm²) of Mussels *Mytilus edulis* and Scyphistomae of *Aurelia aurita* and the Distance (cm) from the Plate Edge

Side of the Plate	Duration of Experiment, days	Equation of regression	Correlation Coefficient, r
Mytilus edulis (x is the distance from the leading edge of the plate)			
Upper	57	$Q_c / S = +5.08 + 6.56 / \sqrt{x}$	−0.947
Lower	57	$Q_c / S = -0.52 + 4.67 / \sqrt{x}$	−0.916
Side	57	$Q_c / S = +0.42 + 3.64 / \sqrt{x}$	−0.914
Aurelia aurita (x is the distance from the trailing edge of the plate)			
Lower	8	$Q_c / S = -1.65 + 17.78 / \sqrt{x}$	−0.769

TABLE 7.3

Empirical Equations of Regression between the Biomass of Fouling (g/m²) on the Inner Wall of a Pipe and the Distance (m) from the Intake

Pipe Diameter, m	Duration of Experiment, days	Equation of Regression	Correlation Coefficient, r
0.022	30	$Q_b / S = -6.42 + 27.24 / \sqrt{x}$	−0.960
0.031	30	$Q_b / S = -14.12 + 57.22 / \sqrt{x}$	−0.980
0.046	30	$Q_b / S = -28.53 + 84.87 / \sqrt{x}$	−0.996
0.083	30	$Q_b / S = -76.83 + 262.60 / \sqrt{x}$	−0.980

Note: The flow rate in the pipe is 0.15 m/s.

Calculated from the data of Yakubenko et al., 1984.

positive rheotaxis, and therefore the initial gradients are not leveled by redistribution. The surfaces considered above belong to this particular type.

In connection with the gradient distribution of biofouling formed on surfaces in a parallel flow, it is necessary to consider the problem of estimating the average abundance and biomass. Indeed, their values vary at different points of a body placed in such a flow. Sampling from arbitrary areas or transects may produce biased estimates of abundance and biomass. Of course, these values could be determined precisely by collecting the fouling from the entire surface and relating the total

TABLE 7.4
Empirical Equations of Regression between the Biomass (g/m²) and Density (ind./m²) of Ship Hull Fouling and the Distance (m) from the Bow (Stern)

Place of Sampling	Duration of Experiment, days	Equation of Regression	Correlation Coefficient, r
		Direct Gradient of Abundance (x is the distance from the bow)	
Board (vessel at sea)	365	$Q_b / S = -230.72 + 1174.16 / \sqrt{x}$	−0.970
		$Q_c / S = 13065.70 + 67082.93 / \sqrt{x}$	−0.964
Board (vessel in dock)	365	$Q_b / S = -345.80 + 1558.42 / \sqrt{x}$	−0.969
		$Q_c / S = 20120.41 + 69147.47 / \sqrt{x}$	−0.956
		Inverse Gradient of Abundance (x is the distance from the stern)	
Bottom (vessel at sea)	365	$Q_b / S = -157.31 + 880.04 / \sqrt{x}$	−0.995
		$Q_c / S = -26871.78 + 193255.66 / \sqrt{x}$	−0.998
Bottom (vessel in dock)	365	$Q_b / S = -166.48 + 1058.93 / \sqrt{x}$	−1.000
		$Q_c / S = -54166.82 + 274124.29 / \sqrt{x}$	−0.987

Note: The length of the vessel is 14 m.

Calculated from the data of Zvyagintsev and Mikhailov, 1980.

abundance or biomass to the total area. However, this approach is reasonable only in the case of small surfaces on which biofouling is not too strongly developed.

It is evident that sample areas or transects must be placed not arbitrarily but in the particular area of the biofouling gradient, where its density and biomass are equal to the mean values. Applying the mean value theorem to the function describing the gradient distribution of biofouling, we obtain from Equation 7.34

$$Q / S = \frac{1}{b-a} \int_{a=0}^{b=l} (K_1 + K_2 / \sqrt{x})dx = K_1 + 2K_2 / \sqrt{l} , \qquad (7.35)$$

where l is length of the plate. Substituting the Q/S value from Equation (7.34) gives

$$Q / S = K_1 + K_2 / \sqrt{x} = K_1 + 2K_2 / \sqrt{l} ; \qquad (7.36)$$

therefore,

$$x = 0.25 \, l . \qquad (7.37)$$

Thus, the biomass and abundance of fouling on plates in a longitudinal flow and on similar surfaces have values that are equal to the averages at one-fourth of their length from the leading or trailing edge, depending on which part is more intensively fouled.

In conclusion, let us discuss to what extent the laminar models considered above can be applied to describing the colonization of hard surfaces under field and experimental conditions.

The water flow in seas, oceans, and rivers is to some extent turbulent (Dodds, 1990; Abelson and Denny, 1997). However, the turbulent boundary layer that exists around bodies in such a flow is always underlain by a viscous sublayer, the flow in which is close to laminar (Schlichting, 1979). The viscous sublayers that exist near the seafloor can be as thick as 6 mm (Caldwell and Chriss, 1979). This value is usually much smaller at greater flow velocities in the water column and near the surface. Nevertheless, it may be assumed that larvae and especially spores of macroalgae and microorganisms, owing to their small size (Fraser, 1962; Raymont, 1983), settle and attach within this viscous sublayer with laminar properties. Therefore, laminar models appear to be adequate for a description of real colonization processes. This is confirmed by the facts considered in this chapter in connection with the models. In addition, it should be mentioned that, at the flow velocity that does not limit feeding and growth, juveniles of the bivalve *Agropecten irradians* showed the same growth rates in the turbulent and the laminar flows (Eckman et al., 1989). The validity of our models is also confirmed by the gradient distribution of biofouling predicted by these models and proved by our field experiments as well as published data. The qualitative differences between the processes of hard-surface colonization in the laminar and turbulent flows may not in fact be as significant as they appear.

8 General Regularities of Biofouling

8.1 CAUSES, MECHANISMS, AND LIMITS OF BIOFOULING CONCENTRATION ON HARD SURFACES

In this chapter I summarize and reconsider the material presented in the preceding chapters. References to specific publications will be given only where especially necessary; otherwise, the reader is advised to review the corresponding preceding chapters. The fundamental problems that are supposed to be solved here are as follows: Why do organisms concentrate on hard surfaces? What are the main mechanisms of this concentration? Why is it especially distinct in the coastal belt, where many man-made structures are situated? What are the limits of abundance and biomass of organisms on hard surfaces, and what are the factors that determine them? It should be noted that, in Chapter 1, the phrase "concentration of organisms" was proposed for designating the high abundance and biomass of organisms reached on hard substrates.

The first of our specified problems may be subdivided further: Why are hard surfaces colonized? Why does the colonization result in the progressively increasing abundance and biomass of organisms? Why don't the organisms leave the colonized surface? The answers to these questions may appear simple and obvious, but only at first sight. For instance, to explain the very phenomenon of colonization, one must know how the minute dispersal form, which ranges in size from 0.001 mm in bacteria to 1 to 2 mm (sometimes up to 5 mm) in large swimming larvae, finds a hard surface (sized from millimeters to centimeters or meters) in the vast ocean (Abelson and Denny, 1997). If we bear in mind that the phenomenon of settlement of invertebrate larvae and macroalgal spores involves not just any surface but a very limited set of natural surfaces that are suitable for their settlement and living, we will realize that colonization and therefore the concentration of biofouling should be regarded as very unlikely events. The settlement of a larva on a hard surface in a water flow also appears difficult to explain at close sight. Indeed, the velocity of the current is usually no less than 10 times as great as that of the larva. To use a demonstrative example, the problem of settling larvae is analogous to that of a man trying to jump off a fast-moving train and succeeding in landing safely onto a platform. Assuming that the man runs at 5 m/s (the record speed being 10 m/s), the train must be moving at 180 km/h to reach a velocity ratio of 10:1. This example will probably reveal the difficulty of the problem, even though the questions that we have posed were considered trivial at first sight.

As was shown in Chapters 2 to 6, the colonization of hard surfaces and the related processes are not stochastic but represent a directed sequence of events. Indeed, the release of propagules into the plankton is based on specific mechanisms, both passive (in immotile microorganisms and spores) and active (e.g., in larvae — see Section 3.1). The amount of propagules released into the plankton by microorganisms (bacteria, diatoms) and mass species of multicellular organisms is enormous. Therefore, despite the heavy losses (which could be as high as 99% of the larvae) resulting from their temporary planktonic existence (Thorson, 1963, cited after Mileikovsky, 1971), the number of surviving organisms is large enough to ensure the mass colonization of natural and artificial substrates.

The probability of microorganisms encountering a hard surface is small, because their settlement mechanisms are not very efficient. Indeed, they are passively transported by the current. Their chemotaxis to a surface can be manifested only in its direct proximity. Therefore, the small probability of settlement and attachment in these forms is compensated for by the high abundance of microorganisms in the plankton (Raymont, 1980, 1983). Even though the first bacterial cells can be found on experimental substrates within 1 h, the rate of their settlement is very slow (see Section 3.2). This is also characteristic of diatoms. The increase in the microorganism population on a hard surface in the ocean is accomplished mainly by the division of their cells (see Section 6.5).

Although the abundance of invertebrate larvae (and motile spores of macroalgae) in the plankton is several orders of magnitude lower than that of microorganisms, their encounter with a hard substrate is still possible, and in many cases it is guaranteed. The probability is increased abruptly due to the behavioral reactions that are peculiar to the larvae: photo-, geo-, and chemotaxis. The relatively high velocity of the larvae also plays a considerable role (Table 3.1). The choice of habitat, based on the gradual narrowing of the search area during the last period of meroplanktonic existence, as was shown to be the case in some species (see Section 4.4), allows the larvae to not only concentrate in a particular horizon but also to settle at a specific depth in the characteristic biotope and in suitable substrates with a specific set of features (Chapters 4 and 5). Various outgrowths (flagella, appendages, mucous filaments, etc.) that are present in the propagules also increase the probability of their encounter with the substrate (see Section 6.2).

Our example illustrating the apparent impossibility of larval settlement on a hard surface in a flow emphasizes the peculiarities of settlement in the moving aquatic medium and the role of adhesion, which can hardly be overestimated. In Chapter 7, I have formulated the basic conclusion that adhesion represents the crucial process of colonization (biofouling) of any hard substrate. Adhesion is carried out by a number of mechanisms, the most important of which is the secretion of adhesives (cements) on the organism surface. Propagules are very "sticky." Everyone who has worked with larvae knows well that they readily attach themselves to pipettes or preparatory needles and cannot be removed easily from the substrate. They are so small that, while in contact with a surface, they become enclosed within the boundary layer (see Section 7.1), where the current velocity and the hydrodynamic forces impeding their attachment are not very strong and their adhesion is easier (Abelson and Denny, 1997).

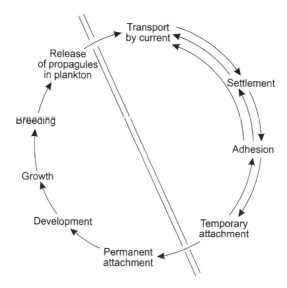

FIGURE 8.1 Colonization cycle of sessile species with motile dispersal stages. Events shown above the double line belong to the reversible phase, and those shown below belong to the irreversible phase.

As a type of link between the meroplanktonic (pelagic) and periphytonic exist-ences, adhesion induces a certain sequence of interrelated and mutually dependent processes, such as firm attachment (which is permanent in sessile species), devel-opment (including individual growth of microorganisms and metamorphosis in inver-tebrates), and individual growth of multicellular organisms (population growth in microorganisms).

Two phases can be distinguished in the cycle of (colonization) biofouling by sessile organisms: the reversible and the irreversible. The former embraces such events as transport by the current, settling, adhesion, and temporary attachment. Having evaluated the surface, the larvae may leave it and go back to the plankton, where they will be transported by the current to the next substrate. This process may occur repeatedly until a habitat is finally selected. The transition from temporary to permanent attachment corresponds to the transition from a reversible to an irrevers-ible phase of colonization and represents the end of the meroplanktonic period of life (Figure 8.1). This scheme also appears to hold true for the motile spores of macroalgae and sessile microorganisms.

Once the swimming propagules of vagile invertebrates and microorganisms have settled on a surface and selected it as the prospective habitat, they finally become attached to it by polymer secretions and adhesive forces (see Sections 6.1 and 6.2). The possibility of their migration to another hard substrate is determined by the reversibility of their attachment (Figure 8.2). As for the propagules that are incapable of active swimming, their colonization cycle will probably be different for sessile (Figure 8.3) and vagile (Figure 8.4) species.

Our analysis has shown that the sequence of colonization processes is not stochastic; moreover, it is directional, since each of the processes involved actually

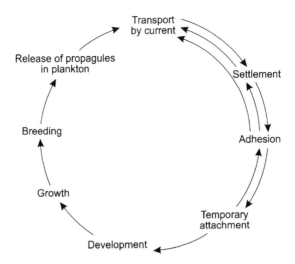

FIGURE 8.2 Colonization cycle of vagile species with motile dispersal stages.

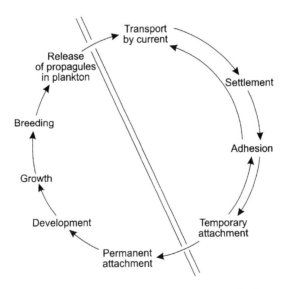

FIGURE 8.3 Colonization cycle of sessile species with non-motile dispersal stages.

activates the subsequent one. Therefore, biofouling is inevitable. The directed nature of (colonization) biofouling is obviously the main cause of biomass concentration on hard surfaces (see Chapter 1). In the course of succession, this leads to the progressive growth of abundance and biomass of organisms. The concentration of organisms is based on the mechanisms that we have just mentioned and which are considered in detail in Chapters 2 to 7.

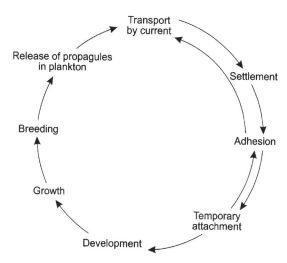

FIGURE 8.4 Colonization cycle of vagile species with non-motile dispersal stages.

An important prerequisite for a large biomass concentration on hard substrates is that the basis of hard-substrate communities is formed by sessile species. Vagile organisms do not leave the colonized surfaces simultaneously and in great numbers, if they do at all. On the other hand, the substrates are exposed to a flow of propagules, the rate and duration of which are determined by larval supply (Todd, 1998). Macroorganisms settle in particular biotopes and on quite specific natural surfaces. If they did not show such selectivity, they would obviously settle in very different habitats, including those that are totally unsuitable for them. Such populations and species would be nonviable and would be eliminated by selection. Propagules have a certain habitat selectivity that was formed in the course of evolution and determined genetically, which allows them to choose microbiotopes with a certain degree of reliability. Failing to find a suitable substrate, the larvae delay their settlement and not infrequently their metamorphosis. Propagules of some species may be eliminated by natural death if they cannot find a suitable habitat (see Section 4.4). Therefore, once a larva (or a macroalgal spore) has finally selected a suitable substrate, it does not leave it, regardless of whether the adult form is motile or firmly attached.

The exchange of propagules between hard substrates or between hard and soft (loose) ground (see Figure 1.3) is accompanied by a partial loss of biomass. However, since the reproduction and dispersion periods of different species do not coincide but are generally separated in time (and may even occur in different seasons), this does not lead to noticeable changes of the total biomass of communities, especially because new organisms are constantly added to them. Important aspects of recruitment in benthic communities have been considered in a number of reviews (Hadfield, 1986; Pawlik, 1992; Thièbaut et al., 1998; Todd, 1998).

Concentration of organisms on hard substrates proceeds at unequal rates in different parts of the world's oceans. It is well known that coastal biofouling is characterized by a greater biomass and higher species diversity as compared to that of the open ocean (Reznichenko et al., 1976; Zevina, 1994). In my opinion, the main reasons for the higher degree of organism concentration on hard substrates in coastal areas and on the shelf are the following: (1) a higher temperature of the surface layer, which has the greatest productivity; (2) a higher concentration of food, biogens, and growth factors; (3) a greater number of hard surfaces; and (4) the proximity of benthic communities, which serve as the most important sources of biofouling recruitment. Therefore, biofouling is very intensive in coastal waters. On the one hand, this indicates good prospects for the aquaculture of invertebrates and algae in these areas. On the other hand, the abundance of propagules of biofoulers indicates very unfriendly conditions for the exploitation of technical objects and requires severe antifouling measures (see Chapter 9).

Another group of reasons for the higher concentration of biofouling near the coast is as follows. Coastal marine areas represent unstable habitats, in which the communities do not reach the climax stage. Their succession is interrupted by various factors (storms, predation, denudation, abrasion), and the communities get "thrown back" to earlier stages of succession, characterized by the prevalence of fast-growing species, such as hydroids, cirripedes, polychaetes, bryozoans, and ascidians, which determine the high overall rate of community growth (see Section 2.2). Thus, unstable habitats are characterized by unfinished, cyclic successions (Oshurkov, 1992).

The high rate of accumulation and growth of foulers results in enormous biomasses developing on hard substrates. The biomass may reach many tens and even hundreds of kilograms per square meter, which is not observed in the benthos (Zevina, 1994). The biomass of the giant barnacle *Megabalanus tintinnabulum*, which has a massive shell, may be as large as one-third of a ton per m^2 (Zevina and Negashev, 1994). This value is probably close to the maximum possible biomass in a fouling community.

The accumulating capacity of hard surfaces, though great, still has its limits. For purely physical reasons, it is limited by the area of the fouled surface, the genetically determined sizes of the organisms inhabiting it, and the layered structure of the community, i.e., the additional surfaces that appear as a result of biofouling. As was shown in Section 6.5, the fouling biomass reaches its maximum at the critical current velocity, equal to 0.2 to 0.5 m/s in many species, and decreases abruptly if this limit is exceeded. There are also biological limitations, related to the stable climax state, when the composition of the community and the abundance and mass ratio of the species become relatively stable and undergo almost no changes for quite a long time.

The theory of functional morphology of algae (Khailov et al., 1992) allows one to make important practical conclusions about the maximum biomass and the productivity of plant communities in mariculture. A fixed volume may contain a different number of artificial construction elements (e.g., cylindrical bars, cones, corrugated surfaces, etc.), which determine the different "concentrations" of surfaces and their "density" in relation to the bottom area. These parameters of artificial constructions most strongly affect the algal biomass output. When the density of the surface in relation to the unit bottom area changes within four orders of magnitude, the respective phytomass changes within one order of magnitude (Khailov et al., 1992). Theoretically, this phenomenon allows one to increase the phytomass production to the maximum possible level under the given physical conditions.

8.2 EVOLUTION OF HARD-SUBSTRATE COMMUNITIES

Recent fouling communities occur on a variety of hard substrates: non-living natural substrates, living organisms, and man-made structures. In addition, micro-foulers are also present on detritus particles, which are usually no more than 1 to 2 mm in size. Detritus provides for microfoulers as solid a base as underwater rock or ship bottom for macrofoulers.

It is quite natural to assume that the evolution of biofouling had proceeded in close association with the hard substrates, their properties, and their position in the water relative to the bottom. At the early stages of evolution, only inert bodies existed; later, multicellular animals and algae appeared; and only in the nearest historical past, artificial materials and objects became available for foulers.

Assuming that life has developed from non-living matter (the abiogenesis concept), it is quite probable that the prebiological (chemical) evolution proceeded mostly on the interfaces between the water and atmosphere in the surface film and between the water and hard surface in shallow waters. It is on the interfaces that substances are concentrated and chemical reactions are accelerated. Similar ideas are included in J.D. Bernal's (1967) hypothesis, which postulates that life originated and initially developed on clay surfaces. These ideas were further elaborated on in the hypothesis of N.P. Yushkin (1995, 1998), concerning the role of minerals in the origin of life.

The most favorable conditions for the abiogenetic origin of life appear to have occurred on interfaces between different media, on hard surfaces in particular. It may be assumed that the increased concentration of nutrients on the water/mineral boundary served as an important prerequisite for the concentration of protobionts, and later microorganisms, on these surfaces.

The initial stages of the modern succession of hard-substrate communities possibly reflect these most ancient mechanisms. The fouling proper is preceded by the prebiological period (biochemical conditioning), including the adsorption of molecules and ions on the hard surface. The biological stage starts with the settlement and attachment of copiotrophic bacteria, which can quickly consume adsorbed organic matter and grow in the oligotrophic marine waters. The copiotrophic forms perform the microconditioning of the surface for the subsequent succession (see Section 2.2).

Some stages can be distinguished in the evolution of communities, related to the evolution of the hard substrates on which they developed. At the first stages of life on the sea bottom, hard-substrate communities were predominantly formed on non-living hard substrates. At the second stage, as large multicellular organisms appeared, biofouling expanded onto living objects inhabiting both the ocean bottom and the water column. At the third (modern) stage, anthropogenic substrates became an important factor in the evolution of hard-substrate communities, apart from the non-living and biological substrates.

It may be assumed that, at the early stage of the biofouling evolution, the majority of forms inhabited the ocean bottom in its coastal areas. The bottom was mostly hard, while soft ground was only in the process of being formed. Due to the low current velocities near the bottom, no special means of remaining on the bottom were required. Therefore, the foulers living in deep waters were probably dominated by motile forms. In shallow waters, due to the higher hydrodynamic activity, selection may have resulted in the appearance of attached forms, which subsequently spread beyond the littoral and upper sublittoral zones. The biogeochemical activity of micro- and macroorganisms (e.g., Vernadsky, 1929, 1998; Lovelock, 1979; Krumbein, 1996; Krumbein and Lapo, 1996) probably facilitated the formation of soft ground and the detritus, thus preparing the conditions for the appearance of species inhabiting soft and hard ground.

Soft-ground communities seem to have appeared simultaneously with hard-substrate communities, or even later, when soft ground was formed on the ocean bottom. Benthic forms inhabiting soft and hard ground may have evolved from motile forms of foulers as a result of adaptive radiation in the ground, as a new ecological medium. This assumption may be confirmed by some similarities in the species composition of the recent communities of hard substrates and soft ground and also by the fact that edifying forms in the benthos of soft ground are motile organisms (see Section 1.1). Thus, this stage of evolution can be regarded as the period of development of soft-ground communities. At that time, detritus became an indispensable component of the water column milieu. The microfouling communities that had been restricted to the hard substrates on the bottom also started to develop on the detritus, including its fraction suspended in the water column. With the development of soft-ground communities, the exchange of propagules between them and the hard-substrate communities was established.

The second stage of evolution of hard-substrate communities is characterized by the appearance of large multicellular organisms in the ocean. Their surfaces provided hard substrates that could be colonized by other organisms, both unicellular and multicellular, thus starting the long evolution of epibiotic and then symbiotic and parasitic relations among organisms. The appearance of foulers on living hard substrates resulted in the development of attached forms that were resistant to the chemical and physical defenses of their basibionts (see Section 10.1). The exchange of propagules between the different benthic communities was accomplished via the plankton; within the latter, the meroplanktonic component appeared as a pool of temporary planktonic dispersal forms: larvae of invertebrates, spores of macroalgae, and microorganisms.

The third, or modern, stage started rather recently by historical standards. The non-living hard surfaces no longer occupy vast and continuous territories. Consequently, organisms concentrated on the bottom have a mosaic rather than a continuous distribution. The role of marine organisms' surfaces as habitats for foulers has increased. The appearance of anthropogenic surfaces in the hydrosphere and the exponential growth of their overall area has become an essentially new factor in the evolution of the concentration of organisms on hard substrates. In the coastal areas, these surfaces are represented primarily by technical objects, the largest total area being that of submerged parts of vessels. In the open ocean, flotsam and oceanic litter are the dominant forms of anthropogenic substrates (see Section 1.3). At the present historical stage, artificial surfaces have become an important factor in the concentration and microevolution of marine organisms. It may be assumed that their role will be even more significant in the future, since human activity in the ocean, including not only its shelf area but also more remote and deeper parts, will be increased.

The concentration of organisms, especially those well manifested on artificial materials, seriously hampers the use of technical objects in the marine environment and requires constant antifouling measures.

9 Protection of Man-Made Structures against Biofouling

9.1 PHYSICAL PROTECTION

Protection of technical and biological objects from biofouling can be based on physical and chemical factors, as well as on their joint effect. In the literature, mechanical factors, such as scrubbing off the fouling, are usually considered as a separate group (Cologer and Preiser, 1984); however, in my opinion, they can be regarded as a type of physical factor. The assemblage of physical (chemical) methods and means, under whose actions colonization by propagules, juveniles, and adult foulers is suppressed, is referred to as physical (chemical) protection. The basic ideas and methods of protection of man-made structures against biofouling are discussed in several reviews (Fischer et al., 1984; Marshall and Bott, 1988; Gurevich et al., 1989; Foster, 1994; Wahl, 1997; Walker and Percival, 2000). Classifying antifouling methods by the acting factors allows one to consider the protection of not only man-made structures (Chapter 9) but also living organisms (Chapter 10) from the same viewpoint.

One of the simplest methods of physical protection against biofouling is creating a mechanical barrier to fence off the settling propagules. Such a barrier can, for example, be realized in the form of a curtain of air bubbles surrounding a ship's hull. To create this curtain, air is released under pressure from openings in a system of air pipes installed on the ship's hull (Rasmussen, 1969a). This type of protection by itself is not sufficiently reliable and can be used mostly to protect vessels with smooth contours. However, it is considerably more efficient when combined with chemical (toxic) factors. For example, biofouling can be suppressed if kerosene containing a dissolved toxin (*bis*-tributyltinoxide) is released together with air bubbles. Such a combined method of protection, named the "Toxion," was used at the beginning of the last century (Gurevich et al., 1989).

An example of physical protection is flaking of paint, in which case paint chips peel off the surface together with the organisms attached to them. B. Ketchum (1952) reported trial data for 378 ship paints, three-quarters of which proved to be exfoliating. However, none of these paints prevented biofouling completely, as it developed again where the coating had peeled off.

Organotin-containing self-polishing copolymer (SPC) coatings, originally designed by International Paint Marine Coatings (United Kingdom), have been in use in many countries since the late 1970s. The matrix of these coatings is formed by vinylic, acrylic, and methacrylic copolymers. The antifouling properties of SPC result from the joint effect of chemical and physical factors. As a result of the hydrolysis of the covalent bond between the biocide (for example, tributyltin) and a polymeric matrix, the former is released into the boundary layer (see Section 7.1), where it can reach a concentration that is lethal for settling propagules. The areas of the polymer that are devoid of any biocide are dissolved. These processes intensify with increasing water flow velocity past the coated surface. In coarser areas, including those on which foulers have settled, the dissolution of coating proceeds faster and the fouling becomes detached. Thus, with long-term exploitation, the roughness of the surface decreases, resulting in a "polishing" effect. In addition to antifouling, SPCs also reduce fuel consumption. Besides reducing costs, this also has a positive ecological impact: it weakens the global greenhouse effect (Wahl, 1997).

Polishing is highly efficient. The roughness of the underwater part of a ship's hull equal to 75 to 170 μm at the time of construction is considered quite satisfactory in many countries (Gurevich et al., 1989). Roughness naturally increases during the course of exploitation, mainly due to biofouling, mechanical damage, and corrosion, and may rise to 0.5 to 0.8 mm in 10 years. Conversely, when SPCs are used, the hull roughness may be diminished to 50 μm after only 9 months of operation, i.e., it may become even less than the original value. SPCs are more expensive than ordinary coatings; however, due to the self-polishing properties, the expenditure is recompensed in 2 years. These coatings can last for 5 years and longer (Clare, 1996; Frost et al., 1999).

The effect of the second generation of self-polishing coatings, the ABC (ablative coatings) class, as well as that of the SPC class, is based on a combination of physical and chemical factors. When these coatings are immersed in water, the organosilicon polymer dissolves slowly, releasing biocide, and the biofouling flakes off (Yuki and Tsuboi, 1991; Tsukerman and Rukhadze, 1996).

An important difference between ABCs and SPCs is that the former have low adhesion, allowing the accumulated biofouling to be removed easily. Probably because of this, a number of authors (for example, Reisch, 2001; Watermann, 2001) consider the development of ablative silicone coatings, including non-biocidal ones, a promising direction and predict that in the future these coatings will compete with biocide-based SPCs. At any rate, ABCs can last as long as 5 years (Ameron, 2000). Their antifouling mechanism appears to be flaking off a film of coating together with the foulers (Tsukerman and Rukhadze, 1996), which may be caused, in particular, by the biodeterioration of the coating (Swain et al., 1998). Tests of the promising silicone-based fouling-release coating (Intersleek, USA) conducted in Pearl Harbor showed it to be efficient against biofouling (Holm et al., 2000). However, according to the justified opinion of some authors (for example, Costa, 2000), silicone coatings are not likely to be used widely in the near future, because there are many difficulties connected with their application, such as the need to paint the ship in dry-dock, the use of quick-drying primers, and the high cost of the coating. In addition, the ways and rates of degradation and the utilization of ABCs in the marine environment are still unknown (Wahl, 1997).

There are other kinds of polymeric coatings that have been designed to reduce the adhesion of foulers and dynamic friction. Such non-stick antifouling materials include oxyethyl cellulose, polyethylene oxide, acrylic resins, polyurethanes, fluorinated graphite, and fluorinated epoxies (Bultman et al., 1984; Gurevich et al., 1989; Wahl, 1997). However, all of them are insufficiently effective in preventing biofouling.

The research performed by E. P. Mel'nichuk (1973) demonstrated the possibility of using antifouling protection based on the formation of a liquid layer on the protected surface to reduce adhesion. Mastics made of paraffin and petrolatum oil did not become fouled during a 1-year exposure in the Black Sea. The best protective properties were displayed by compounds that had a petrolatum oil content of 13 to 30%. According to Mel'nichuk (1973), their antifouling effect resulted from the syneresis mechanism, i.e., the bleeding of petrolatum oil onto the paraffin surface. The liquid film formed on the surface of the coating hampered the attachment of propagules. Further research in this direction (for example, Itikawa, 1983), however, produced no appreciable practical outcomes.

Ultrasonic methods of protecting vessels (Fischer et al., 1984; Edel'kin et al., 1989; Shadrina, 1995) and objects of mariculture (Lin et al., 1988) from biofouling are being developed. The antifouling effect of ultrasound is based on the mechanical destruction of firmly attached organisms by acoustic vibrations at frequencies ranging approximately from 20 to 200 kHz and higher and pulse radiation power up to 1 kW. Oscillations are fed directly onto the ship's hull. Organisms with hard skeletons, for example, hydroids (Burton et al., 1984), are destroyed more easily. In the freshwater bivalve *Dreissena polymorpha*, ultrasound treatment breaks the attachment of the byssus threads to the surface (Lubyanova et al., 1988).

According to the data of M. A. Dolgopol'skaya (1973), who studied the effects of ultrasound in marine conditions, it is the intensity of the elastic vibrations, rather than their duration, that is of crucial importance in suppressing biofouling. However, even high intensity does not ensure complete protection from fouling by invertebrates and macroalgae. In Dolgopol'skaya's opinion (1973), in addition to the standing waves, which are generated by the oscillating plate and locally damage the foulers, a progressive wave must be generated in order to destroy fouling over the entire surface.

The ultrasonic method by itself is not presently used for antifouling protection of vessels, because of its poor efficiency and high cost (Fischer et al., 1984; Gurevich et al., 1989). However, when it is used in combination with antifouling paints (Shcherbakov et al., 1972) or with the electrolysis of sea water (Edel'kin et al., 1989), it provides more reliable protection against marine biofouling. In these cases, the physical effects can be supplemented and considerably augmented by the chemical, biocidal effects.

Protection against biofouling using low-frequency vibrations has been attempted (Jackson and Gill, 1990; Rittschof et al., 1998). The idea of the method is based on the fact that these oscillations are perpendicular to the surface and therefore may, to some extent, hamper the attachment of animal larvae and algal spores. However, protection of ship hulls with infrasound proved to be effective only when it was combined with a biocide-containing coating. In addition, is should be noted that the generation of low-frequency vibrations in an aquatic medium requires quite large-sized devices, the use of which is undesirable for technological reasons.

The damaging effect of radiation on various organisms is well known. This property serves as a basis for attempting to use radioactive isotopes for protecting against biofouling. For example, an isotope of technetium-99, which emits β nuclides, is rather efficient and at the same time promising. According to short-time tests under marine conditions (Makarova, 1990), this isotope ensures protection for up to 2 years, while the actual lifetime of technetium-based coatings appears to be many times longer. Despite the high efficiency of this method, it is unlikely to become widely used, because of its high health and environmental hazards.

Apart from those surveyed above, there are other approaches to physical protection against biofouling. They are based on such factors as temperature, magnetic and electric fields, currents, hydrodynamic forces, and even blast waves (Fischer et al., 1984; Gurevich et al., 1989). However, these techniques are still in the experimental stages. It is possible that some of them will find practical uses in the future.

Based on the above, it is possible to name several general approaches to developing physical protection against biofouling. The most radical approach is to isolate (separate) the protected object from the flow of settling propagules. This can be realized in a variety of ways: for example, by using propagule-free filtered water in cooling systems; by creating a mechanical barrier to fence off the propagules; or by maintaining a liquid layer over the protected surface to prevent attachment. The other two general approaches involve removing the foulers that have already settled on and attached to the surface. In the first case, the fouling is removed along with the part of the surface (ablative paints, self-polishing coatings), etc., whereas, in the second case, only the fouling itself gets detached (ultra- and infrasound).

Consideration of various methods of physical protection shows that the most efficient among them are self-polishing coatings and radioactive protection using β nuclides, though the latter method is hazardous for the environment. In view of the advantages of many physical methods over chemical ones (the adjustable dose and duration of action, the possibility of localized application, safety for the environment), further development of physical protection appears rather promising.

9.2 COMMERCIAL CHEMOBIOCIDAL PROTECTION

Chemical protection against biofouling represents a collection of protection methods and techniques based on the action of chemical factors on dispersal, juvenile, and adult forms of foulers. According to this definition, the methods of chemical protection include antifouling coatings, chlorination, ozonation, treatment with copper sulphate, anodic protection, and plating of the surface (Fischer et al., 1984; Gurevich et al., 1989).

Commercial chemical protection is carried out mainly with the use of copper, zinc, and lead oxides; organotin compounds; chlorine; and ozone. Mercury oxide and organoarsenic compounds were common protective agents in the recent past. Considering the high toxicity of these compounds for foulers, chemical protection can quite rightfully be called *chemobiocidal*.

The base of the commonly used antifouling coatings is formed by paints and enamels that contain copper, organotin, and other biocides that kill the propagules (Gurevich et al., 1989). To protect metal surfaces that come in contact with sea water, for example, a ship's hull, antifouling coatings are applied over the undercoat and anticorrosion paints in one or several layers, depending on the type of paint and operating conditions. There are two types of antifouling paints and enamels (which we refer to collectively as paints): those with a soluble matrix, or film-forming base, and those with an insoluble matrix. They act to create a concentration of biocide within the laminar (boundary) layer (see Section 7.1) that is high enough to kill the propagules entering this layer. The necessary concentration, exceeding some critical value, is maintained due to the continuous leaching of the biocide from the paint.

In soluble paints, the matrix and the biocide are dissolved simultaneously (Frost, 1990). The protective effect will last longer if the two processes have the same rate. An example of paints of this type are the self-polishing coatings, whose polishing mechanism was considered in the previous section (9.1). They are carboxyl-containing organotin polymers, for example, acrylic ones, that dissolve slowly in water. In self-polishing coatings, the dissolution of the polymer matrix is preceded by a hydrolysis stage, during which the biocide — tributyltin (TBT) or copper — is released. Therefore, the dissolution rates of the matrix and the biocide are practically equal. Dissolution of polymers in self-polishing paints and the necessary rate of biocide leaching are attained mainly when the vessel is in motion, whereas, while at their berths, the vessels painted with self-polishing coatings are still subject to fouling. For this reason, in order to ensure the efficiency of such coatings in various operating regimes, they are manufactured with the addition of biocides, such as copper, that are not bound with the polymer (Gurevich et al., 1989).

The leaching rate of organotin compounds increases with the increasing movement rates of the vessel, i.e., in the very situation when biofouling becomes less probable. This is one of the drawbacks of self-polishing coatings. To compensate for this deficiency, many companies produce various kinds of antifouling coatings with controlled polishing rates, depending on the movement rate of the vessel.

In paints with an insoluble matrix, the biocide, for example, copper, is released on the surface through pores and capillaries that are formed as a result of the washing-out of soluble ingredients and the biocide itself into the laminar water sublayer (Gurevich et al., 1989). Thus, the leached layer of paint gradually becomes thicker. To reach the laminar water sublayer, the biocide must diffuse through the entire width of the leached layer of the coating matrix. Consequently, the biocide release rate decreases exponentially during the course of exploitation, so that the paint loses its biocidal properties long before the biocide source is depleted. Removing this drawback is one of the ways to increase the efficacy of paints with an insoluble matrix (Frost, 1990).

The principal biocides that are used in ship coatings are compounds of copper and tin. Coatings that include arsenic and mercury, which were widely used in the past (Gurevich et al., 1989), were prohibited in many countries between 1950 and 1980 because of their high ecological and technological hazards.

Copper is present in antifouling paints mostly in the form of the more highly toxic cuprous oxide, Cu_2O, whereas the low-toxic cupric oxide CuO is hardly used at all. Cuprous thiocyanate is used for antifouling protection in the variable waterline area, where algae develop. Rather toxic, especially to animals, are trialkyltin and triaryltin compounds, for example, tributyltin fluoride and triphenyltin chloride (Figure 9.1). Organoarsenic compounds also have high antifouling capabilities. In

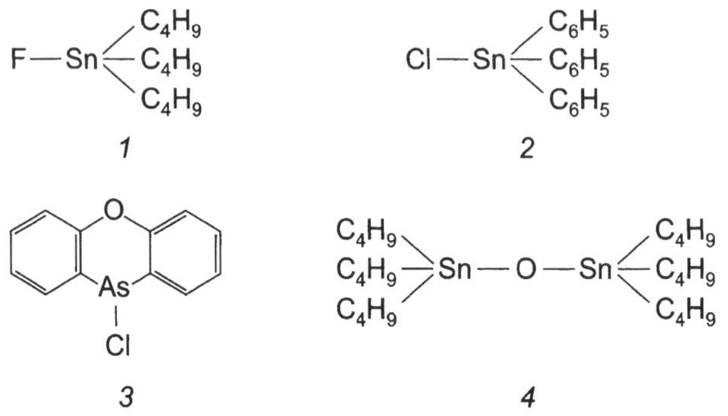

FIGURE 9.1 Tin and arsenic compounds previously used in ship paint. (1) Tributyltin fluoride, (2) triphenyltin chloride, (3) chlorophenoxarsine, (4) *bis*(tributyltin) oxide.

particular, chlorophenoxarsine (Figure 9.1) suppressed both animal and algal macrofouling approximately equally, although it was less efficient than the organotin compounds (Izrailyanz et al., 1976).

Many paints contain not one but several basic biocidal agents, which enhances their protective effect. In practical uses, *bis*(tributyltin) oxide, tributyltin chloride, and triphenyltin chloride proved to be the most efficient of the organotin compounds. However, using of any of them in paints with an insoluble matrix prevents biofouling only for a period of 6–8 or 12–15 months (Gurevich et al., 1989). Adding cuprous oxide in the paint increases the period of protection to 1.5 to 2 years and more.

It should be borne in mind that copper compounds are more toxic to animals than to macroalgae (see, e.g., Gurevich and Dolgopolskaya, 1975). To overcome this deficiency, zinc oxide, which is a good algicide, is added to ship paints. This additive also increases the dissolution rate of copper and thus enhances its antifouling effect. To improve the operation characteristics of paints, manufacturers also add such biocides as derivatives of carbamino acid, carboxylic acids (especially salicylic acid), and thio- and isothiocyanates (Gurevich et al., 1989). They quite often display synergism with the principal biocides, enhancing or expanding their action or improving other operational characteristics.

From the biological point of view, the antifouling effect of ship coatings results from the biocidal action on propagules, juveniles, and adult forms of foulers. Information on the toxicity of the principal and auxiliary biocides of ship paints is summarized in reviews on larvae (Deslous-Paoli, 1981–1982), adult animals, and macroalgae growing on technical objects (Polishchuk, 1973; Patin, 1979; Polikarpov and Egorov, 1986; Filenko, 1988; Khristoforova, 1989). Propagules, as a rule, are more sensitive to toxicants than adults, whereas juveniles usually occupy an intermediate position.

However, there are some known exceptions to this rule. In the mussel *Mytilus edulis*, toxic resistance to copper increases in the following sequence: adults, juveniles, veligers (Beamont et al., 1987; Hoare and Davenport, 1994). In addition, the veligers display a high lethality threshold to copper ions. Cyprid larvae of barnacles of the genus *Balanus* are also less sensitive to copper ions than juveniles and adults (Dolgopolskaya et al., 1973). The known phenomenon of *Balanus amphitrite* cyprids

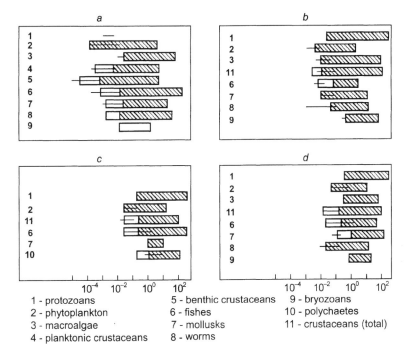

FIGURE 9.2 Ranges of toxic (rectangles) and threshold (bold lines) concentrations of dissolved compounds for different groups of marine organisms. Unshaded areas — ranges of toxic concentrations for early ontogenetic stages. (a) Mercury, (b) copper, (c) lead, (d) zinc. Abscissa: decimal logarithm of concentration, mg/l. (After Patin, 1979. With permission of the Russian Publishing House Pischevaya Promyshlennost, Moscow.)

and some other foulers (oysters, the polychaetes *Hydroides*, the hydroids *Tubularia*, the bryozoans *Membranipora*) settling on copper-based toxic paints (Rudyakova, 1981), the so-called copper tolerance, also appears to be at least partly determined by their resistance to high concentrations of copper ions.

Species of the same taxon (frequently the same genus) may differ considerably in their toxic resistance to heavy metals and other pollutants of the aquatic environment (Stroganov, 1976; Filenko, 1988). For example, the sensitivity of cirripedes to the main biocides of ship paints decreases in the following sequence: *Verruca*, *Balanus perforatus*, *B. amphitrite*, and *B. improvisus* (Gurevich et al., 1989).

Analysis of the data obtained by S. A. Patin (1979) (Figure 9.2), in my opinion, allows us to divide marine aquatic organisms into the following groups according to their toxic resistance: more sensitive (phytoplankton, crustaceans), less sensitive (mollusks), and, finally, resistant (macroalgae, protists, polychaetes, and bryozoans). A similar though more cautious conclusion was made by Patin himself (Patin, 1979). According to the data of R. A. Polishchuk (1973), red algae are more sensitive than green algae to the influence of mercury, copper, silver, and zinc.

Summarizing the published data concerning the effect of biocides contained in antifouling paints on marine organisms, one may conclude that their toxicity is reduced in the sequence: tin, copper, lead (zinc), and arsenic (Patin, 1979; Deslous-Paoli,

1981–1982; Polikarpov and Egorov, 1986; Filenko, 1988; Khristoforova, 1989; Korte et al., 1992). In this sequence, tin and arsenic are supposed to be in the form of their organic compounds. Inorganic tin, for example, as stannic oxide, would occupy the last position in the toxicity sequence, whereas mercury oxide would lead the list.

The information on suppression of settlement, adhesion, and attachment by biocides is no less important. In Chapter 7 it was concluded that the most effective protection against biofouling should be aimed at suppressing these processes. Copper sulphate at a concentration of 0.03 mM, or 4.8 mg/l, blocks locomotion and attachment of spores of all species of red, green, and brown algae studied in this respect (Polishchuk, 1973). The settlement of mollusk larvae is prevented at a leaching rate of copper equal to 1 to 2 $\mu g/cm^2 \cdot day$, and the corresponding value for cirripede cyprid larvae is 10 $\mu g/cm^2 \cdot day$ (Evans, 1981). For the more toxic *bis*(tributyltin) oxide, the leaching rate necessary to suppress the settlement of cyprids is about 1 $\mu g/cm^2 \cdot day$. In fact, these data have been used to determine the desired leaching rate of copper and tin from ship coatings.

It should be noted that the specified rate of biocide leaching is not sufficient to suppress the development of microfoulers — some bacteria, diatoms, and protists, which are more resistant to copper, tin, and other toxins (Gorbenko, 1963, 1981; Robinson et al., 1985; Callow, 1986; Watanabe et al., 1988). Therefore, a number of resistant microorganisms are always present on the surface of antifouling coatings. Toxic-resistant species are known among macroorganisms as well. These are green algae of the genera *Enteromorpha* and *Ulothrix*; brown algae of the genus *Ectocarpus* (Evans, 1981; Hall, 1981; Hall and Baker, 1985; Callow, 1986); cirripedes, in particular *Balanus amphitrite* (Rudyakova, 1981; Gurevich et al., 1989); bivalves of the genera *Mytilus* and *Pecten* (Beamont et al., 1987); polychaetes of the family Serpulidae; and the ascidian *Ciona intestinalis* (Lenihan et al., 1990). It should be emphasized that the macrofouler species whose dispersal forms are resistant to toxins usually dominate in the biofouling communities on engineering objects.

Given the ideal conditions of manufacture, application, and drying, the best copper-based paints with an insoluble matrix will protect vessels from biofouling for 2 to 3 years (Frost et al., 1999). Self-polishing organotin coatings of the soluble type last longer, about 3 to 5 years and more. Their lifetime is largely determined by the coating thickness, with the leaching rate of the biocide being constant. However, the service time of vessels is quite often prolonged up to 20 years (Lyublinskii and Yakubenko, 1990). Therefore, in order to prevent a decrease in performance due to biofouling, the vessels must be periodically dry-docked, cleared of fouling, and repainted.

Despite the above-mentioned deficiencies of antifouling coatings, they are likely to remain an important method of protection in the future. This is because of their high efficiency, profitability, comparative simplicity, the possibility of renewal, and also the fact that no special attendance personnel are needed aboard the vessel (Frost, 1990).

In the following, we shall briefly consider other approaches to the chemical protection of technical objects. The plating method consists of the airless application of a toxic metal, for example, copper or its alloys, from melt onto a surface pretreated with an anticorrosion coating (Gurevich et al., 1989). Copper is dissolved in sea water and protects the surface from colonization by foulers.

The plating methods include the use of copper-nickel alloys, which were introduced in the 1980s, though on a rather restricted scale, for the protection of small vessels, sonars, and some other objects (Cassidy, 1988). The copper:nickel ratio in the alloy varies from 7:3 to 9:1. The alloy is applied onto ship hulls in the form of sheets, foil, or scales. Although copper-nickel cladding is more expensive than copper-based paints, it has a number of important advantages: smoothness (reducing fuel consumption), corrosion and impact resistance, and high toxicity (Grimmek and Sander, 1985). Therefore, its use becomes economically justified after 3 years of service, whereas it may last as long as 10 years. However, copper-nickel alloys were inefficient against microfouling (Srivastava et al., 1990).

The so-called fouling-resistant concrete (Usachev and Strugova, 1989) can be regarded as an analog of antifouling coating. This is structural concrete with the addition of biocides, the best known of which are catapins (organocopper biocides), trialkylstannate compounds of the Lastanox group (Chemapol, Czech Republic), and catamine (alkylbenzyl dimethylammonium chloride). Constructions made of fouling-resistant concrete contain sufficient amounts of the biocide to ensure long-term protection against biofouling. The protective mechanism in this case is basically the same as that of paints with an insoluble matrix. The antifouling effect of concrete is related to the leaching of biocide into the boundary layer around the surface. In Russia, this method of protection was applied in the Kislogubskaya tidal power plant (the Barents Sea). The walls and supporting structures made of concrete with the addition of organocopper biocides did not become fouled for 6 to 9 years, and those with the addition of organotin compounds did not become fouled for more than 10 years (Usachev and Strugova, 1989).

Methods of cathodic and anodic protection are applied in industry to prevent corrosion (Lyublinskii, 1980), but the anodic method is also used for antifouling defense of ship systems, pipelines, heat exchangers, drilling platforms, and power buildings (Yakubenko, 1990). The anodic protection is based on the electrochemical dissolution of the metal anode, which is routinely made of copper, cadmium, zinc, or other metals. As these metals dissolve, the ions that are toxic to dispersal forms are released into the water.

Since vessels are most intensively fouled when at they are at their berths, the anodic protection is switched off while they are on the move. This is one of the advantages of the anodic method, and its fundamental difference from antifouling coatings, which function continuously. Another difference is that, in the case of anodic protection, the toxic agent (e.g., copper solution) is produced outside the defended surface and brought to it by the water flow.

From the viewpoint of toxicology and chemistry, but not technology, anodic protection generally resembles treatment with copper sulphate. The latter method is applied mostly in the freshwater or seawater supply systems of industrial enterprises (Gurevich et al., 1989). Copper sulphate solution is released into the pipes to protect their inner surface against biofouling. Use of this biocide in a concentration of 5 to 15 mg/l for several hours once a week or more frequently is usually sufficient to prevent the development of both microorganisms (Il'ichev et al., 1985) and macroorganisms (Zevina and Lebedev, 1971).

The electrochemical chlorination method is widely known in the protection of technical objects — in particular, cooling systems, power buildings, and also vessels — from marine biofouling (Rasmussen, 1969b; Yakubenko et al., 1981, 1983; Smith and Kretschmer, 1984; Shcherbakova et al., 1986; Usachev and Strugova, 1989; Yakubenko, 1990; Walker and Percical, 2000). This method is based on the electrolysis of seawater by a direct electric current. As a result of the electrochemical processes, hydrogen is released on the cathode and active chlorine on the anode, which is described by the following equation:

$$2NaCl + 2H_2O \rightarrow Cl_2 + 2NaOH + H_2 . \qquad (9.1)$$

With the participation of active chlorine and hydroxyl, further chemical changes proceed according to the equation

$$Cl_2 + H_2O \rightarrow HClO + HCl. \qquad (9.2)$$

Thus, electrolysis of sea water produces active chlorine and hypochloric and hydrochloric acids, which not only kill the propagules, juveniles, and adult foulers but also remove corrosion products (Edel'kin et al., 1989). Therefore, this method offers reliable and efficient protection.

An evident advantage of the electrochemical chlorination method over antifouling coatings and plating is the possibility of controlling the protection and even switching it off. In addition, the effect of chlorine on dispersal forms of foulers occurs in a great volume of water, i.e., even before their contact with the hard surface. The use of rather toxic chlorine, the high efficiency of the equipment used for the electrolysis of water, and competent maintenance ensure a long period of protection of technical objects, which, according to some estimates (Usachev and Strugova, 1989; Lyublinskii and Yakubenko, 1990), may be up to 10 to 15 years or more in the Arctic and boreal waters. In tropical waters, continuous electrochemical chlorination yields actual protection for at least 1 year, whereas a daily 30-min chlorination protects the surface for only 4 months (Smith and Kretschmer, 1984). To protect cooling systems and pipelines in the Black Sea, it was sufficient to treat them periodically with active chlorine in a concentration of 1.0 to 1.5 g/m^3 for 1 to 4 h, with 3- to 5-h intervals (Yakubenko et al., 1983; Shcherbakova et al., 1986). Such diverse regimes of chlorination nevertheless produce similar concentrations of residual chlorine in water, which are sufficient for its disinfection. In the sea water cooling systems of hydroelectric stations, gaseous chlorine dioxide (ClO_2) is used, since it is less hazardous for the environment than the products of electrochemical chlorination (Ambrogi, 1993; Geraci et al., 1993).

In order to reduce the microbial population, which may include pathogenic forms (Camper and McFeters, 2000), potable water is chlorinated or ozonized (Razumov, 1969; Walker and Percical, 2000). Chlorine is more toxic than ozone, and products of its reactions with organic substances (chlororganic compounds) present health and environmental hazards (Patin, 1979). Ozonation is a more gentle method of treating water, sparing the metal tubes. It is well known that ozonation is used to destroy microorganisms in cooling systems and in the treatment of potable water.

There are also patent developments of using ozone or a mixture of ozone with air to protect ship hulls against macroorganisms (Rasmussen, 1969a; Zainiddinov, 1981).

The analysis of industrial chemical methods of antifouling protection shows that all of them are based on the principle of biocidal elimination of propagules, juveniles, and adult foulers located on or near a hard surface. Therefore, one can distinguish between bulk (volume-based) and superficial (surface-based) protection of technical systems (Lyublinskii and Yakubenko, 1990). Other conditions being equal, bulk protection is certainly more efficient, as the foulers in this case are subject to the action of biocides during a longer period and even before their penetration into the laminar-flow layer near the surface. On the other hand, the release of biocides in the water volume around the protected object is exuberant and leads to two consequences. First, maintenance of high concentration of toxicants in a large volume requires extra energy and resource expenditures and cannot be applied generally. Second, the surplus release of such toxic agents as copper and active chlorine is dangerous to the environment.

Antifouling coatings, which are a type of surface-based protection, also have some shortcomings. Their action is uncontrollable. Such protection is also exuberant, since it works continuously, independently of the operational conditions of an engineering object. Therefore, it would be very desirable to develop a physically controllable surface-based protection against biofouling (Lyublinskii and Yakubenko, 1990) that would combine the advantages of both superficial and bulk protection.

9.3 ECOLOGICAL CONSEQUENCES OF TOXICANT APPLICATION

The popular comprehension of the threat of an ecological catastrophe is based, on one hand, on the estimations of the scale of environmental pollution and, on the other hand, on studies of the absorptive capacity of the environment, the paths of accumulation and transformation of hazardous materials, and the resistance of individual species and communities to the action of toxicants.

Heavy metals (mercury, tin, copper, lead, cadmium, zinc, etc.), arsenic, and free chlorine, i.e., the main chemical agents of industrial protection against biofouling and biodeterioration, which were used for this purpose in the past and still are used now, also constitute the major contaminants of aquatic environments, the most hazardous of them being heavy metals and oil products (see, e.g., Tushinsky and Shinkar, 1982). The hazards of heavy metals include their high toxicity, circulation in food chains, and accumulation in organisms. It is necessary to note that the use of tinorganics in ship paints is supposed to be banned as of 2003. Yet the introduction of the ban in different countries will probably take several years. Therefore, we will consider tin based coatings together with copper-based and other coatings.

The principal transformations that heavy metals undergo in aquatic environments are reduction and methylation (Korte et al., 1992). Mercury, tin, arsenic, and lead are easily methylated by microorganisms (Filenko, 1988). Binding with a methyl radical considerably increases the toxicity of these metals and facilitates their transport along food chains and accumulation in organisms (Korte et al., 1992). For example, methylation of arsenic produces extremely toxic di- and trimethylarsines (Filenko, 1988). Methylmercury and methyltin are especially hazardous (Khristoforova, 1989). Studies

TABLE 9.1
Mean Accumulation Factors of Elements in Marine Organisms

Chemical Element	Benthic Algae	Phytoplankton	Zooplankton
Mercury	—	1700	—
Tin	—	6000	450
Copper	100	30000	6000
Lead	700	40000	3000
Zinc	410	15000	8000
Arsenic	2000	—	—

Note: Dashes indicate the absence of data.

Compiled from Polikarpov and Egorov, 1986. With permission of *Energoatomizdat.*

of the global cycle of mercury (Mason et al., 1995) showed that the mercury ion can be transformed into metal mercury in surface oceanic waters. The vapor pressure of mercury is so great that part of it evaporates into the atmosphere. Chlorine easily binds with dissolved organic material to form toxic chlororganic compounds, which are not easily utilized by microorganisms (Korte et al., 1992).

Heavy metals, chlorine, and products of their transformations are distributed in marine environments in a variety of ways. They can be dispersed in the medium, accumulated by aquatic organisms, and transferred along food chains. It is well known (Polikarpov and Egorov, 1986; Khristoforova, 1989) that heavy metals, such as tin, mercury, copper, lead, and zinc, as well as arsenic and other toxins, can be accumulated in marine organisms in concentrations exceeding those in ambient sea water by two, three, and even four orders of magnitude (Table 9.1). As a result, the effect of antifouling chemicals can be observed far from the site of their application.

In areas where a large number of industrial objects is concentrated, antifouling agents kill not only the foulers; they also exert an extremely negative effect on neustonic, planktonic, and benthic organisms located both nearby and at a considerable distance from the protected objects.

Invertebrates, algae, and microorganisms have metallothioneins — special cellular systems that detoxify heavy metals (e.g., Luk'yanova and Evtushenko, 1982; Yoshikawa and Ohta, 1982; Korte et al., 1992; Bebianno and Langston, 1995; Ivanković et al., 2002). These low-molecular proteins, including thiol (sulfur-containing) domains, bind a number of heavy metals (mercury, lead, zinc, copper, and cadmium) and thus inactivate them. This system is effective only at a low level of intoxication.

As concentrations of heavy metals in marine environments exceed their respective threshold values, their toxic effect becomes evident. It shows first of all at the biochemical level. Mercury and copper are strong inhibitors of some enzymes (Korte et al., 1992). The physiological effect of heavy metals on algae is observed in the suppression of photosynthesis and a decline in the primary production, that on invertebrates, in the suppression of respiration and growth (e.g., Kraak et al., 1999; Reinfelder et al., 2000; Ong and Din, 2001). Marine organisms are more sensitive to various contaminants than are freshwater organisms (Patin, 1979).

The comparative toxicity of heavy metals, chlororganics, and other compounds to marine organisms was characterized by S.A. Patin (1979). Unfortunately, tin compounds were studied by him in less detail and consequently were not included in the general scheme (Figure 9.2). Despite major differences in the sensitivity of various groups of organisms, the threshold toxicity values vary to a lesser extent and fall within a rather narrow range. For example, these values are 0.1 to 10 µg/l for mercury, 1 to 10 µg/l for copper, 1 to 100 µg/l for cadmium, and 10 to 100 µg/l for lead and zinc. The range of threshold values for the organotin compounds tributyltin and triphenyltin, which are used in antifouling paints, is 0.1 to 10 µg/l (Burridge et al., 1995), which approximately corresponds to the toxicity of mercury. Freshwater organisms are more resistant to organotin compounds than are marine organisms (Stroganov, 1976). In marine environments, the heavy metals used presently and in the past for protection against biofouling can be arranged in the following sequence, in ascending order of toxicity: lead (zinc) oxides, arsenorganic compounds, copper oxides, organotin compounds, mercury oxide (Rybal'skii et al., 1989).

Chlorine and especially chlororganic compounds, which are formed in organic-rich sea water, are no less hazardous to the environment than are heavy metals (Korte et al., 1992). Experiments have shown a high sensitivity of various invertebrates to active chlorine. Its biocidal concentrations during industrial protection against biofouling are 1.0 to 1.5 g/m^3 (Yakubenko et al., 1983; Shcherbakova et al., 1986). At the same time, treatment with chlorine for 2 h at a concentration of 100 mg/l is sufficient to suppress fouling by barnacles (Shadrina, 1989). Even lower concentrations (20 mg/l, in a regime of 15-h continuous chlorination) are also toxic, suppressing the locomotion activity in cyprid larvae. The toxic action of chlorine on animals can be detected at still lower concentrations. The concentration of residual chlorine in water equal to 5.5 mg/l causes a decrease in many biochemical parameters in the hydroid polyp *Gonothyraea loveni* (Beregovaya, 1991). In particular, the content of carotenoids, glycogen, and free nucleotides changes by several times. The pattern of these changes is quite similar to that observed under the influence of tributyltin oxide. The twofold decrease in the carotenoid content demonstrates the absence of biological mechanisms of detoxification and adaptation to chlorine. The toxic effect may not manifest itself immediately. Residual chlorine at a concentration of only 0.05 mg/l causes death in freshwater mollusks *Corbicula fluminea* within 2 weeks (Ramsay et al., 1988).

Especially strong is the effect of antifouling agents (heavy metals, chlorine, and products of its transformations) on the early developmental stages of macroalgae, invertebrates, and fish (Stroganov, 1976; Patin, 1979; Filenko, 1988). To date, many developmental anomalies have been described, most of which are caused by mercury and organotin compounds. Mollusks, which are used as monitors and indicators of an environment, have been more thoroughly investigated in this respect (e.g., Khristoforova, 1989; Kuhn, 1999; Nehring, 1999). At the most heavily polluted sites, such as port areas and foreshore navigable regions, mollusks may develop misshapen shells or no shells at all (Minichev and Seravin, 1988).

In the sea urchin *Strongylocentrotus intermedius*, fertilization and larval development are disrupted at concentrations of mercurous chloride as low as 1 µg/l and are completely suppressed at 32 µg/l (Vashchenko et al., 1995). Similar concentrations of mercury cause developmental anomalies and retarded growth in the larvae of the mussel *Mytilus galloprovincialis* (Beiras and His, 1995).

The number of successfully germinating embryos of the brown alga *Phyllospora comosa* is considerably reduced as the concentration of tributyltin increases (Burridge et al., 1995). Under the influence of organotin compounds from ship paints, female gastropods in natural populations develop a condition known as imposex. Such individuals have secondary sexual characters of males, whereas the penis length shows a direct relation to the shipping intensity. Both the frequency and the degree of these anomalies rise as the tributyltin concentration in the water increases (see, e.g., Oehlmann et al., 1991; ten Hallers-Tjabbes et al., 1994; Horiguchi et al., 1995; Rilov et al., 2000). Such anomalies of the genital system hamper the process of reproduction and can lead to high mortality.

Even low concentrations of copper (0.4–4.0 µg/l) cause abnormal development of embryos of the bivalves *Mytilus edulis*, *Crassostrea gigas*, *Mizuhopecten yessoensis*, and other species (Malakhov and Medvedeva, 1991). It is manifested in the underdevelopment of the shell and internals at the veliger stage, whereas higher concentrations of copper prevent the deposition of calcium in the shell and result in the evertion of the shell gland. Such larvae have low viability and die early. Zinc is less toxic for embryos of mollusks. In a number of species, this metal also causes evertion of the shell gland, but at concentrations that are 25 times as high as that of copper. In areas polluted with copper, the number of larval anomalies in the mussel *Mytilus edulis* is positively correlated with the contamination level (Hoare et al., 1995). Settlement and metamorphosis in the larvae of the chiton *Ischnochiton hakodadensis* are completely suppressed at copper and zinc concentrations of 10 and 20 µg/l, respectively (Tyurin, 1994). The number of normally developing embryos of the sea urchin *Strongylocentrotus intermedius* is reduced twofold at a copper concentration of 5 to 10 µg/l (Durkina, 1995). Zinc is also less toxic for the larvae of these animals. A threefold decrease in the number of normal late gastrulae is observed at a concentration of zinc that is an order of magnitude greater than that of copper.

The described anomalies of the genital system and the disruption of fertilization and development affected by the basic biocides from antifouling paints reduce the survival rate of individuals and the biotic potential of sensitive species, and can result in the disappearance of the mass species, which play an important role in communities.

Consequently, heavy pollution of marine environments by toxicants released as a result of the antifouling protection of engineering objects from biofouling leads to a noticeable decrease in species diversity and the stability of communities (Prater and Hoke, 1980; Kelly et al., 1990; Weis and Weis, 1995). For example, H.S. Lenihan and co-workers (1990) reported a noticeable decrease in the abundance of sponges, mollusks, and bryozoans in hard-substrate communities on the concrete foundations of coastal buildings and on flotsam in the areas of San Diego Bay (California) used as an anchorage by a large number of vessels (Figure 9.3). According to these authors, this phenomenon was caused by toxins from ship paint, mostly by tributyltin. The abundance of these invertebrates was several times higher in the parts of the water area that hosted fewer vessels. Serpulid polychaetes were, on the contrary, more numerous in the polluted areas, which can be explained not only by their high

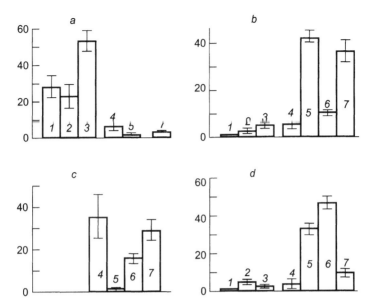

FIGURE 9.3 Effect of the number of vessels on the abundance of invertebrates in San Diego Bay. (a) Polychaetes of the family Serpulidae, (b) sponges, (c) mussels *Mytilus edulis*, (d) bryozoans. *1–3* – water areas used by many vessels, *4–7* – water areas used by few vessels. Ordinate – fraction of surface occupied by mass groups of invertebrates, %. (After Lenihan et al., 1990, with modifications. With permissions of the *Marine Ecology Progress Series* and Prof. J. S. Oliver.)

resistance to the toxicants of ship coatings, but also by the reduced competition of other common groups.

These data show that the use of heavy metals and chlorine in industrial protection against biofouling, which are freely released in the water, has an adverse effect not only on the level of organisms or populations, but also, and not to a smaller extent, on the ecosystem level.

Since the 1970s, marine transport and pleasure vessels have been regarded as one of the principal sources of the growing copper and tin concentrations in near-shore areas, where navigation is most intense. The inflow of copper from ship paints by the 1980s became approximately the same as its release in water with municipal wastes (Gerlach, 1985). The monitoring of copper performed in Arcachon Bay (France) and Chesapeake Bay (U.S.) showed that the principal source of copper was the antifouling coatings of vessels (Alzieu et al., 1987; Scott et al., 1988; Cosson et al., 1989, Russell et al., 1996, etc.). Similar conclusions were made about offshore areas and harbors in other regions polluted with copper, tributyltin, chlorine, and other biocides used for the protection of vessels and commercial plants (see, e.g., Prater and Hoke, 1980; Veglia and Vaissiere, 1986; Bertrandy, 1988; Lenihan et al., 1990; Evans et al., 1995; Hoare et al., 1995).

In view of the ecological hazard of tinorganics, the leading seafaring countries introduced limitations on the use of tributyltin ship coating in the 1980s and 1990s. As a result of this, the concentration of tributyltin in many offshore areas was reduced to the normal value by the end of the last century (see, e.g., Evans et al., 1995; Minchin et al., 1995; Russell et al., 1996).

Nevertheless, in view of the increasing intensity of shipping traffic, vessel tonnage, and the continuing marine pollution with heavy metals hazardous to man, the Marine Environment Protection Committee of the International Maritime Organization has banned the use of tributyltin in ship paint since January 1, 2003. Other biocides, including copper, will probably be banned by 2008 (Anderson and Hunter, 1999).

Thus, the chemobiocidal methods of antifouling protection of man-made structures has actually come to a standstill. Although there are publications concerning environment-friendly coatings, based mainly on non-biocidal silicone polymers (e.g., McGregor and Marr, 1998; Ameron, 2000; Kempf, 2001; Watermann, 2001), the development of commercial antifouling protection on this basis in the near future is problematic enough, since a number of technological (Costa, 2000) and ecological (Wahl, 1997) problems still remain unsolved.

Purely empirical attempts at designing ecologically safe protection appear inefficient. The analysis of colonization processes undertaken in this book may serve as a tool for a purposeful search and development of alternative approaches (Chapter 10), based on the general mathematical model of antifouling protection (Chapter 11).

10 Ecologically Safe Protection from Biofouling

10.1 DEFENSE AGAINST EPIBIONTS

The total surface area of living organisms in marine environments is really enormous. It appears to be comparable with, or even exceed, the area of non-living hard substrates on the shelf. This seems quite probable if one takes into account not only the population of benthos, including the hard-substrate communities (see Section 1.1), but also plankton and nekton. Many "living" surfaces are populated by certain organisms. The extent of epibiosis becomes evident from the following example. Out of 2254 pairwise interactions between species of multicellular algae, invertebrates, and ascidians inhabiting underwater rocks in New England (U.S.), 59% represent active interactions and are the result of the overgrowing of one organism by others (Sebens, 1985b). Competition for space is especially severe on natural substrates in coastal areas and on the shelf.

The dispersal forms of micro- and macroorganisms of benthos that settle on attached or vagile animals and macroalgae (basibionts) become epibionts. The following situations are theoretically possible. The basibiont surface or a part of it may be either chemically inert or attractive for epibionts or, conversely, repellent, toxic, or biocidal. According to many workers (Goodbody, 1961; Jackson, 1977a; Gauthier and Aubert, 1981; Bakus et al., 1986; Pawlik, 1992; Wahl, 1989, 1997; Slattery, 1997; Targett, 1997), negative epibiotic interrelations between macroorganisms are more common than are positive or neutral interrelations. This phenomenon may serve as a prerequisite for developing ecologically safe antifouling protection, based on epibiotic defense mechanisms.

Let us consider in greater detail the published data on the protection of sea organisms from epibionts. Some attached macroalgae and animals, despite being surrounded by hundreds of potential foulers, have almost no macroorganisms on their surface. Such resistant species have special mechanisms of protection, which have been considered in a number of reviews (Jackson, 1977a; Gurin and Azhgikhin, 1981; Gauthier and Aubert, 1981; Bakus et al., 1986; Elyakov and Stonik, 1986; Wahl, 1989, 1997; Sammarco and Coll, 1990; Pawlik, 1992; Clare, 1996; Slattery, 1997; Steinberg et al., 1997; Targett, 1997). Similar to commercial antifouling protection (see Chapter 9), biological defense against epibiosis can be divided into a chemical and a physical one on the basis of the acting factors; in this case, mechanical defense will be considered as a type of physical defense.

The most widespread means of direct physical (mechanical) protection of basi-bionts from epibionts are the following: release of mucus; peeling of outer teguments; molting; filtering-off of dispersal forms; and the presence of needles and other skeletal structures hampering the fouling. For relatively motile animals (for example, dolphins and fish), the speed with which they are able to swim should be added to this list, along with the mucus-rich integuments that may release toxicants (Jackson, · 1977a; Wahl, 1989, 1997; Duffy and Hay, 1990; Pawlik, 1992; Slattery, 1997; Targett, 1997; Wahl et al., 1998). Means of indirect physical protection may include a high growth rate, allowing one species to avoid fouling by another, more slowly growing species (see Section 6.5). Such situations are commonly observed in those algae and invertebrates that grow preferentially along a hard surface, such as cal-careous coralline algae, sponges, corals, bryozoans, and compound ascidians.

Release of mucus, peeling, and molting as a means of tegument renewal are routine mechanisms used to remove fouling from the outer surface of sea plants and animals. They make it possible to periodically cast off the biofouling and the dead cover tissues. Such methods of physical protection are widespread in nature (Wahl, 1989, 1997; Targett, 1997). In more than 20 investigated species of coralline algae, epithelial cells were shown to peel off (Johnson and Mann, 1986; Keats et al., 1997). Peeling provides a short-term antifouling effect but does not completely solve the problem of defending against epibiosis. In algae, it reduces the abundance of micro- and macrofouling only by several times, whereas animals usually are more efficiently protected by tegument peeling or molts. The sponge *Halichondria panicea*, a typical inhabitant of temperate waters, regularly renews its tegument about every 3 weeks (Barthel and Wolfrath, 1989), which protects it from biofouling to a certain extent. In a similar way, peeling, along with other means of physical and chemical protec-tion, is quite efficient in maintaining the low level of biofouling in the ascidian *Polysyncraton lacazei* (Wahl and Lafargue, 1990; Wahl and Banaigs, 1991). Similar mechanisms were described in corals (Sammarco and Coll, 1990) and other inver-tebrates (Wahl, 1989, 1997).

Peeling is a rather slow process. It occurs with certain periodicity but is not continuous. Therefore, it cannot be regarded as a radical method of protection from epibiosis. This may be said even more categorically about molting, which occurs less frequently than peeling. In addition, molting is usually accompanied by the interruption of other protective mechanisms, such as the removal of dispersal forms from the plankton in the process of feeding, which play a significant role in defending against epibiosis. Molting at the adult stage is characteristic of the animals whose growth is limited by chitinous integuments (Wahl, 1989). It is observed in cirripedes, ascidians, and some other sessile forms. Physical defense by itself appears to be insufficient for completely protecting a basibiont from epibionts. Therefore, physical protection is considered to be the main method of epibiosis control in those species that are resistant to even a considerable degree of biofouling.

One of the means of physical defense against epibionts is low surface energy, i.e., hydrophoby of the basibionts' teguments, which suppresses larval attachment to them (e.g., Wahl, 1989, 1997; Targett, 1997; Rittschof et al., 1998). Quite often this may be linked to the properties of the biofilms on their surface. The attachment of bacteria is hampered or rendered impossible at free energy values of about 20 to

30 mN/m (see, e.g., Dexter, 1978). Similar critical values of surface energy have been established for the attachment of the cirripede cyprids (Rittschof et al., 1998). However, it should be noted that different parts of the body of a living organism may, generally speaking, have different wettability; and, in addition, this parameter may change in ontogenesis. Therefore, such a method of physical protection from epibiosis appears to have a restricted distribution in nature.

The efficiency of physical protection in basibionts can be augmented by chemical agents. In general, chemical protection from epibionts (and predators) is more reliable and successful than physical protection, since it works continuously. It is quite widely spread and represents a more sophisticated adaptation of organisms to the changing environment. For example, damage to the thallus of *Fucus distichus* increases the production of phenolic compounds that are used by this brown alga for protection against feeding by the gastropod *Littorina sitkana* (van Alstyne, 1988). The phenol content in the plant increases by 20% within 2 weeks, not only in the damaged part, but also in the adjacent branches. As a result, the rate of consumption of such fucoids by the mollusks decreases twofold. The induction of chemical defense by physical damage has been observed in 17% of the 42 algal species studied (Cetrulo and Hay, 2000).

The term "allelochemical action" will be preferentially used in the following to designate the general chemical influence of one plant or animal species over another (Gilyarov et al., 1986), whereas negative interactions involving toxic effects will be referred to as "allelopathy" (Rice, 1984). Studies performed in the 1960s to 1980s (Goodbody, 1961; Khailov, 1971; Kucherova, 1973; Jackson and Buss, 1975; Bak et al., 1981; Targett et al., 1983; Rittschof et al., 1985; de Ruyter et al., 1988, etc.) demonstrated the existence of chemical defense against epibiosis in sea plants and animals. It was found that the allelochemical action of macroalgae was based on the release of secondary metabolites of quite various natures. It should be noted that secondary metabolites include the compounds that are not directly associated with basic metabolism, i.e., processes of growth and reproduction, but perform regulatory, signal, protective, and some other functions.

The objects of many studies were marine algae, which have little or no fouling at all. For example, Z.S. Kucherova (1973) studied the biological activity of exudates of the Black Sea macrophytes. She found that the green alga *Enteromorpha linza*, the brown alga *Padina* sp., and the red algae *Corallina officinalis* and *Callithamnion* sp. released into sea water some compounds that, under experimental conditions, caused the cessation of movement in the larvae of such macrofoulers as the mussels *Mytilus galloprovincialis* and barnacles of the genus *Balanus*. Prolonged exposure to these compounds killed the larvae. In addition, these exudates suppressed the development of cultures of some bacteria and diatoms.

A widespread group of secondary metabolites released in water by marine macroalgae is that of phenolic derivatives — tannins and tannin-like compounds (Slattery, 1997; Targett, 1997). Hydrolysable tannins represent esters of sugars and gallic acid or gallic and hexaoxydiphenic acids (Figure 10.1(1, 2)). Condensed tannins are the product of the oxidative polymerization of catechols (Figure 10.1(3)) and belong to the group of polyphenolic compounds. They are considerably more resistant to microbial destruction than hydrolysable tannins (Barashkov, 1972). Phlorotannins,

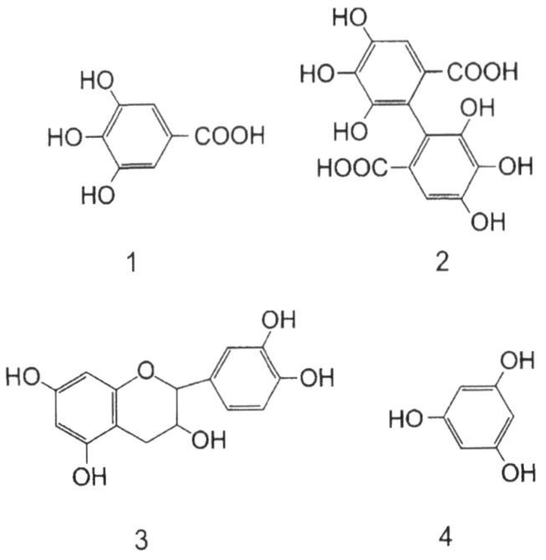

FIGURE 10.1 Phenolic components of tannins. (1) Gallic acid, (2) hexaoxydiphenic acid, (3) catechol, (4) 1,3,5-trihydroxybenzene (phloroglucinol).

which are commonly present in brown algae, are polymers of 1,3,5-trihydroxyben-zene, known by the common name phloroglucinol (Figure 10.1(4)). The content of polyphenolic compounds in brown algae can reach 10 to 25% by dry weight (van Alstyne, 1988; Targett et al., 1995) and is especially high in young growing parts of the thallus — at the tips of branches, where the epibionts are routinely scarce or absent (Sieburth and Conover, 1965).

Tannins are highly toxic and kill the mollusks feeding on the algae even at low concentrations (Duffy and Hay, 1990). Tannins of the brown alga *Sargassum natans* are toxic to various marine invertebrates, such as hydroid polyps, triclads, nematodes, sea spiders, and copepods (Sieburth and Conover, 1965), causing loss of motility followed by death. According to the data of the same authors, a paint containing tannic acid was resistant to fouling by cirripedes and algae. Other studies (Lau and Qian, 1997, 2000) showed that tannic acid, phloroglucinol, and their polymers (phlorotannins) inhibit the settlement of the larvae of the polychaete *Hydroides elegans* and the barnacle *Balanus amphitrite amphitrite.*

In addition to polyphenols, algae may secrete other high-molecular compounds as well. The green alga *Ulva reticulata* suppresses larval settlement and metamor-phosis in the polychaete *Hydroides elegans* (Harder and Qian, 2000). The antifouling factor of still unknown composition (a polysaccharide, protein, glycoconjugate, or a mixture thereof) has a molecular weight of more than 100 kD.

Rather high biological activity is characteristic of halogenorganic compounds, such as trichloroethylene, bromopentane, iodoethane, and others, secreted by various species of green, brown, and red algae (Gschwend et al., 1985; Abrahamsson et al., 1995; Steinberg et al., 1998; Wright et al., 2000). These compounds are toxic to micro- and macroorganisms and protect the algae from epibionts.

The red alga *Plocamium hamatum* exerts a contact toxic effect on the coral *Sinularia cruciata* and the sponges inhabiting the coral reef (de Nys et al., 1991). Animal tissues undergo necrosis under the action of a specific monoterpene (Figure 10.2). Terpenoids of green algae are effective against microorganisms, sea urchins, and fish (Paul and Fenical, 1987).

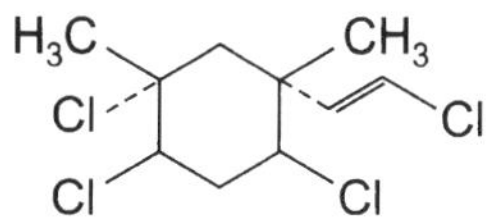

FIGURE 10.2 Monoterpene chloromertensine.

Secondary metabolites of macroalgae also serve as a defense against phytoph-ages. For example, phlorotannins of the brown alga *Ecklonia stolonifera* protect it from being eaten by the gastropod *Haliotis discus* (Taniguchi et al., 1991). Diterpe-noids rather efficiently protect green algae of the genus *Halimeda* from grazing by fish (Paul and van Alstyne, 1988). Terpenic compounds are secreted not only by green but also by red and brown algae (Barashkov, 1972). A mechanism that is common to 39 species of green, brown, and red algae, which protects them from being eaten by invertebrates, involves the release of acrylic acid and acrylate as a result of the metabolic transformation of dimethyl-sulfoniopropionate (van Alstine et al., 2001).

It is interesting to note that allelochemical interactions between different flow-ering plants, and also between flowering plants and animals, are based on similar principles, while the protective compounds used belong to the same classes as in marine macroalgae (Harborn, 1993). This indicates both the ancient origin and the universality of the corresponding biochemical mechanisms, which have been pre-served during the long evolution from the lower to the higher plants.

Among marine animals, allelochemical protection from epibiosis has been stud-ied most extensively in sponges, corals, and ascidians, which is reflected in several reviews (Wahl, 1989; Pawlik, 1992; Clare, 1996; Slattery, 1997; Targett, 1997). Compounds secreted by many sponges are toxic to microorganisms, animals, and plants. Therefore, keeping sponges in aquaria together with other organisms quite often results in the death of the latter. The antimicrobial activity of sponges is a widespread phenomenon. It was described by J.E. Thompson and his colleagues (1985), who studied 40 species of sponges and found that, for 28 of the species (i.e., 70%), the extracts of sponges suppressed the growth of bacterial and yeast cultures. More than 40 different antimicrobial substances, mostly belonging to terpenes, have so far been isolated and identified in sponges.

Defense of sponges against macrofouling, though common, is still not a general rule; for example, it was revealed in only 6 of 20 studied species of Caribbean sponges, i.e., in 30% (Engel and Pawlik, 2000). The marine sponge *Aplisina fistularis* protects itself from fouling by bryozoans and polychaetes (Thompson, 1985). In experiments, the settlement of larvae of the bryozoan *Philodophora pacifica* and the polychaete *Salmacina tribranchiata* was impeded in water in which the sponge had been kept for 1 h. The suppression of metamorphosis and death of juveniles of the mollusk *Haliotis rufescens* also were observed. After several days, the water sur-rounding the sponges became toxic to mollusks and starfish. Two heterocyclic compounds, identified as aerothionin and homoaerothionin (Figure 10.3 [1,2]), were isolated from exudates of *Aplisina fistularis*. They proved to be responsible for protecting this sponge from epibionts (Walker et al., 1985).

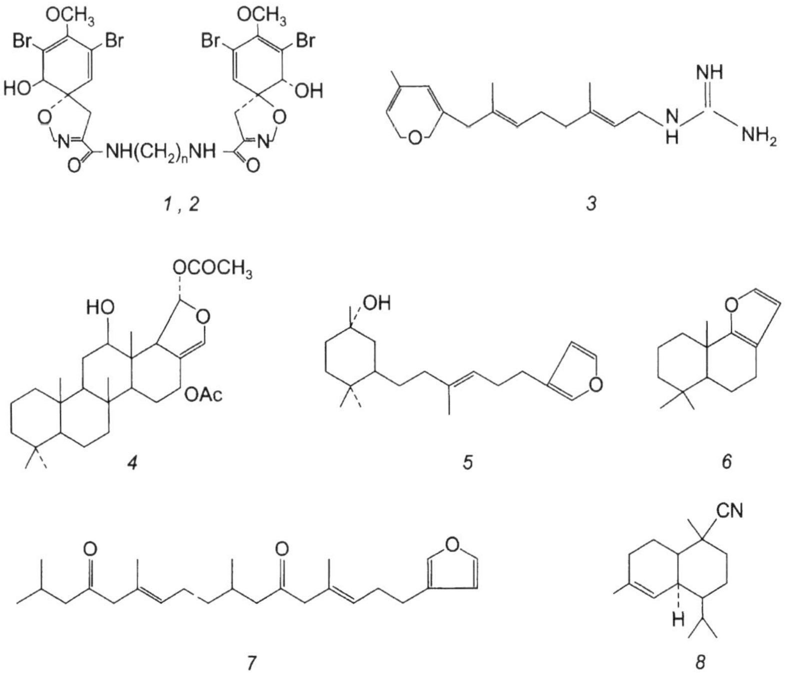

FIGURE 10.3 Terpenes of sponges with a broad action spectrum. (1) Aerothionin (n = 4), (2) homoaerothionin (n = 5), (3) siphonodictidine, (4) heteronemin, (5) ambiol A, (6) pallescensin A, (7) idiadione, (8) δ-cadinen cyan.

Observations of *Siphonodictyon* spp., which settle on the coral *Montastrea cavernosa* near the Caribbean islands, show that a sterile zone is formed around these sponges (Jackson and Buss, 1975). Their toxic agent is siphonodictidine (Figure 10.3 [3]), which consists of a furan ring and a sesquiterpene (Sullivan et al., 1983).

The sponges *Aplisina fistularis*, *Haliclona cinerea*, *Dysidea amblia*, *Euryspongia* sp., *Axinella* sp., and some others, which are capable of the simultaneous suppression of three to six species of bacteria, have practically no macrofouling on their surface (Thompson et al., 1985). The antimicrobial metabolites of sponges show a wide spectrum of actions: in particular, they suppress the growth of a red alga, the locomotion of a limpet and a starfish, and the feeding of a hydroid and a bryozoan. They also inhibit settlement and development in the propagules of the brown alga *Macrocystis pyrifera*, the polychaete *Salmacina tribranchiata*, the bryozoan *Philodophora pacifica*, and the abalone *Haliotis rufescens*. Other terpenes secreted by sponges (Figure 10.3) also have a wide spectrum of actions (Elyakov and Stonik, 1986; Clare, 1996).

Sponges, like macroalgae, can defend themselves from grazing by means of secondary metabolites that they release in water (Wright et al., 1997), for example, halogenorganic compounds (Assmann et al., 2000) and triterpene glycosides (Kubanek et al., 2000).

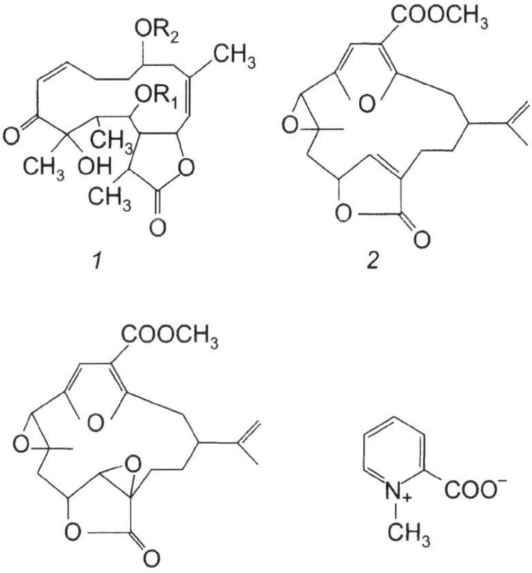

FIGURE 10.4 Antifoulants of corals. (1) Renillafoulins ($R_1 = R_2 = C_2H_5$; $R_1 = C_2H_5$, $R_2 = C_2H_5CO$; $R_1 = C_2H_5$, $R_2 = C_3H_7CO$), (2) pukalide, (3) epoxypukalide, (4) homarine.

Coral polyps produce a number of toxic aliphatic, heterocyclic, nitrogen-containing compounds (including pyridines), and terpenes. Especially noticeable among them is palitoxin (Orlov and Gelashvili, 1985; Gleibs and Mebs, 1998), which is structurally close to saponins, the triterpene derivatives known to exist in echinoderms. Palitoxin has been isolated from the coral *Palythoa toxica* that lives in the Caribbean Sea. This is the most powerful toxin known in marine organisms. Its lethal dose for mice is only 0.15 µg/kg of body mass, which is 3,000 times lower than that of curare and 60,000 times lower than that of potassium cyanide.

Terpenes, many of which are also toxic, are especially common in corals (Gurin and Azhgikhin, 1981; Elyakov and Stonik, 1986; Clare, 1996). Large quantities of terpenes are present in soft-bodied corals (Alcyonaria); they protect the corals from predation by fish; from other corals in the course of interspecific competition; and also from fouling by filamentous algae, bryozoans, sedentary polychaetes, and cirripedes (Sammarco and Coll, 1990; Puglisi et al., 2000).

One of the representatives of sea fans (Gorgoniacea), *Renilla reniformis*, secretes diterpenes known as renillafoulins (Figure 10.4 [1]), which suppress larval settlement in the barnacle *Balanus amphitrite amphitrite* by acting as biocides (Keifer and Rinehart, 1986). The diterpenes pukalide and epoxypukalide (Figure 10.4 [2,3]) from the gorgonian *Leptogorgia virgulata* also inhibited settlement in the larvae of *B. amphitrite* (Gerhart et al., 1988). However, according to the data of these authors, the mechanism of suppression does not appear to be biocidal.

Secondary metabolites of corals can suppress the growth of microorganisms, bacteria, and diatoms via a non-toxic mechanism (Wilsanand et al., 1999, 2001). The corals *Leptogorgia virgulata* and *L. setacea* were found to contain homarine

(Figure 10.4 [4]), a pyridine derivative that is effective in protection from epiphytic diatoms (Targett et al., 1983).

One of the methods of allelochemical protection from epibiosis in ascidians involves releasing substances with a low pH value onto the tunic surface (Wahl, 1989). In 13 of the 35 ascidian species from the families Ascidiidae, Didemnidae, Polycytoridae, and Polyclinidae, living near the Bermudas, the pH of the excreta was less than 2.0 (Stoecker, 1980). Such a high concentration of hydrogen ions was caused by the release of sulfuric acid from the vacuoles of special cells — vanadocytes. Another method of toxic protection from epibionts and predators in ascidians is the high content of vanadium, which is also accumulated in the vanadocytes. Such a method of protection was found in 10 of the 35 ascidian species occurring on the Bermudas. As a result of chemical and possibly physical protection, or their combination, 60% of the species studied were completely free of macroscopic epibionts (Stoecker, 1980).

The compound ascidian *Polysyncraton lacazei* has only one species of multicellular epibionts — a small entoproct *Loxocalyx* sp. — even though hundreds of potential epibiont species are present in the nearest environment (Wahl and Lafargue, 1990). Another very rare epibiont is the diatom *Navicula* sp. Besides physical protection, which is mainly related to its filtering activity, this ascidian possesses means of chemical protection. Special studies (Wahl and Banaigs, 1991) have shown that the ascidian releases two secondary metabolites that suppress the reproduction of unicellular algae. One of them is a still unidentified lipid, while the other is probably a protein. These metabolites also suppressed the development of the sea urchin *Paracentrotus lividus* and were toxic to its larvae.

The surface of the ascidian *Cystodytes lobatus* was shown to be practically free of microorganisms (Wahl et al., 1994). The density of bacteria on it is about 10 to 100 cells/cm², whereas on other species this value may reach up to 10^5 cells/cm². Extracts and secretions of this ascidian reduce the number of attaching bacteria of various species by several times. The fractions suppressing the settlement of microorganisms did not influence their growth, even though the growth inhibitors were present in the extracts of this ascidian. Thus, the suppression of settlement was not caused by any expressed toxic effect on the bacteria. Similar results were obtained for *Aplidium californicum*, *Archidistoma psammion*, *Didemnum* sp., and *Trididemnum* sp. (Wahl et al., 1994).

Different ascidians possess protective properties to an unequal extent. The study of 12 species showed that only some of them, such as *Aplidium proliferum*, *Botryllus schlosseri*, and *Morchellium argus*, possessed a noticeable defense against a hydroid polyp and two species of bryozoans; *M. argus* also was protected against the polychaete *Spirorbis spirorbis* (Teo and Ryland, 1994). The highest mortality rate in this polychaete and the bryozoans was caused by the ascidian *Clavelina lepadiformis*. The same species revealed a distinct antibacterial biocidal activity.

The surfaces of the ascidians *Eudistoma olivaceum* and *E. glandulosum* are protected from biofouling due to the release of special compounds — eudistomins (Figure 10.5 [1,2]) — which are alkaloids that have high biological activity (Davis et al., 1991).

Ascidians, like sponges and corals, also secrete deterrents that protect them from such predators as crabs and fishes (Teo and Ryland, 1994).

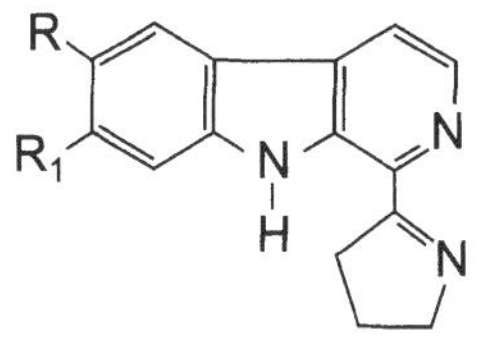

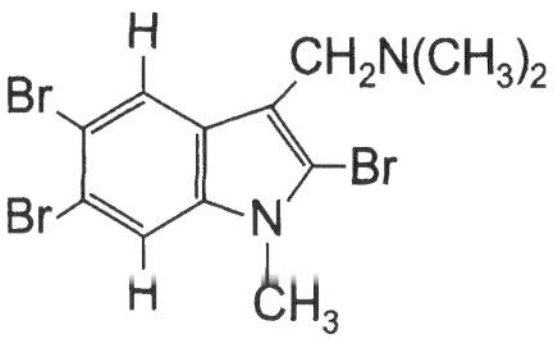

1, 2   3

FIGURE 10.5 Some antifoulants from ascidians and bryozoans. (1) and (2) Eudistomins G and H (in 1, R = H, R_1 = Br; in 2, R = Br, R_1 = H); (3) 2,5,6-tribromo-1-methylgramine.

Other groups of animals have been less thoroughly studied as potential sources of antifoulants. Of special interest are the data on the suppression of settlement in *Balanus amphitrite* cyprid larvae by 2,5,6-tribromo-1-methylgramine (Figure 10.5 [3]), isolated from the bryozoan *Zoobotryon pellucidum* (Kon-ya et al., 1994). This compound was efficient in concentrations smaller than the working concentrations of the common biocide tributyltin oxide, which is used in industrial antifouling protection (see Section 9.1). It should be emphasized that settlement was suppressed at non-toxic concentrations of the agent. Another example is the work in which exudates of two bryozoan species *Bugula pacifica* and *Tricellaria occidentalis* were shown to have a broad spectrum of antibacterial activities and to suppress the bacterial film on these animals (Shellenberger and Ross, 1998).

The data on the possible suppression of biofouling by the microbial communities developing on hard surfaces (including the integuments of basibionts) are directly related to understanding the mechanisms of defense against epibiosis. In particular, films of the diatoms *Stauroneis constricta* and *Nitzschia closterium* were shown to be toxic to the prothalli of the red alga *Gigartina stellata* and to suppress its growth (Huang and Boney, 1985). The combined negative effects of the two species of diatoms proved to be more expressed than the separate influence of each species. Such data are especially important for understanding the protection from epibiosis in a natural environment, where multispecific communities of microorganisms, consisting preferentially of bacteria and diatoms, develop on the surfaces of macroalgae, invertebrates, and ascidians.

Different epiphytic bacteria have different effects on the settlement of larvae. Their effects can be toxic, biocidal, neutral, or stimulating (e.g., Thompson et al., 1985; Maki et al., 1988, 1990; Rittschof and Costlow, 1989; Holmstrom et al., 1992; Dobretsov and Qian, 2002; see also Section 5.5). Of special interest in connection with defense against epibiosis are data on the bacterial suppression of the settlement and attachment of larvae. Toxic epiphytic bacteria are rather common and comprise about 25% of all cultures isolated from natural marine substrates.

Comparison of the toxicities of some bacteria allows one to arrange them in the following sequence, in order of their increasing biocidal activity: *Vibrio campbelli*, *Pseudomonas atlantica*, *Deleya (Pseudomonas) marina*, and *Vibrio vulnificus* (Maki et al., 1988, 1990). According to some evaluations (Rittschof and Costlow, 1989), *D. marina* reduced the settlement of the cyprid larvae of *Balanus amphitrite* and the cyphonautes of *Bugula neritina* by 14 and 17 times, respectively. An exopolymer that suppresses the settlement and attachment of barnacle larvae was isolated from the culture of this bacterium (Maki et al., 1990). When adsorbed on polystyrene, this high-molecular substance reduced the number of attached juvenile barnacles by more than 10 times, whereas on glass it was reduced by only 3 times. Such differences were accounted for by the structural features of the adhesive polymer, which appears to have a greater affinity to a hydrophobic surface than to a hydrophilic one (Maki et al., 1990). According to other data (Holmström et al., 1992), the settlement and attachment of barnacles can be suppressed by a bacterial factor with an MW of only 500 D. The toxin includes a carbohydrate component and probably is a carbohydrate.

Consideration of the published data allows some general conclusions to be made. Defense against epibiosis, which is widespread in nature, reduces the abundance of epibionts on marine algae and animals. This may partly explain the fact that the biomass of foulers is usually higher on non-toxic artificial substrates than on the surfaces of living objects (Zevina, 1994). As in the case of industrial antifouling protection, chemical factors prove to be the most efficient against epibiosis. They work continuously and therefore provide permanent protection. Some classes of substances and individual compounds are efficient enough to be considered as potential antifoulants in industrial protection systems. In my opinion, this group should include, first of all, the phenolic and halogenorganic secondary metabolites of sponges with a broad spectrum of biocidal activities, and also the terpenes of corals, since they have distinct toxic properties. These substances or their analogs will probably be used for this purpose in the future. Non-biocidal protection from biofouling, which will be surveyed in Sections 10.3 and 10.4, appears quite promising, especially from an ecological point of view

10.2 NATURAL AND INDUSTRIAL ANTICOLONIZATION PROTECTION

The concentration of foulers on natural and artificial hard substrates (see Section 1.2) is largely determined by their colonization pressure, which may be rather intensive (Wahl and Mark, 1999). However, during the course of evolution, basibionts have acquired certain mechanisms of chemical and physical defense against colonization by epibionts (see Section 10.1). In industry, special protection methods have been designed to counteract the colonization pressure (see Section 9.2).

Despite the wide use of biocides (and toxicants) by basibionts, the natural defense against biofouling is ecologically safe. However, the use of biocides for protecting man-made systems has rather negative ecological consequences (see Section 9.3). The fact is that the chemical nature of antifouling agents is completely different in the two cases. Natural defenses against epibiosis involve secondary organic exometabolites: phenolic (polyphenolic) and halogenorganic compounds, terpenes, heterocyclic compounds, and other compounds. In the protection of man-made systems, heavy metals (copper, zinc, and lead oxides), low-molecular organotin compounds, chlorine, ozone, and their derivatives are used as biocides.

The essentially different chemical natures of the two groups of biocides leads to important ecological consequences. Biocides produced by basibionts are rather easily destroyed and utilized by microorganisms. The contribution of these secondary metabolites to the general matter and energy flows in ecosystems is insignificant and does not lead to any functional disturbances. Conversely, biocides used for the protection of vessels and other man-made structures either do not decompose to less toxic compounds, or decompose very slowly, or their transformation may produce even more toxic compounds (see Section 9.3). Therefore, industrial biocides cause anomalies of development (teratogenic effect), concentrate in planktonic and benthic organisms, reduce the abundance of sensitive species that do not participate in biofouling, are transported along food chains, and eventually lead to the destabilization and degradation of large ecosystems (see Section 9.3).

The following also should be noted about the differences between natural and industrial protection from biofouling. The group of potential colonists of a particular basibiont includes only a restricted number of species. This is related both to the selectivity of settlement and to the presence of defense mechanisms in basibionts. Therefore it is possible to think that not all but only some biocides excreted in water by basibionts are simultaneously effective against many species of foulers. The same reasoning also appears to hold true for natural antiadhesive and repellent compounds (but see Sections 10.3 and 10.4). Conversely, industrial objects, for example, unlimited-range vessels, must be protected simultaneously from many species of foulers, which means that the biocides used must be universal. This is usually achieved by using not one but two or several biocides, and also by their high concentration in the protected volume.

Industrial coatings implement the principle of a biocide, for example, TBT, immobilized on the surface of a polymeric matrix (SPC). This principle has not, however, been detected in biological systems of defense against epibiosis. The use of immobilized biocides (repellents, antiadhesives) in protection systems has important advantages: local action of antifoulants, low consumption rate (or no consumption at all) of the biocide, and, consequently, reduced costs of the coating.

Natural chemical protection from epibiosis works continuously, although its efficiency may drop with aging, which is probably related to the reduced rates of metabolism. The lifetime of protective ship coatings is limited by their thickness and the amount of biocides stored and usually does not exceed 5 to 10 years. At the same time, electrochemical protection, which is based on the electrolysis of seawater (see Section 9.2), can in theory operate on a continuous basis.

The efficiency of industrial protection from macrofoulers is extremely high but drops when the coating becomes depleted or damaged. Natural protection from epibionts is also rather effective, especially in macroalgae, sponges, corals, and ascidians. However, many species of these and other groups are not completely free of epibionts. In a number of cases, the protection used only inhibits the colonization process to some extent, but it does not destroy the epibionts.

Natural defenses against fouling essentially differ from industrial protection in that they include not only biocidal methods but also repellent, antiadhesive, and antilocomotory protection. Substances possessing a deterrent effect (repellents), those reducing or suppressing attachment (antiadhesives), and those that block locomotion (anesthetic and narcotizing agents) work in natural systems at concentrations that do not kill the organisms and in many cases do not cause any apparent toxic effects (Clare, 1996; Wahl, 1997; Rittschof, 2000). There are also agents that suppress metamorphosis and growth in the foulers (see the above reviews and Section 10.1). Although the structure and mechanism of action of natural non-biocidal antifoulants are poorly studied, one may suppose that they are released in rather small amounts, have an organic nature, and can be metabolized by microorganisms.

Acting against foulers, anesthetic and narcotizing agents would interrupt the flux of motile propagules, repellents would suppress their settlement, and antiadhesives would hamper their attachment to a surface. The substances that block metamorphosis and growth would halt the further development of colonization. Such defenses against biofouling that are aimed at suppressing the process of colonization can be called *anticolonization protection*. It may involve the suppression of one, several, or all colonization processes. Protection from epibiosis may obviously be assigned to this type. According to some authors (e.g., Wilsanand et al., 1999, 2001; Armstrong et al., 2000), macrofouling can be delayed or suppressed, at least in part, by blocking the development of microfouling communities.

Industrial protection from biofouling may be quite reasonably based on the above-stated principles. The possible lines of development of such anticolonization protection may be distinguished based on the specific targets of suppression, which are as follows:

1. The microfouling stage
2. Active (or passive) movement of the propagules to the surface being protected;
3. Settlement
4. Attachment
5. Development (metamorphosis) of foulers on the surface
6. Growth

Apparently, it must be assumed that a radical approach to such protection will involve disruption of the colonization process at a reversible phase (see Chapter 8), before the foulers pass to periphytonic existence, i.e., to permanent life on a hard surface. This can be realized by suppressing the initial stages of colonization: flux of foulers to the surface, their settlement, and attachment.

Industrial protection from biofouling is based on a similar design, because the biocides that are used in it completely suppress the same stages of colonization. However, industrial antifouling protection is ecologically hazardous (see Section 9.3 for details).

It is evidently necessary to develop such methods of anticolonization protection that would be based on principles other than industrial biocidal antifouling protection. The basic requirements of ecologically safe antifouling protection that are already known from the literature (e.g., Seravin et al., 1985; Railkin et al., 1990; Clare, 1996; Wahl, 1997; Rittschof, 2001, etc.) can be defined as follows:

1. Local protective action, spatially restricted to the preferentially defended object
2. Effect directed at the dispersal forms of foulers (microorganisms, larvae, spores)
3. Multiple protective action, i.e., suppression of not one but several colonization processes
4. Biological or spontaneous destruction of the chemical protective agents or products of their reactions
5. Non-toxicity or low toxicity of the protective agents and their derivatives to man, macroalgae, crustaceans, mollusks, fish, mammals, food objects, and objects of aquaculture
6. Absence of hazardous effects, such as carcinogenic, mutagenic, and teratogenic, in the chemical protective systems

Let us discuss in greater detail the main approaches to ecologically safe antifouling protection, aimed at the suppression of the reversible phase of colonization.

10.3 REPELLENT PROTECTION

The ideas of using repellents in antifouling protection were put forward repeatedly over at least the last 30 years (see, e.g., Gurevich and Dolgopol'skaya, 1975; Tsukerman, 1983; Mitchell and Kirchman, 1984; Seravin et al., 1985; Braiko, 1987; Tsukerman and Rukhadze, 1987; Maki et al., 1988; Railkin et al., 1990; Zevina and Rukhadze, 1992; Zevina, 1994; Clare, 1996; Wahl, 1997, etc.). In the studies of defense against epibiosis and the action on larvae of metabolites of microorganisms, algae, and animals, as well as synthetic compounds, vast data have been accumulated that show a possibility of non-biocidal suppression of settlement (e.g., Thompson et al., 1985; Maki et al., 1988; Sears et al., 1990; Davis et al., 1991; Railkin, 1995a, 1995b; Railkin and Dobretsov, 1994; Wahl et al., 1998; Dobretsov, 1999a, 1999b; Dobretsov and Qian, 2002). This was reflected in several reviews (Clare, 1996; Slattery, 1997; Targett, 1997; Wahl, 1997; Rittschof, 2000). However, the acting factors were not precisely identified in many works. Therefore, in the following we shall consider only those publications in which the chemical structure of antifoulants has been elucidated.

In particular, it was found that tannic acid could suppress, probably in a non-biocidal way, the settlement of the polychaete *Hydroides elegans* (Lau and Qian, 1997) while phloroglucinol suppresses the settlement of the cyprid larvae of *Balanus amphitrite amphitrite* (Lau and Qian, 2000). The compounds that were repulsive to the larvae of the bryozoan *Phidolophora pacifica* were 12-epi-deoxoscalarin, which is secreted by the sponge *Leoselia idia*, and isonitriles of the sponge *Axinella* sp. The latter were also effective in suppressing the settlement of the polychaete *Salmacina tribranchiata* (Thompson et al., 1985). A pukalide (Figure 10.4 [2]) derivative, α-acetoxypukalide, isolated from the gorgonarian *Sinularia* sp., suppressed the settlement of the larvae of *B. amphitrite* (Mizobuchi et al., 1994). The sponge *Phyllospongia papyracea* produced free fatty acids, which repelled the blue mussel *Mytilus edulis* (Goto et al., 1992), and also a furanoterpene furospongolide, which suppressed the settlement of *B. amphitrite* (Goto et al., 1993).

As is justly stated in the literature (e.g., Wahl, 1997), the rejected surfaces may simply be unattractive to a fouler or possess some other properties (not necessarily repellent ones) that hamper the settlement of propagules. The true repellent effect may be revealed only by special behavioral tests, and unfortunately there are very few works in which this was taken into account. It must be noted that, in many studies, the repellent function of various substances or their mixtures, extracts, or exudates of microorganisms, plants, or animals is supposed rather than experimentally proved.

In view of the above, it is easy to understand why the term "repellent" is frequently treated very broadly in the literature devoted to antifouling protection, namely, as a substance in the presence of which there is no settlement. Accordingly, not only non-toxic but also low-toxic substances that reduce settlement are referred to as repellents. It should be noted that many authors (Crisp, 1984; Lindner, 1984; Elfimov et al., 1995, etc.; see also Section 4.1) regard attachment as part of settlement. From this viewpoint, substances that suppress attachment should also be considered as repellents. In view of this, we shall discuss the definition of the term "repellent" in greater detail, since not only biologists but also chemists, physicists, technologists, and engineers are engaged in the field of protection from biofouling (and biodeterioration), and the ambiguous treatment of terms could hamper their collaboration.

According to the Russian *Biological Encyclopedic Dictionary* (Gilyarov et al., 1986), repellents are "natural and synthetic substances repulsing animals. Repellents act upon distant or contact chemoreceptors. Substances inducing negative chemotaxis in unicellular organisms are also referred to as repellents" (p. 536). In another biological dictionary (Reymers, 1980), a repellent is similarly defined as "a substance of natural or synthetic origin, scaring off animals. In nature, an agent of allelopathy, in economy, one of the types of pesticides" (p. 148). According to the *Webster University Dictionary of the English Language* (1987), "Repel — to exert a force tending to move (a body) further away; to drive back. Repellent — 1. Repelling, driving back. 2. Preparation for repelling insects or pests" (p. 844).

These definitions correctly reflect the essence of repellency, as repulsion from the source of a repellent cue. Nevertheless, they have their drawbacks. First, not only chemical preparations but also factors of other natures, for example, ultrasonic oscillations, may act as repellents, i.e., scare off organisms that are sensitive to them. For example, there were attempts to repel birds from airports by broadcasting the cries of avian predators via loudspeakers (Ilichev et al., 1987). Second, repellents may affect the behavior of not only adult animals but also their larvae, swimming spores of macroalgae, and microorganisms, i.e., any motile organisms, not just insects or pests. Like multicellular organisms, unicellular organisms also possess chemoreceptors, including those that participate in the chemotactic response. Finally, besides negative chemotaxis, the organisms (unicellular, multicellular, adults, and larvae) may exhibit chemokineses. The definition given in N. F. Reymers' dictionary includes not only behavioral but also toxicological criteria connected with allelopathy and pesticides. In fact, repellents have nothing in common with either term, even though they are certainly used to control insect pests.

The following more precise definition may be proposed. Repellents are cues inducing a negative motor response, taxis, or kinesis in organisms at a certain stage of development, which causes them to move away from the source of these cues. Chemotaxis is locomotion directed toward the source of chemical cues in the case of an attractant and away from it in the case of a repellent. Chemokinesis is also a locomotor response orienting an organism relative to the source of chemical stimulation and resulting in it approaching the source (positive kinesis) or moving away from it (negative kinesis). Chemokinesis is a modification of the intensity of motor reactions (the speed of movement or the frequency of turning) that depends on the intensity of the chemical stimulus (Fraenkel and Gunn, 1961). If the locomotion of organisms slows down or the frequency of their turning increases as they approach the source of a chemical cue, then in due course they will amass near such a source, i.e., will be attracted to it; otherwise, they will move away from the source.

Therefore, the substances reported in the literature as repellents cannot be considered as such unless special behavioral tests have been performed. These tests may confirm or refute their repellent nature. What is known about repellents for marine foulers? With the use of capillaries it has been demonstrated that bacteria isolated from biofouling show a negative motor response, similar to chemokinesis, to a number of organic substances: phenylthiourea, indole, N,N,N',N'-tetramethylethylenediamine, tannic acid, and benzoic acid (Chet and Mitchell, 1976). The amount of bacteria entering a capillary with nutrient broth dropped by 20 and more times when the capillary was filled with a solution of one of these repellents. It is important to note that these solutions were not toxic to microorganisms. The addition of the most efficient of them (tannic or benzoic acids) in a non-toxic varnish reduced the amount of bacteria on its surface by more than a million times — from 5×10^{12} to 1×10^6 cells/cm^2 — but did not suppress bacterial fouling completely (Chet and Mitchell, 1976).

Further examinations showed that solutions of benzoic and tannic acids placed in capillaries repelled unicellular green algae of the genus *Dunaliella* (Mitchell and Kirchman, 1984). Benzoic acid and indole, applied in concentric circles on a sheet of filter paper, prevented the escape of adult gastropods *Monodonta neritoides* from the center of the circle (Ohta et al., 1978). Larvae of the oyster *Ostrea* avoided capillaries with solutions of benzoic, tannic, and especially alginic acids (Mitchell and Kirchman, 1984). Alginic acid included in a paint coating protected it from biofouling to some extent. These data show that various micro- and macroorganisms can have common repellents.

Exposure tests performed in a coastal area of California (U.S.) showed that the addition of tannic acid to a paint coating provided 80% protection from algal fouling for 3 months (Mitchell and Kirchman, 1984). According to other data (Sieburth and Conover, 1965), tannic acid added to a varnish suppressed settlement on plates not only of macroalgae but also of cirripedes.

This line of research was continued by the author and his colleagues (Railkin et al., 1993a; Railkin, 1995a, 1995b; Railkin and Dobretsov, 1994) in laboratory and field experiments, in which the objects of study were not only microfouling but also macrofouling communities. Field experiments were carried out in the White Sea

TABLE 10.1
Repellent Effect of *N,N,N′,N′*-Tetramethylethylenediamine and Benzoic Acid on Planulae of Hydroids and Postlarvae of Mollusks

Substance	Concentration, mM	Turning from the Capillary				
		Angle α,°	Distance, mm	α > 90°,%	t	t $_{0.05}$
Dynamena pumila						
N,N,N′,N′-Tetramethyl-	0	10 ± 2	1	0	—	—
ethylenediamine	43	8 ± 2	0.98 ± 0.03	0	0.71	2.02
	172	68 ± 8	0.41 ± 0.04	40	7.03	2.02
	516	105 ± 11	0.70 ± 0.07	55	8.50	2.02
Benzoic acid	0	10 ± 2	1	0	—	—
	116	110 ± 11	1.30 ± 0.20	85	8.99	2.02
Gonothyraea loveni						
N,N,N′,N′-Tetramethyl-	0	6 ± 3	1	0	—	—
ethylenediamine	43	45 ± 9	1.00 ± 0.07	20	5.11	2.10
	172	90 ± 11	1.13 ± 0.10	70	8.18	2.05
Mytilus edulis						
N,N,N′,N′-Tetramethyl-	0	8 ± 2	1	0	—	—
ethylenediamine	86	90 ± 11	1.80 ± 0.10	85	7.44	2.02
Benzoic acid	0	8 ± 2	1	0	—	—
	744	90 ± 15	0.60 ± 0.10	45	5.45	2.02

Based on the data of Railkin, 1995a.

coastal area (Kandalaksha Bay, Chupa Inlet) with the principal design similar to that described in Section 7.3. The substances to be tested were added in non-toxic concentrations to non-toxic vinyl-rosin varnish, with which 5×10-cm plates were coated (for details, see Railkin and Dobretsov, 1994). In control tests, the varnish contained no additions. To provide optimal conditions for colonization, the plates were exposed in a horizontal position at a depth of 1 m on a hydrovane (see Figure 7.2).

In the laboratory, the repellent effect was studied using glass capillaries several centimeters long and 0.2 mm in diameter, filled with solutions of the substances being tested. In addition, chemotactic chambers made of Plexiglas and measuring $36 \times 40 \times 80$ mm were used for the same tests. Each chamber consisted of three sections, separated with 0.92-µ nucleopore filters. Hydroid larvae were placed in the middle section; one of the side sections contained the test solution and the other (control) contained sea water.

The laboratory experiments showed that planulae of the hydroids *Gonothyraea loveni* and *Dynamena pumila* and postlarval young mussels *Mytilus edulis* (shell lengths 1–2 mm) were indifferent to the capillary filled with sea water (Table 10.1, zero concentration). The hydroids swam, and the mollusks crawled past the tip of the capillary and did not respond to it. However, their behavior changed drastically if the capillary was filled with benzoic acid or *N,N,N′,N′*-tetramethylethylenediamine. At optimal

TABLE 10.2
Suppression of Macrofouling by Repellents Added to Coatings

Coating	Relative Fouling, % of Control			
	10–19 days	22–25 days	45 days	70 days
Varnish + benzoic acid	29 ± 12	45 ± 7	48 ± 18	18 ± 19
	(2.6)	(11.0)	(1.9)	(0.9)
Varnish + N,N,N',N'-tetramethylethylenediamine	—	17 ± 1	32 ± 4	—
		(13.5)	(3.2)	
Antifouling ship enamel KHV-5153	8 ± 0.1	9 ± 3	15 ± 5	31 ± 6
	(10.5)	(15.3)	(4.0)	(3.8)

Notes: (1) Fouling of the control vinyl-rosin coatings without additions is taken to be 100%; (2) calculated *t*-test values are given in parentheses; (3) in Tables 10.2, 10.3, 10.5, and 10.6, the values significantly ($p < 0.05$) different from the control are underlined; (4) dashes indicate the absence of data.

Based on the data of Railkin and Dobretsov, 1994.

concentrations of these substances, all animals avoided the capillaries with them, stopping at a distance equal to their body length. The planulae turned the front end of the body, and the mollusks turned the foot under the shell, and moved away. In one-half of the cases and sometimes more frequently, the turning angle was 90 to 180°. The negative chemotactic response was augmented as the concentration of the substance in the capillary was increased. The experiments with chemotactic chambers (Railkin, 1995b) showed that, as the substance entered owing to diffusion through the membrane into the section containing the planulae, the larvae gradually moved away from the membrane, i.e., were repelled.

It should be added that the substances tested at the concentrations providing the maximum repellent effect were not toxic to the test organisms — pediveligers of *M. edulis* and postlarval juveniles of the starfish *Asterias rubens* (Railkin and Dobretsov, 1994). The above data leave no doubts that both substances tested are true repellents.

In the field experiments, plates coated with varnish with or without repellents were exposed in the sea during the period of mass settlement of mussel larvae. The density of these mollusks on control coatings reached 1,000,000 ind./m². Second in abundance were hydroids, whose density constituted up to 20,000 zooids/m². Thus, the colonization pressure of foulers was rather high. Despite this, however, both repellents were efficient enough against macrofouling. The protective action of benzoic acid lasted for about 1 month, and that of N,N,N',N' tetramethylethylenediamine lasted for no less than 1.5 months (Table 10.2). Such a short period of protection was probably conditioned by the sufficiently fast washing-out of repellents from the coatings, because of which their concentration in the outer coating layer dropped below the effective value.

These repellents were also effective enough against microfouling (Table 10.3). The repellent suppression of all major groups of microfoulers (bacteria, diatoms,

TABLE 10.3
Suppression of Microfouling by Repellents

	Relative Fouling, % of Control					
	Bacteria		Diatoms		Heterotrophic Flagellates	
Coating	14 days	21 days	14 days	21 days	14 days	21 days
Varnish + benzoic acid	68 ± 11	60 ± 15	63 ± 15	8 ± 2	52 ± 52	2.0 ± 0.4
	(2.5)	(2.2)	(1.4)	(12.8)	(0.7)	(12.3)
Varnish + N,N,N',N'-	—	57 ± 12	—	4 ± 2	—	1.0 ± 0.4
tetramethylethylenediamine		(3.1)		(13.4)		(12.4)
Antifouling ship enamel	244 ± 21	72 ± 19	61 ± 8	7 ± 2	335 ± 57	2.0 ± 0.5
KHV-5153	(4.9)	(1.3)	(1.1)	(13.4)	(3.97)	(5.1)

Note: See Table 10.2.

After Railkin et al., 1993a. With permission of the *Russian Journal of Marine Biology.*

and heterotrophic flagellates) did not differ significantly ($p < 0.05$) from the antifouling effect of a commercial ship paint.

The data presented (literary and original) show that benzoic and tannic acids, as well as N,N,N',N'-tetramethylethylenediamine, have a broad spectrum of repellent effect, repulsing bacteria, unicellular algae, and larvae of hydroids, cirripedes, and two species of mollusks. Thus, it is quite possible that the same substance may act as an efficient repellent against the dispersal forms of many species, in which case it will suppress the development of a multispecific fouling community. This may to some extent remove the objections against repellent protection (Zevina, 1994), based on the notion that it is difficult to accomplish because of repellent specificity.

10.4 ANTIADHESIVE PROTECTION

The analysis of colonization processes performed in Chapter 7 brought us to an important conclusion: that the crucial event in the whole biofouling cycle was the adhesion (see Chapter 6) of dispersal forms on their contact with a hard surface. It is adhesion and the subsequent firm attachment with the help of biopolymers that makes the colonization of a surface irreversible (see Figure 8.1 and Figure 8.3).

It is evident that antiadhesive protection from biofouling can be based on the suppression of the adhesion of propagules to the hard surface. This idea has been partly implemented in low-adhesion organosilicon polymeric ship coatings, often referred to as fouling-release coatings (see Section 9.1). Strictly speaking, being biocide-free, these coatings do not prevent the process of biofouling. However, organisms accumulated on such coatings can be removed easily even after several years of service.

Apparently, antiadhesive protection can be developed on the basis of not only physical principles but also chemical factors. Despite the great fundamental and practical significance of the processes of adhesion and attachment, the effect of non-biocidal antiadhesive substances has been studied only on some species of foulers.

Consequently, chemical antiadhesive protection is still in the early stages of its development.

Phenolic secondary metabolites of plants are known to have a broad action spectrum (see Section 10.1). Sulphate of p-coumaric acid (zosteric acid), isolated from *Zostera marina*, prevented the attachment of marine bacteria as well as a number of algae and invertebrates (Todd et al., 1993). These facts indicate the possibility of a universal chemical antiadhesive protection.

A widely used method of evaluation of the antiadhesive action of chemical compounds involves counting the byssus threads formed by the bivalve *Mytilus edulis* on coatings containing the substances tested and the respective controls (see, e.g., Ina et al., 1989). A new approach (Hellio et al., 2000), which appears promising, involves the estimation of activity of the enzyme phenoloxidase catalyzing byssus formation in the presence of the substances tested. However, to introduce this method on a larger scale, the data obtained by both methods will have to be compared.

Using *M. edulis* as a test object, the following facts were established (Ina et al., 1989; Etoh et al., 1990). If terpenes of some flowering plants are chemically bound to a polymeric matrix (polyvinyl resin), the number of byssus threads formed by blue mussels on such a coating decreases. The most prolonged effect (up to 3–4 months) was shown by coatings containing the terpene 6-methylthiohexyl isothiocyanate.

Antiadhesive action on *M. edulis* also was exerted by a mixture of fatty acids isolated from the sponge *Phyllospongia papyracea* (Goto et al., 1992) and the sesquiterpene β-bisabolene isolated from the gorgonarian *Sinularia* sp. (Shimidzu et al., 1993). The detachment of the limpet *Megathura crenulata* and the starfish *Pisaster giganteus* from the aquarium wall in the presence of secondary metabolites of some sponges (Thompson et al., 1985) probably can be accounted for by their antiadhesive action on the attachment organs — the foot and the tube feet. Some adrenoceptor compounds such as medetomidine and clonidine inhibited the attachment and metamorphosis in cyprid larvae of *Balanus improvisus* (Dahlström et al., 2000).

Research performed by me and my colleagues (Railkin et al., 1993a; Railkin, 1995b; Railkin and Dobretsov, 1994) at sea and in the laboratory showed that some repellents and anesthetic or narcotizing materials represented quite effective non-toxic antiadhesion agents.

The effect of solutions of N,N,N',N'-tetramethylethylenediamine, benzoic acid, and two barbiturates (5,5-diethylbarbituric and 5-p-diethylamino anilinomethylene barbituric acids) was tested on juvenile mussels *Mytilus edulis* with shell lengths of 1 to 2 mm. All of these substances caused a similar sequence of physiological responses: temporary immobilization, detachment of the sole from the substrate (the Petri dish bottom), and partial or complete relaxation of the abductor muscle. This sequence was observed most distinctly at low concentrations of the agents, when the suppression of attachment and contractility was slower. The speed of suppression depended on the substance and its concentration (Table 10.4). The effects observed were completely reversible. After rinsing in clear sea water, the mollusks restored all of the responses in a reversed order.

The substances mentioned above are likely to exert the same action on pedive-ligers during their settlement on the coatings containing these substances, because at the moment of settlement they possess completely formed mechanisms of adhesion

TABLE 10.4
Reversible Suppression of Locomotion, Adhesion, and Contractility in Juvenile Mussels *Mytilus edulis*

Substance	Concentration, mM	Time of Suppression			Time of Locomotion Recovery, h
		Locomotion	Adhesion	Foot Muscle Contractility	
*N,N,N',N'-*Tetramethylethylenediamine	0.9	12 h	Not suppressed in 12 h	24–48 h (42 ± 2) h	12–24
	3.4	5–15 min	5–15 min	1–20 h (6 ± 1) h	6–24
Benzoic acid	28	Not suppressed in 24 h		Not suppressed in 12 h	—
	83	5–15 min	5–15 min	30–60 min	0.5–1
	248	0–5 min	0–5 min	15–30 min	1–3
	744	0–5 min	0–5 min		1–6
5,5-Diethylbarbituric acid	22	Not suppressed in 72 h	6–16 h	Not suppressed in 72 h	—
	65	6–16 h	5–30 min	48–60 h (50 ± 1) h	24–48 (without rinsing)
	194	5–15 min			1–3
	581	1 min	1 min	4–12 h (8 ± 1) h	3–6
5-*p*-Diethylamino anilinemethylene barbituric acid	12	Not suppressed in 36 h			—
	37	0–5 min	0–5 min	12–39 h (30 ± 1) h	8
	111	1–5 min	1–5 min	1–24 h (8 ± 2) h	3–12
	332	0–1 min	0–1 min	0.5–12 h (2 ± 1) h	20–24

Note: Mean values and their errors are given in parentheses.

After Railkin, 1995a. With permission of the *Hydrobiological Journal.*

and locomotion on a hard substrate. The suppression of fouling by repellents possessing an antiadhesive effect was considered in the previous section.

Additional experiments showed that barbiturates at the concentrations tested were not toxic to juvenile mussels (Railkin and Dobretsov, 1994). The substances were added to a vinyl-rosin varnish that was applied to non-toxic polymeric plates. The coatings were tested in the Kandalaksha Bay coastal area (the White Sea) at a depth of 1 m, during the season of mass settlement of larvae of mussels and hydroids (for details, see Section 10.3).

The experiments showed that, on control coatings (vinyl-rosin varnish without additions), the density of mussels reached 1,000,000 ind./m^2 and that of hydroids reached 20,000 zooids/m^2. At the same time, coatings with an addition of barbiturates considerably reduced macrofouling (Table 10.5) as well as microfouling (Table 10.6). The effect of the coatings with respect to macrofouling lasted for at least 1.5 months. It ceased by the end of that period, apparently owing to the rapid washing-out of barbiturates from the varnish. The stronger antifouling effect of benzoic acid and N,N,N',N'-tetramethylethylenediamine (see Table 10.2), as compared to that of barbiturates, indicates that, in addition to the antiadhesive effect (Table 10.4), they also have a distinct repellent action (Table 10.1).

The above data on antifouling protection using antiadhesive substances show that the same substances can suppress in a non-biocidal way locomotion, adhesion, and attachment not only in invertebrates but also in microorganisms. Such compounds can be used to develop antiadhesive chemical protection from sea biofouling that will be safe for the environment.

10.5 BIOCIDAL PROTECTION

Consideration of the general approaches to ecologically safe anticolonization protection from biofouling (discussed in Section 10.2) has shown that the hazards of industrial protection are not determined by the very fact that it involves the use of biocides. The actual and more essential cause is that the biocides used are rather toxic compounds, which are very slowly transformed or not transformed at all into non-toxic ones. Therefore, they exert long-term effects not only on foulers (i.e., target organisms) but also on any other organism that is within their range of influence, which leads to rather negative ecological consequences.

It is evident that, for biocides to be used for ecologically safe antifouling protection, they must comply to certain conditions. These conditions, in particular, include: ability of biological or spontaneous degradation; low toxicity to man and commercial marine organisms; and finally the absence of delayed hazardous effects, for example, carcinogenic, mutagenic, and teratogenic.

Natural biocidal antifoulants used in chemical defense against epibiosis completely meet these requirements. Antifoulants of macroorganisms include various terpenic, heterocyclic, and halogenorganic compounds; furan, lactone, and phenol derivatives (including polyphenols); pyridines; alkaloids; and other compounds. The second group comprises the biocides produced by microorganisms (see, e.g., Huang and Boney, 1985; Holmström et al., 1992; Clare, 1996; Armstrong et al., 2000). Their chemical nature was characterized only in a few cases. For example,

TABLE 10.5

Suppression of Macrofouling by Antiadhesive Substances

	Relative Fouling, % of Control			
Coating	10–19 days	22–25 days	45 days	70 days
Varnish + 5,5-diethylbarbituric acid	57 ± 9	63 ± 19	27 ± 9	50 ± 19
	(1.6)	(9.6)	(3.2)	(1.4)
Varnish + 5-*p*-diethylamino anilinomethylene barbituric acid	—	26 ± 3	51 ± 7	—
		(10.8)	(2.2)	
Antifouling ship enamel KHV-5153	8 ± 0.1	9 ± 3	15 ± 5	31 ± 6
	(10.5)	(15.3)	(4.0)	(3.8)

Note: See Table 10.2.

Based on the data of Railkin and Dobretsov, 1994.

TABLE 10.6

Suppression of Microfouling by Antiadhesive Substances

	Relative Fouling, % of Control					
	Bacteria		Diatoms		Heterotrophic Flagellates	
Antiadhesives Added to Non-Toxic Varnish	14 days	21 days	14 days	21 days	14 days	21 days
5,5-Diethylbarbituric acid	66 ± 14	118 ± 24	120 ± 18	12 ± 1	119 ± 62	0 ± 0.6
	(2.2)	(0.6)	(0.8)	(12.5)	(0.3)	(12.2)
5-*p*-Diethylamino anilinomethylene barbituric acid	—	437 ± 107	—	34 ± 5	—	45 ± 31
		(3.1)		(1.1)		(1.5)
Antifouling ship enamel KHV-5153	244 ± 21	72 ± 19	61 ± 8	7 ± 2	335 ± 57	2.0 ± 0.5
	(4.9)	(1.3)	(1.1)	(3.4)	(4.0)	(5.1)

Note: See Table 10.2.

After Railkin et al., 1993a. With permission of *the Russian Journal of Marine Biology.*

ubiquinone secreted by the bacterium *Alteromonas* sp. was shown to suppress the settlement of *Balanus amphitrite* cyprids (Kon-ya et al., 1995). The third group includes various natural or synthetic biologically active substances as well as extracts and exudates of marine organisms: lipids, fatty acids, sterols, steroids (including saponins), alkaloids, antibiotics, and other compounds (see reviews: Pawlik, 1992; Clare, 1996). The incorporation of some of them into the coatings has demonstrated that ecologically safe protection from biofouling is possible. I shall give only a few examples. When added to a coating composition, the extracts of some microorganisms (the bacteria *Vibrio, Bacillus, Escherichia* and the yeast *Saccharomyces*) were efficient against biofouling (Ina and Takahashi, 1990). The addition of some enzymes (for example, proteases) to the antifungal antibiotic natamycin enhanced its protective antimicrobial effect in a varnish (Noël, 1984). The bacteria isolated from non-fouled surfaces of the brown alga *Fucus serratus* and the nudibranch *Archidoris pseudoargus* were used as a source of a biologically active fraction that protected a polymeric coating from bacteria (Armstrong et al., 2000). A total of 51 species of sponges with little or no fouling on them were collected off Curaçao Island. When their extract was added to a non-toxic paint, the plates coated with the paint got half as much fouling as the plates coated with clear paint (Laban, 1993). Extracts of starfish added to a coating were also effective against biofouling (Tanimizu, 1964), possibly due to the biological action of the saponins contained in them. Some monoterpenes and diterpene-sugar esters that were isolated from two eucalyptus species suppressed the attachment of adult mussels to the hard substrate (Etoh et al., 1990). The same effect was shown by isothiocyanate derivatives containing the SCN group (Ina et al., 1989), which appears to be responsible for their antifouling action. Complex salts of hydrocyanic acid, containing the CN group, have been suggested as safe antifouling agents (Gansloser and Nissenbaum, 1982).

These examples show the possibility of using organic matters for biocidal antifouling protection. Although their biological transformation into non-toxic or low-toxic compounds has not been specifically studied, one may assume that ecologically safe protection based on such compounds can theoretically be developed.

Oxides of copper, lead, and zinc are used in industrial antifouling protection. However, the range of potential inorganic antifoulants is not limited to these compounds. In my opinion, a promising direction is studying the possibility of using reactive oxygen species in antifouling protection systems (Railkin et al., 1987). These forms include hydrogen peroxide, H_2O_2; ozone, O_3; singlet oxygen, $^1O_2$; hydroxyl radical, $OH^.$; hydroperoxy radical $HO_2^.$; and oxygen anion-radical, $O_2^{.-}$, also called superoxide or dioxide. Their chemical properties and their role in biological systems are considered in several reviews (e.g., Afanas'ev, 1979; Myers, 1980; Forman and Boveris, 1982; Mason, 1982; Shinkarenko and Aleskovskii, 1982; Vladimirov et al., 1991; Rice-Evans and Burdon, 1994).

Reactive oxygen species, in particular radicals, can damage important biological macromolecules (polysaccharides, lipids, proteins, enzymes, DNA, RNA, and others) and cause peroxidation of lipids of the cell membranes. In other words, they can inflict damage that in the long run can prove fatal to the cells (e.g., Vladimirov et al., 1991; Winston and di Giulio, 1991; Chaudière, 1994).

Having high reactivity, active forms of oxygen show a non-selective biocidal effect, which means that they can be used against any species of foulers. The most toxic of them, such as the radicals and singlet oxygen, are very quickly (in fractions of seconds or minutes) transformed in water into non-toxic products and have a short range of diffusion (Afanas'ev, 1979; Shinkarenko and Aleskovskii, 1982), which suggests a local action. In view of these facts, reactive oxygen species may theoretically be used for ecologically safe antifouling protection. In my opinion, the most promising in this respect are ozone, singlet oxygen, and superoxide, whose toxicity is intermediate between the most toxic hydroxyl radical and the least toxic hydrogen dioxide. Chemical properties of these forms are known well enough, and they can be produced as individual compounds in large quantities.

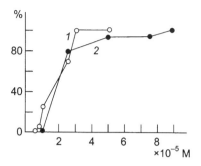

FIGURE 10.6 Toxicity of some porphyrins to *Paramecium caudatum*. Abscissa – concentration of porphyrin solution, M; ordinate – mortality, %. (1) Disodium salt of cobalt hematic acid IX complex, (2) tetrasodium salt of cobalt tetra-(*p*-sulphophenyl)-porphin complex. (After Railkin et al., 1984. With permission of *Doklady Akad. Nauk.*)

Reactive oxygen species are produced during metabolism in prokaryotic and eukaryotic cells as a result of redox reactions. Intracellular protection from these matters is ensured by enzymes, such as glutathione peroxidases, catalases, and superoxide dismutases (SODs), which catalyze transformation of reactive oxygen species into less toxic or non-toxic ones (Winston and di Giulio, 1991). Protection from oxygen anion-radical at the cell level is performed by SODs. In bacteria, this enzyme is present in two forms (Gregory et al., 1973): Mn-SOD occurs in mitochondria and Fe-SOD in the peryplasmic space. The former performs the function of intracellular protection from oxygen, and the latter protects mostly from oxygen coming from external sources. Eukaryotic unicellular and multicellular organisms possess only an intracellular Cu-SOD, which differs in a number of properties from the corresponding bacterial enzymes (McCord and Fridovich, 1969). It probably does not provide sufficient enough protection against exogenous oxygen. Therefore, protection from eukaryotic foulers with the use of superoxide may be quite efficient.

Superoxide can be obtained by radiolysis of water as well as by electrochemical, photochemical, enzymatic, and non-enzymatic methods (Merzlyak and Sobolev, 1975). For example, in model systems, it can be produced by a catalytic one-electron reduction of oxygen by cobalt porphyrin complexes (Enikolopyan et al., 1983). As a result of the coordination of an oxygen molecule with the cobalt ion, an oxocomplex is formed, within which an electron is transferred to oxygen. The complex then disintegrates, producing an oxygen anion-radical.

Model experiments showed (Railkin et al., 1984, 1987) that cobalt porphyrins were toxic to a eukaryotic unicellular organism — the ciliate *Paramecium caudatum*, used in toxicological tests (Figure 10.6). In increasing order of their toxic effect, the substances studied can be arranged in the following sequence: disodium salt of cobalt hematic acid IX complex, tetrasodium salt of cobalt tetra-(*p*-sulphophenyl)-porphin complex, and cobalt meso-(tetra-N-methylpiridine)-porphin complex.

In solutions of these substances, the swimming velocity of the ciliates decreases gradually, and subsequently the protists die. At the first stage, the toxic effect is reversible. The addition of an equimolar concentration of SOD not only restores cell motility but also keeps the ciliates alive (Figure 10.7). The positive test with reduction of Nitro

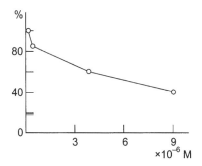

FIGURE 10.7 Suppression of the toxic effect of tetrasodium salt of cobalt tetra-(p-sulphophe-nyl)-porphin complex at a concentration of 6.4×10^{-5} M, by superoxide dismutase. Abscissa – concentration of superoxide dismutase, M; ordinate – mortality of *Paramecium caudatum*, %. (After Railkin et al., 1984. With permission of *Doklady Akad. Nauk.*)

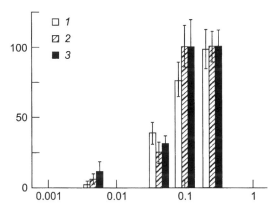

FIGURE 10.8 Toxicity to *Paramecium caudatum* of cobalt-tetraphenylporphin (1) and its complexes with polyvinylpirrolidone (2) and acrylic acid (3). Abscissa – concentration of the complexes, mM; ordinate – mortality, %.

Blue tetrazolium chloride into formazan (Nishikimi et al., 1972) confirms that the death of the ciliates is connected with the appearance of electrons in the medium, whereas the antitoxic effect of SOD indicates that the superoxide-based mechanism of toxicity is highly probable in this case.

The mortality of ciliates depends on the porphyrin concentration. The protists may die in some minutes, hours, or after a longer period. The average daily lethal dose of one of the biocides, cobalt tetra-(p-sulphophenyl)-porphin complex, is 1.1×10^{-6} M. A group of chemists from Moscow and Samarkand succeeded in immobilizing this biocide on a firm matrix by means of copolymerization of por-phyrin with N-vinylpirrolidone, and also with acrylic acid. It is very important that the biological activity of the biocide was preserved at the initial level after such an immobilization (Figure 10.8).

How does such a model coating work? The immobilized chemical agent, for example, a cobalt complex of porphyrin, activates the dissolved oxygen by donating an electron. As a result, a superoxide is formed and the complex itself acquires a positive charge. The reduction of the complex proceeds due to the electron donors that are routinely present in the water medium. Therefore, the cobalt porphyrin complex can be involved in a cyclic process. Donating an electron, it activates oxygen, and, receiving an electron, it becomes ready to activate the next oxygen molecule. In its turn, superoxide dismutates spontaneously, reacting with water to produce hydrogen dioxide and singlet oxygen:

$$2\ O_2^- + 2H_2O \rightarrow H_2O_2 + {}^1O_2 + 2HO^-. \qquad (10.1)$$

The main toxic effect will be determined by the rate of superoxide and singlet oxygen generation and by their concentrations in the outer layer of the coating. The half-life of superoxide at a pH value typical of seawater is about 1 min (Afanas'ev, 1979). Hydrogen dioxide is low-toxic. Conversely, singlet oxygen is highly toxic to both unicellular and multicellular organisms (Shinkarenko and Aleskovskii, 1982). However, in water this form of oxygen is very quickly (in 10^{-5} s) transformed into its common triplet state, i.e., into the oxygen that we breathe. The toxic action of reactive oxygen species is likely to be observed in the near proximity of the surface protected and to have no negative ecological effects.

Experiments with polymeric matrices containing metal porphyrin complexes that were carried out under White Sea conditions showed that some of these compounds, including cobalt tetra-(p-sulphophenyl)-porphin complex, provided sufficiently effective protection from microfouling (Railkin et al., 1987). The swelling capacity of matrices in water and, accordingly, the rate of washing-out of porphyrins from them increased in the sequence: poly(methyl methacrylate), polyglycidyl metacrylate, butyl metacrylate. At the same time, the antifouling effect of cobalt porphyrins decreased in the same sequence. This suggests that the protective action of cobalt porphyrins is at least partly determined by the catalytic activation of oxygen and not only by their toxicity to microfoulers.

What are the prospects of developing antifouling methods that employ reactive oxygen species? These compounds are not xenobiotics, i.e., materials of non-natural origin. On the contrary, they are secreted in water by microorganisms and plants (Skurlatov et al., 1994) and also some invertebrates (Thomason et al., 1996; Peskin et al., 1998) as a by-product of redox processes. Reactive oxygen species play an important role in the self-purification of natural waters, since they participate in the oxidative destruction of organic matters (alkanes, phenols, etc.). As compared to organic derivatives of chlorine, organic ozone derivatives are less toxic (see, e.g., Sokolova and Markov, 1985). All of the above suggests that reactive oxygen species can be used to develop biocidal ecologically safe protection from marine biofouling.

What are the specific features of hypothetical coatings based on immobilized compounds that serve as a source of active oxygen, for example superoxide? What distinguishes them from copper-containing paints? First, the mechanism of action of such coatings is not connected to the diffusion of the biocide from the coating, as is the case with industrial chemobiocidal protection. Their action is conditioned by the activation of dissolved oxygen in the direct proximity of the surface being protected, within the boundary layer, i.e., exactly where the dispersal forms of foulers settle and attach. In addition, superoxide and products of its transformations will have a rather restricted range of biocidal action, and they will not be accumulated in the environment and in organisms. Thus, there is good reason to assume that reactive oxygen species, including superoxide, are promising agents of ecologically safe protection from marine and freshwater biofouling.

10.6 PROSPECTS OF DEVELOPING ECOLOGICALLY SAFE ANTICOLONIZATION PROTECTION

Antifouling protection aimed at the suppression of colonization processes employs a variety of ideas, approaches, and methods. Many of them have been considered in this chapter, but the prospects and possible directions of development of ecologically safe anticolonization protection from biofouling have not yet been discussed in detail. However, this must be done, especially in view of the ban imposed in 2003 on rather efficient organotin-based industrial coatings.

The most radical approach to antifouling protection would obviously involve preventing the contact of organisms with the hard surface. This can be achieved by different methods, depending on the situation. For example, in closed or semi-closed water circuits, such as in industrial cooling systems or potable water treatment plants, it is possible to use water from which the larvae have been removed by filtration or sterilization (e.g., Razumov, 1969; Turpaeva, 1987b; Walker and Percical, 2000). Visits to a freshwater port by ocean-going vessels prevent their fouling by marine species (Zevina, 1994; Wahl, 1997). There is a patent (Duddrige and Kent, 1985) on a method of antifouling protection that isolates the protected surface with a special disposable coating. In practice, however, situations in which the protected object can be completely isolated from contact with foulers are quite rare. Much more frequently, the protected object is subject to a continuous or periodic flux of dispersal forms of foulers, which determines the need for constant or periodic antifouling protection.

Fouling of surfaces of man-made constructions consists of a series of colonization processes. For sessile species, this sequence becomes irreversible at the stage of permanent attachment (Figures 8.1 and 8.3). Therefore, protection from biofouling can be most efficiently carried out by suppressing transport, settlement, adhesion, and temporary attachment, i.e., the processes preceding permanent attachment.

A fairly radical method involves killing all of the propagules of foulers around the protected object with the use of biocides. It is implemented, in particular, when water is treated with chlorine dioxide in potable water supply systems (Ambrogi, 1993; Geraci et al., 1993) or with chlorine and its derivatives during the electro-chemical control of marine biofouling (e.g., Smith and Kretschmer, 1984; Usachev and Strugova, 1989). Antifouling coatings containing organotin and copper compounds are also rather effective in similar situations. These methods of protection actually destroy the flux of living foulers onto the protected surface, i.e., conditionally speaking, they block their transport. However, industrial protection using biocides is ecologically hazardous.

The flux of propagules can be also interrupted by non-biocidal chemical methods. This can be achieved, for example, by using anesthetic and narcotizing agents, in particular chloroform, menthol, carbon dioxide, chloral hydrate, hydroxylamine, magnesium chloride (Kaplan, 1969), nickel or cobalt ions (Railkin and Seravin, 1989), analogs of diterpenes (Clare et al., 1999), and other compounds. In model experiments, a curtain of carbon dioxide bubbles offered short-term protection of a surface from fouling by cirripedes in the Black Sea (Terent'ev et al., 1966). Calcium channel blockers efficiently suppress locomotion in both microfoulers (Cooksey 1981, Geesey et al., 2000) and macrofoulers (Railkin and Seravin, 1989); their action appears to be reversible and, as a rule, is not connected with any toxic effect. Thus, the compounds mentioned could be considered as potential antifoulants, given the overall or sufficiently wide spectrum of their action (Tsukerman, 1983; Railkin et al., 1990; Rittschof, 2000). However, it should be borne in mind that antilocomotor protection can work only against those species whose dispersal forms are motile; it would obviously fail in the case of immotile spores of macroalgae and many micro-foulers.

The idea of antifouling protection using chemical repellents (e.g., Seravin et al., 1985; Railkin et al., 1990; Wahl, 1997; Rittschof, 2000) is rather productive. Its advantages arise from the fact that the repellent effect, as a rule, does not involve toxicity. Therefore, repellent-based protection is very likely to be safe for the environment. There are, however, two aspects that can impede the development of such protection. First, among the variety of known antifoulants, excluding biocides, compounds with a broad action spectrum are difficult to find. Second, behavioral tests of antifoulants using a wide variety of objects are highly labor-consuming. The latter problem can be solved in part by exposing to the ocean the coatings containing the materials to be tested. However, this does not exclude the need for behavioral tests in the laboratory. At the same time, the results presented in Section 10.3 show that some materials, such as benzoic and tannic acids and some other compounds, are repellent to a number of macrofoulers and, apparently, to many microfoulers as well. Unfortunately, this does not mean that such compounds will efficiently repel all principal sessile foulers. On the other hand, the data just mentioned allow one to address the problem of the existence of similar ways in which repellent cues are received in various organisms (from bacteria to mollusks). This problem, although poorly studied with regard to negative chemotaxis (chemokinesis), is very significant from the evolutionary and practical points of view. It should be emphasized that, in addition to the case of antilocomotor protection, repellent protection can be used only against foulers that have motile dispersal or settling stages. Therefore, it must be supplemented by other means of protection that are directed against immotile propagules.

During the colonization of a hard surface, all organisms, both relatively motile and immotile ones, necessarily come in contact with the surface, which is accompanied by physical adhesion and subsequently by attachment to it. The attachment may be temporary, controlled by both physical and biological mechanisms (Chapter 6). The physical mechanisms of adhesion are essentially the same in microorganisms, animal larvae, and macroalgal spores. Therefore, one may assume that antiadhesive protection is more likely to yield an all-purpose effect rather than repellent protection. The standard approach to antiadhesive protection is to decrease the surface energy. It is successfully realized in the modern silicone fouling-release coatings. However, the opportunities for antiadhesive protection are not limited to these coatings. Adhesion and attachment can be blocked not only physically or mechanically but also chemically. Examples of this have been considered in Section 10.4. The possibility of suppressing the adhesion and attachment of mollusks by barbiturates that possess narcotizing properties (Railkin, 1995b) may indicate the similarity or even the identity of some mechanisms underlying the narcosis and suppression of adhesion in bivalves. The protective effect of barbiturates under the marine conditions, where many species of potential foulers occur, suggests that universal chemical antiadhesive protection is possible in principle.

The chemical antifouling protection aimed at suppressing metamorphosis, feeding, and growth of the organisms populating hard substrates is also discussed in the literature (e.g., Fusetani, 1987; Paul and Fenical, 1987; Railkin et al., 1990; Teo and Ryland, 1994; Clare, 1996; Wahl, 1997; Wright et al., 1997; Cetrulo and Hay, 2000; Rittschof, 2000). Theoretically, such protection should be less efficient than the methods directed at suppression of the early colonization processes. It is logical to assume that suppression of metamorphosis, feeding, or growth of the settled and attached organisms will result in their elimination. However, their bodies will create an insulating layer and screen the chemical protective action of the coating (Rudyakova, 1981; Rittschof, 2000).

The ideas of protection from macrofouling by suppressing microfouling or regulating its composition have gained some popularity (e.g., Clare, 1996; Steinberg et al., 1997; Wilsanand et al., 1999; Armstrong et al., 2000). Such an approach is based on the idea that, during the initial succession, the stage of microfouling precedes, and in a certain way prepares for, that of macrofouling (Chapter 2). There are also data (e.g., Robinson et al., 1985; Dobretsov and Railkin, 1994) showing a high correlation between the abundance of microfoulers and mass species of macrofoulers on the substrate occupied by both groups. At the same time, it is known that the larvae of invertebrates (and spores of algae) can colonize on surfaces with weakly developed microfouling or no fouling at all (see Section 5.5 and Section 5.8). During the long life in the plankton, the substrate selectivity of larvae decreases (see Section 4.4). In view of all of this, suppression of microfouling can hardly be considered a reliable protection from macrofouling. However, protection from microorganisms by itself may impede corrosion and improve the operation characteristics of industrial objects, for example, heat exchangers (Chapter 1).

Biocidal protection based on natural antifoulants or their analogs has indisputable prospects, because it meets the basic criteria of ecologically safe protection (see Section 10.2). First, antifoulants are effective against principal foulers. Second, their teratogenic, mutagenic, or carcinogenic effects are not known. Third, they are biodegradable and consequently cannot accumulate in water or in organisms and have hazardous effects on the ecosystem. Therefore, they may be assumed to be ecologically safe.

More than 100 of the identified antifoulants of various chemical natures have presently been isolated from microorganisms, algae, and animals (see reviews: Elyakov and Stonik, 1986; Paul and Fenical, 1987; Pawlik, 1992; Clare, 1996; Slattery, 1997; Targett, 1997). However, it is difficult to choose from among them the specific materials to be used in ecologically safe antifouling protection. The problem is that laboratory and field tests of antifoulants are carried out using a variety of techniques and biological objects, so that a comparative analysis of their antifouling effect is difficult. It obviously would be practical to unify both the procedures used and the test organisms. Important recommendations on this problem are given in the reviews (Wahl, 1997; Rittschof, 2000). It should be noted that the main question, namely, that of using the results of short-term laboratory or field tests to predict the possible biological and ecological consequences of long-term applications of antifoulants in the protection systems, is practically not covered in the literature. The principles and methods of predicting the ecological impact of antifoulants remain poorly studied, which evidently impedes the development of ecologically safe protection.

Nevertheless, based on the published data (Fischer et al., 1984; Houghton, 1984; Seravin et al., 1985; Railkin et al., 1990; Wahl, 1997; Rittschof, 2000), we can define some general conditions that must be met by antifoulants for them to be used in antifouling protection systems:

1. Versatility (= multiple activity), i.e., a wide spectrum of protective actions against the dispersal forms of all (or at least the principal) groups of foulers
2. Multilevel effect, i.e., the suppression of not one, but two or several colonization processes simultaneously
3. The suppression of processes constituting the reversible phase of colonization, such as transport, settlement, adhesion, and temporary attachment
4. Antifouling effect at low concentrations

It should be noted that secondary metabolites of the basibionts, for example, macroalgae, may act differently on different species of foulers, exerting a toxic effect or suppressing settlement in some cases and, on the contrary, stimulating settlement in others (Walters et al., 1996). Taking this and other facts into account, the most promising antifoulants to be used in protection systems appear to be natural phenolic and terpenic biocides, furans, lactones, and antimicrobial metabolites of sponges (see Section 10.1) or their synthetic analogs. They are to be found in many marine algae and animals and have a wide spectrum of biocidal actions.

The simplicity of some ideas of biological antifouling protection is appealing. For example, it has been proposed that preventing the colonization of a surface by foulers may be achieved by introducing their competitors or predators; for example, nudibranch mollusks could be used to eliminate hydroids (Seravin et al., 1985; Minichev and Seravin, 1988). Despite a certain degree of attractiveness, the proposed methods are not practically feasible and, apparently, may have a rather restricted application, for example, in mariculture.

In my opinion, general ideas and concrete technologies of surface immobilization of antifoulants have good prospects; these antifoulants may be both biocidal (Hüttinger, 1988) and non-biocidal (Seravin et al., 1985; Railkin et al., 1990). One of the relevant examples was considered in Section 10.5, in connection with the experimental development of antifouling protection using reactive oxygen species. The activation of dissolved oxygen and the formation of an oxygen anion-radical owing to the electron donor immobilized on a surface have significant advantages over the protection by chemical agents that are being washed away from a coating. Such advantages include local action, near-zero consumption rate, and a long period of protection.

It should be borne in mind that different man-made objects, owing to their operational specificity, will probably require different approaches to their protection. In particular, fouling of vessels must be suppressed completely for several years, and such objects can be protected by environmentally safe biocides. On the contrary, culture cages for scallops (a commercially important bivalve) obviously must be protected by non-toxic compounds. The inner walls of pipes in industrial cooling systems may in principle be protected by fairly strong toxins that must not, however, enhance their corrosion and pollute the environment. Since there are no clear criteria to evaluate the prospects of the various directions of anticolonization protection, one has to conclude that all of these directions may be explored further. However, in the near-term outlook, efforts toward the development of ecologically safe biocidal and antiadhesive protection appear to be justified. It is in these directions that the most impressing practical results have already been achieved. It may be promising to combine the two approaches to design protective coatings. In this connection, I believe that the immobilization of some natural biocidal antifoulants on silicone fouling-release coatings would make it possible to develop ecologically safe protection with a prolonged effect that could be applicable for various types of man-made objects.

It is to be hoped that harmless methods for protecting man-made structures against biofouling that are similar to those existing in nature will be developed. The general ideas and principles considered in this book, and especially the formal-logic approach to the problem presented in the following chapter, may serve as guides in this intricate process.

11 The General Model of Protection against Biofouling

Until recently, the development of new methods of protection against biofouling has proceeded mostly on an empirical basis. With the development of the basic quantitative theory of colonization (Chapter 7) and the concept of anticolonization protection (see Chapter 10), there appears to be a way to analyze the theoretical foundations of protection against biofouling.

Let us consider the mathematical model of accumulation (see Chapter 7). On the basis of Equation 7.7, the biomass B of biofouling on a hard surface can be expressed as follows:

$$B = \alpha K_a b \left(f(U_\infty, x, y) + V \right) t , \qquad (11.1)$$

where α is the coefficient of settlement selectivity, K_a is the adhesion coefficient, b is the biomass of dispersal forms in the plankton per unit volume, $f(U_\infty, x, y)$ is the distribution function of the flow rate within the boundary layer, V is the speed of settlement of dispersal forms, and t is time. It should be borne in mind that α denotes the fraction of propagules settling from the plankton onto the hard surface in unit time. The coefficient K_a describes the fraction of propagules that adhere to the surface, out of the total number of organisms coming in contact with the surface during time t. The function $f(U_\infty, x, y)$ determines the flow rate within the boundary layer at a point with the coordinates (x, y), with U_∞ being the velocity of free (outer) current. Thus, Equation 11.1 describes the fouling of any surface irrespective of its shape and dimensions.

The necessary and sufficient conditions of complete protection from biofouling can be found by putting to zero the left part of Equation 11.1, which describes the fouling biomass. Let us assume that $B = 0$, i.e., fouling is completely absent. From Equation 11.1, this can be the case, in particular, when any of the following conditions are met:

1. $\alpha = 0$.
2. $K_a = 0$.
3. $b = 0$.
4. $f(U_\infty, x, y) = 0, \quad V = 0$.
5. $V = -f(U_\infty, x, y), \quad V > 0$.

The case $t = 0$ is trivial, since it means that the hard surface (object) has not been immersed in water. Other apparent solutions of Equation 11.1 at $B = 0$ represent combinations of those shown above and may be analyzed by the reader. To suppress the fouling of a hard surface, including that of an engineering object, it is sufficient to meet any of the above conditions.

Let us analyze in greater detail the consequences of the five solutions of Equation 11.1 at $B = 0$. They can be interpreted as general directions of prevention of biofouling, protection from it, and extermination of the existing biofouling. Null solutions may exist in any of the three situations: prevention of biofouling; suppression of biofouling, or protection proper; and elimination of biofouling that has already formed. Taking these three situations into account, one may consider the theoretical basis of biofouling control as comprising 15 directions. Each of them does not represent a particular method or a technique of protection but, on the contrary, may include a variety of methods. An attempt at formal analysis undertaken here allows one to offer a general model of antifouling protection. It describes what happens to the dispersal forms during the prevention, protection, and elimination of biofouling, leaving room for creative imagination as to how a particular result can be achieved. It has not been been possible in all cases to find published examples of a particular direction of protection. This shows that the proposed model, apart from the ones that we know about, includes essentially new and unexpected approaches to the problem of protection, and therefore it has a certain prognostic value.

Cases 1–3: $\alpha = 0$. The methods are aimed at modifying or suppressing the behavioral (locomotor) reactions of the propagules (microorganisms, larvae of invertebrates, and spores of macroalgae) by creating such conditions around the protected surface under which they do not move toward the surface or do not settle on it.

Suppression of movement toward the protected surface (case 1) means that some distant-acting measures have been taken to prevent the propagules from approaching the surface. The simplest examples are the distant biocidal action of chlorine in the case of volume-based protection (Yakubenko, 1990) and the immobilizing action of carbon dioxide (Terent'ev et al., 1966). Besides purely chemical protection, physical protection or a combination of the two methods are possible. For example, the barrier may exist in the form of a screen of air bubbles (Rasmussen, 1969a) or by air bubbles combined with underwater spraying of a potent toxin, such as *bis*(tributyltin) oxide, and a surfactant mixed with kerosene (Gurevich et al., 1989). These substances can be applied onto the submerged part of the hull, killing the dispersal forms and preventing their settlement and attachment. Suppression of settlement with the use of biocides is a fairly common phenomenon in epibiotic relations (see Section 10.1).

Case 2 implies active rather than passive protection and characterizes the effect (or effects) aimed at suppressing movement toward the surface. The best example of this type is the use of repellents (see Section 10.3).

Case 3, in my opinion, should be interpreted as movement away from the surface after immediate contact with it or after settlement. This may happen in the natural course of events, when an improper substrate is rejected in the substrate selection phase (see Chapter 4). By analogy with contact-action settlement inductors, one may assume the existence of materials that would suppress settlement after contact chemoreception by the dispersal forms. These hypothetical materials would, in the final reckoning, prevent the substrate selection and settlement and thus protect the surface from colonization. They can be assigned to neither repellents nor antiadhesives. Even though such materials have not been described yet, their existence can be predicted by the general model of antifouling protection and is theoretically quite probable.

Cases 4–6: $K_u = 0$. Measures are taken to prevent or suppress adhesion of dispersal forms to the surface. Adherence can be prevented (case 4), for example, if the surface is insulated from dispersal forms. It has been proposed (Duddrige and Kent, 1985) that the inner walls of sea water conduits can be protected mechanically by means of disposable insulating lining, on which biofouling would be accumulated.

Suppression of adherence (case 5) means that either the surface or the integuments of the dispersal forms are not sticky. Industrial ship coatings with low adhesion (the so-called fouling-release coatings) are well known. Another possibility would be to create a rapid flow of water around the protected object, with the current velocity exceeding 1 to 2 m/s. This would violate the main condition of biofouling (see Section 7.1): namely, the shear force and lift force combined would exceed the adhesion force.

One of the ways to eliminate adhesion (case 6) involves removing the dispersal forms that have adhered to the surface. This can be done by a variety of methods, including mechanical cleaning. A common natural mechanism is peeling of old or dead parts of tegument, together with the foulers attached to them. This is observed in macroalgae and also in animals, in particular during molts (see Section 10.1).

Cases 7–9: $b = 0$. Measures are taken to prevent the dispersal forms from appearing near the surface, to eliminate them, or to suppress their development on the surface. Equally, one may consider the removal of the surface from the area where dispersal forms are present.

A number of examples of fouling prevention (case 7) are given below. In order to protect the piping system of Azovstal Iron and Steel Works, E. P. Turpaeva (1987b) proposed using sea water from which macroorganisms have been removed, for example, by filtration. Contact of the protected surface with water containing dispersal forms of foulers can be avoided by prevention measures, for example, by visits of vessels to freshwater ports, such as the port of St. Petersburg (Zevina, 1990).

Suppression of biofouling by creating a propagule-free water layer near the protected surface is theoretically possible. This could be accomplished by drawing off water together with the propagules from the boundary layer and ejecting it beyond the vessel's contours. Besides antifouling protection, this would increase the speed of the vessel. In fact, such a method of protection is unlikely to be realized in the near future, owing to the high energy demands.

Direct elimination of dispersal forms from the protected surface in such a way that they would simply disappear ($b = 0$), for example, dissolve, cannot yet be illustrated by appropriate examples.

Cases 10–12: $f(U_\infty, x, y) = 0$, $V = 0$. Water flow around the protected object and the movement of dispersal forms are simultaneously prevented or suppressed.

The difficulty of finding methods of protection that comply with the above conditions is that the water flow velocity and the movement velocity of the dispersal forms must be simultaneously equal to zero. In principle, this can be achieved by freezing the water around the protected object, so that the vessel would move in an envelope of ice. Incidentally, its outer surface would remain constantly smooth, reducing fuel consumption and possibly compensating for part of the energy spent on creating the ice envelope.

Cases 13–15: $V = -f(U_\infty, x, y)$, $V > 0$. In my opinion, we are dealing here with prevention or suppression of the movement of propagules relative to the surface, probably within the boundary layer, at a non-zero flow velocity.

At present, it is difficult to propose how this method could be realized. Under the above conditions, the movement of propagules must be directed against the water flow. Moreover, the speed of their locomotion must vary at different sites on the surface, while at the same time being exactly equal to the flow velocity within the boundary layer. In other words, the propagules in this case are not carried in any direction within the boundary layer despite the water flow in it. Although this looks like an unsolvable controversy, such paradoxes often give rise to novel creative ideas (Altshuller, 1986).

Further analysis of the biofouling processes described by the colonization models (see Chapter 7) allows one to propose a total of 15 independent directions of prevention and elimination of biofouling. Each of these directions may be realized in a variety of methods and specific techniques. This indicates vast prospects of further theoretical and practical work. In addition, this demonstrates once more the value of the general approach that I have followed in this book: namely, consideration of biofouling, i.e., the process of colonization of hard surfaces by hydrobionts, as a regular sequence of more elementary events.

12 Conclusion

The colonization of marine hard substrates is a chain of natural events by which the highly productive communities of coral reefs, underwater rocks, hard ground, macroalgae, and animals are formed. Colonization results in the concentration of huge biomasses on submerged bodies, which substantially exceed the corresponding values for soft ground. This fundamental property of organisms, namely, their ability to concentrate on hard surfaces, has long been used in the mariculture of mollusks and algae and, over the last decades, also in the development of artificial reef communities. The concentration of organisms on the surface of technical objects is a negative phenomenon that seriously hampers their operation.

Despite a certain degree of variability, hard bodies have many common properties in regard to the organisms that inhabit them. These are, first of all, soundness, physical strength, relative stability in time, and integrity. These properties, combined with environmental conditions (primarily the current), largely determine the similarity of the principal life forms inhabiting hard substrates.

Dominant among these life forms are attached organisms adapted to holding on to the surface of hard bodies in the current and able to obtain the food carried by the flow. Accordingly, the communities inhabiting hard substrates form a single ecological group (Chapter 1). This thesis has been reliably confirmed by a detailed consideration of the colonization processes and cycles (Chapters 3–8) as well as the succession of hard substrate communities (Chapter 2). Thus, our attention has been focused on the organisms and communities living on hard substrates, and also on the colonization processes that determine the formation and development of such communities.

Two global problems were covered in this book: how and why a high abundance and biomass of organisms is reached in the marine environment of hard natural and artificial substrates, and how man-made structures can be protected from colonization by these organisms in an environment-friendly way. The concentration of foulers on hard substrates (see Section 1.2) and its suppression in the course of antifouling protection are diametrically opposed processes. In the first case, organisms are accumulated on hard substrates, so that their biomass exceeds that of soft ground by tens and hundreds of times. In the second case, the concentration of foulers becomes completely suppressed as a result of special measures. It has been possible to consider and analyze these two problems in one book because the formation and development of communities on underwater rocks, hard ground, coral reefs, macroalgae, animals, man-made structures, etc., are based on the same mechanisms and processes, the most essential of which are settlement, attachment, development, and growth. However, the intensity of these processes on natural and artificial substrates with different properties is not the same.

The analysis of organism concentration on hard bodies performed in this book allowed us to characterize the general causes of this phenomenon; however, many mechanisms remain poorly studied. It is obvious that the concentration of organisms on hard substrates is based on colonization processes the sequence of which resembles a one-way conveyor. It is powered by the current, which constantly brings microorganisms, and also macroorganisms in warm waters, to the hard substrates. In temperate and cold water, the colonization of hard substrates by macroorganisms occurs periodically and is enhanced during the periods of settlement of the dispersal forms.

Planktotrophic larvae, which spend considerable time in the plankton before settlement, are transported by the current, the velocity of which exceeds that of the larva's own locomotion by an order of magnitude or more. The interaction of the larvae with the flow requires further studies, especially with regard to the conditions of turbulent mixing, which is quite distinct not only near the surface but also in the near-bottom layer. It is the consideration of turbulent flow events that allows us to better understand the real processes and mechanisms of hard-substrate colonization in the ocean.

Despite a considerable wealth of data concerning the colonization processes, some of their aspects remain obscure. For example, the mechanisms of stimulation and induction of settlement of the dispersal forms of macroorganisms are insufficiently studied. Only 11 mechanisms of the 18 theoretically possible have been described (see Chapter 5). The mosaic pattern of hard substrates with their various signal properties, and also the signal role of light, gravity, and, for some species, also the current, seem to determine the corresponding mosaic pattern of settlement on natural and artificial substrates. Nevertheless, the hierarchy of the cues produced by the substrate and its environment is very poorly studied.

Our knowledge of taxes, attractants, and repellents is scanty even for the dominant fouling species. In many cases, the reasons for their mass settlement on some substrates and complete absence on other substrates cannot be explained completely. It should be noted that behavioral tests are seldom used in the studies of chemotaxis. However, in order to study colonization and to develop repellent-based protection from biofouling, it is absolutely necessary to apply, develop, and unify the corresponding techniques of behavior study (Section 10.3).

Despite the immense importance of attachment for the colonization of hard substrates (see Chapter 6), we are still far from understanding many of its biological mechanisms. Still, using the available data on the structure, properties, and synthesis of byssus in bivalves, it was possible to develop biological glues providing fast and reliable bonding in moist atmosphere and under water. Such glues are really needed in underwater construction, attendance of vessels and other objects, and until recently this problem has had no satisfactory technical solution.

On the other hand, understanding the intricate mechanisms of adhesion and attachment would allow ecologically safe antiadhesive protection from biofouling to be developed (Section 10.4). Modern silicone fouling-release coatings with low adhesion can be cleaned of fouling very easily even after several years of marine exposure. Such coatings are evidently very promising. In my opinion, progress in the field of antiadhesive protection is connected both with further development of the theory of physical adhesion and with improvement of our knowledge of the biological mechanisms of attachment.

The quantitative models of hard-surface colonization (Chapter 7) that roughly describe the initial stages of this process provide the possibility of controling the accumulation and growth. On the one hand, these models explain the general mechanisms of organism concentration on hard substrates (Chapters 7–8); on the other hand, they allow one to consider the possible methods of ecologically safe anticolonization protection from biofouling (Chapters 10–11). In this connection, the possible application of mathematical models of colonization to mariculture of macroalgae and invertebrates should be mentioned. This aspect has been referred to repeatedly in the book. However, these and similar problems should be the subject of future research.

I believe that our understanding of the specific nature of chemical cues of settlement, attachment, and metamorphosis of the dominant species of hard-substrate communities will in the future allow us to create communities with predefined properties. The use of hard substrates, such as artificial reefs, in coastal areas may be very helpful in recovering disturbed communities and regulating their composition and production, and also play an important role in the decontamination of these water areas. It is quite possible that, in the future, using the extended knowledge of the colonization processes and their mechanisms, we will learn how to manage the coastal marine ecosystems on a strictly scientific basis.

References

Abelson, A. and Denny, M., Settlement of marine organisms in flow, *Annu. Rev. Ecol. Syst.,* 28, 317, 1997.

Abrahamsson, K. et al., Marine algae — a source of trichloroethylene and perchloroethylene, *Limnol. Oceanogr.,* 40(7), 1321, 1995.

Adamson, W.L., Liberatire, G.L., and Taylor, D.W., Control of microfouling in ship piping and heat exchanger systems, in *Marine Biodeterioration: An Interdisciplinary Study,* Costlow, J.D. and Tipper, R.C., Eds., Naval Institute Press, Annapolis, MD, 1984, 95.

Afanas'ev, I.B., Oxygen anion-radical O_2^- in chemical and biochemical processes, *Usp. Khim.,* 48(6), 977, 1979.

Aizatulin, T.A., Lebedev, V.L., and Khailov, K.M., *Okean, aktivnye poverkhnosti i zhizn'* (Ocean, Active Surfaces, and Life), Gidrometeoizdat, Leningrad, 1979, 192.

Alexandrov, B.G., Daily dynamics of abundance of *Balanus improvisus* Darwin larvae in the Black Sea coastal zone, deposited at VINITI, *25.02.86, N 3960-B86,* Kiev, 1986, 15.

Alexandrov, B.G., Hydrobiological Fundamentals of the Black Sea Coastal Ecosystems Management, Doctoral (biology) dissertation, Institute of Biology of the Southern Seas, Sebastopol, 2002, 466.

Alexandrov, B.G. and Yurchenko, Yu., Relationship between structural and functional properties of marine animal fouling and geometry of hard substrates, in *Ecological Safety of Coastal and Shelf Zones and Complex Using Continental Shelf Resourses,* Ivanov, V.A., Ed., Akademie Nauk Ukraine, Sebastopol, 2000, 351.

Alexandrov, B.G., Minucheva, G.G., and Strikalenko, T.V., Ecological aspects of using of tyre rubber as a substrate for artificial reefs, *Biol. Morya (Russ. J. Mar. Biol.),* 28(2), 131, 2002.

Alibekova, I.I., Bagirov, R.M., and Pyatakova, G.M., Fouling of vessels in the Caspian Sea, *Izv. Akad. Nauk Azerb. SSR Ser. Biol.,* 4, 47, 1985.

Allee, W.C., *Animal Aggregations: A Study in General Sociology,* University of Chicago Press, Chicago, 1931, 431.

Altshuller, G.S., *Naiti ideyu. Vvedenie v teoriyu resheniya izobretatel'skikh zadach* (Find an Idea: Introduction to the Theory of Invention), Nauka, Novosibirsk, 1986, 209.

Alzieu, C., Barbier, G., and Sanjuan, J., Evolution des teneurs en cuivre des huitres du bassin d'Arcachon: influence de la legislation sur les peintures antisalissures, *Oceanol. Acta,* 10(4), 463, 1987.

Amato, I., Stuck on mussels, *Sci. News,* 139, 8(1), 1991.

Ambrogi, R., Fouling control systems for water treatment, *Oebalia,* 19, 355, 1993.

Ameron, B.V., Geldermalsen, Schiff, and Hafen, Seewirt, *Kommandobrucke,* 52(10), 62, 2000.

Anderson, C. and Hunter, J., Antifouling coatings and the global regulatory debate — an industry under scrutiny, *Asia-Pacific Fishing' 99: The Second Asia-Pacific Fishing Conference, Cairns, July 6–7, 1999,* Baird Publishers, Melbourne, 1999, 177.

Anderson, M.J., A chemical cue induces settlement of Sydney rock oyster, Saccostrea commercialis, in the laboratory and in the field, *Biol. Bull.,* 190, 350, 1996.

Andreyuk, E.I. et al., *Microbial Corrosion and Its Pathogens,* Nauk Dumka, Kiev, 1980, 288.

Andreyuk, E.I. et al., Microflora of the KhV-5153 antifouling coating under marine conditions, *Mikrobiol. Zh.,* 47(4), 3, 1985.

Armstrong, E., Boyd, K.G., Pisacane, A., Peppiatt, Ch.J., and Burgess, J.G., Marine microbial natural products in antifouling coatings, *Biofouling,* 16, 215, 2000.

Aroujo, J.T.C., Coutinho, C.M.L.M., and Aguiar, L.E.V., Sulphate reducing bacteria associated with biocorrosion: a review, *Mem. Inst. Oswaldo Cruz.,* 87, 329, 1992.

Assmann, M. et al., Chemical defenses of the Caribbean sponges *Agelas wiedenmayeri* and *Agelas conifera, Mar. Ecol. Prog. Ser.,* 207, 255, 2000.

Avelin, M., Marine microfouling algae: the diatoms, in *Fouling Organisms in the Indian Ocean: Biology and Control Technology,* Nagabhushanam, R. and Thompson, M.F., Eds., Oxford and IBH Publishing, New Delhi, 1997, 221.

Avelin, M. and Vitalina, V., Microfouling bacteria, in *Fouling Organisms in the Indian Ocean: Biology and Control Technology,* Nagabhushanam, R. and Thompson, M.F., Eds., Oxford and IBH Publishing, New Delhi, 1997, 189.

Babkov, A.I. and Golikov, A.N., *Hydrobiocomplexes of the White Sea,* Zoological Institute, Leningrad, 1984, 104.

Bagaveeva, E.V., Polychaetes in the biofouling of artificial substrates in aquaculture of *Laminaria* and Japanese scallop, in *Biologicheskie issledovaniya bentosa i obrastaniya v Yaponskom more* (Biological Studies of Benthos and Biofouling in the Sea of Japan), Fadeev, V.I., Ed., Dal'nevost. Nauchn. Tsentr Akad. Nauk SSSR, Vladivostok, 1991, 111.

Bagaveeva, E.V. and Zvyagintsev, A.Ju., The introduction of polychaetes *Hydroides elegans* (Haswell), *Polydora limicola* Annenkova, and *Pseudopotamilla occelata* Moore to the northwestern part of the East Sea, *Ocean Res.,* 22, 25, 2000.

Bagaveeva, E.V., Kubanin, A.A., and Chaplygina, S.F., The role of ships in the introduction of hydroids, polychaetes and bryozoans into the Sea of Japan, *Biol. Morya,* 2, 19, 1984.

Baier, R.E., Initial events in microbial film formation, in *Marine Biodeterioration: An Interdisciplinary Study,* Costlow, J.D. and Tipper, R.C., Eds., Naval Institute Press, Annapolis, MD, 1984, 57.

Bak, R.P.M., Sybesma, J., and van Duyl, F.C., The ecology of the tropical compound ascidian *Trididemnum solidum.* II. Abundance, growth and survival, *Mar. Ecol. Prog. Ser.,* 6(1), 43, 1981.

Baker, J. and Evans, L.V., The ship fouling alga *Ectocarpus.* I. Ultrastructure and cytochemistry of plurilocular reproductive stages, *Protoplasma,* 77(1), 1, 1973.

Bakus, G.J., Targett, N.M., and Schulte, B., Chemical ecology of marine organisms: an overview, *J. Chem. Ecol.,* 12(5), 951, 1986.

Bamforth, S.S., The variety of artificial substrates as used for microfauna, in *Artificial Substrates,* Cairns, J., Jr., Ed., Ann Arbor Science, Ann Arbor, MI, 1982, 115.

Barashkov, G.K., *Sravnitel'naya biokhimiya vodoroslei* (Comparative Biochemistry of Algae), Pishchevaya Promyshlenost', Moscow, 1972, 336.

Barnes, D.K.A. and Dick, M.H., Overgrowth competition in encrusting bryozoan assemblages of the intertidal and infralittoral zones of Alaska, *Mar. Biol.,* 136(5), 813, 2000.

Barnes, H. and Powell, H.T., The growth of *Balanus balanoides* (L.) and *B. crenatus* Brug. under varying conditions of submersion, *J. Mar. Biol. Assoc. U.K.,* 32(1), 107, 1953.

Barthel, D., On the ecophysiology of the sponge *Halisarca panicea* in Kiel Bight. I. Substrate specificity, growth and reproduction, *Mar. Ecol. Prog. Ser.,* 32, 291, 1986.

Barthel, D. and Wolfrath, B., Tissue sloughing in the sponge *Halichondria panicea*: a fouling organism prevents being fouled, *Oecologia,* 78(3), 357, 1989.

Bashmachnikov, I.L. et al., Interaction of a mussel bed and laminaria thickets with water flow, *3-ya nauchnaya sessiya Morskoi Biologicheskoi stantsii SPbGU* (3rd Scientific Session of the Marine Biological Station of St. Petersburg State University), St. Petersburg, 2002, 51.

Bayne, B.L., The responses of the larval *Mytilus edulis* to gravity, *Oikos*, 15(1), 162, 1964.

Bayne, B.L., The biology of mussel larvae, in *Marine Mussels; Their Ecology and Physiology*, Bayne, B.L., Ed., Cambridge University Press, Cambridge, 1976, 81.

Beamont, A.R., Tserpes, G., and Budd, M.D., Some effects of copper on the veliger larvae of the mussel *Mytilus edulis* and the scallop *Pecten maximus* (Mollusca, Bivalvia), *Mar. Environ. Res.*, 21(4), 299, 1987.

Bebianno, M.J. and Langston, W.J., Induction of metallothionein synthesis in the gill and kidney of *Littorina littorea* exposed to cadmium, *J. Mar. Biol. Assn. U.K.*, 75, 173, 1995.

Begon, M., Harper, J.L., and Townsend, C.R., *Ecology: Individuals, Populations and Communities, 3rd ed., Vol. 2*, Blackwell Scientific, Oxford, 1996, 945.

Behning, A.L., To the study of nature of the Volga, in *Monografii Volzhskoi biologicheskoi stantsii* (Monographs of the Volga Biological Station), No. 1, Volzhskoe Knizhnoe Izdat., Saratov, 1924, 210.

Behning, A.L., *Das Leben der Wolga*, Ost-Europa-Verlag, Berlin, 1929, 141.

Beiras, R. and His, E., Effects of dissolved mercury on embryogenesis, survival and growth of *Mytilus galloprovincialis* mussel larvae, *Mar. Ecol. Prog. Ser.*, 126(1–3), 185, 1995.

Bell, E.C. and Goslin, J.M., Mechanical design of mussel byssus: material yield enhances attachment strength, *J. Exp. Biol.*, 199(4), 1005, 1996.

Bell, S.S. and Delvin, D.J., Short-term macrofaunal recolonization of sediment and epibenthic habitats in Tampa Bay, Florida, *Bull. Mar. Sci.*, 33(1), 102, 1983.

Belorustseva, S.A. and Marfenin, N.N., Affect of variable phases of tide cycle on reproduction of *Laomedea flexuosa* (Hydroidea, Thecaphora), *Zh. Obsch. Biol.*, 63(1), 50, 2002.

Benayahu, Y. and Loya, Y., Substratum preferences and planulae settling of two red sea alcyonaceans *Xenia macrospiculata* Gohar and *Parerythropodium fulvum fulvum* (Forskål), *J. Exp. Mar. Biol. Ecol.*, 83, 249, 1984.

Beregovaya, N.M., Changes in the biochemical parameters of *Obelia loveni* and their use in bioindication of active chlorine in seawater, in *Sbornik tezisov dokl. Vsesoyuz. shkoly po tekhnich. sredstvam i metodam issledovaniya Mirovogo okeana* (Proc. All-Union Workshop on Technical Means and Techniques of Studying the World Ocean), Inst. Okeanol. Akad. Nauk SSSR, Moscow, 1991, 144.

Berg, H.C., Physics of bacterial chemotaxis, in *Sensory Perception and Transduction in Aneural Organisms*, Colombetti, G. et al., Eds., Plenum Press, New York, 1985, 19.

Berger, V.Ya. et al., Morphofunctional and ecological aspects of byssus development in the mussel (*Mytilus edulis* L.), in *Ekologiya obrastaniya v Belom more* (Biofouling Ecology in the White Sea), Berger, V.Ya. and Seranin, L.N., Eds., Zoological Institute, Leningrad, 1985, 67.

Bergquist, P.R., *Sponges*, University of California Press, Los Angeles, 1978, 268.

Bergquist, P.R., Sinclair, M.E., and Hogg, J.J., Adaptation to intertidal existence: reproductive cycles and larval behavior in Desmospongiae, in *Biology of the Porifera*, Fry, W.G., Ed., Academic Press, New York, 1970, 247.

Berking, S., Control of metamorphosis and pattern formation in *Hydractinia, Bioessays*, 13(7), 323, 1991.

Bernal, J.D., *The Origin of Life*, Weidenfeld and Nicolson, London, 1967, 345.

Berrill, N.J., Studies in tunicate development. II. Abbreviation of development in the Molgulidae, *Philos. Trans. R. Soc. London Ser. B*, 219(468), 281, 1931.

Bertalanffy von, L., A qualitative theory of organic growth (inquiries on growth laws. II), *Hum. Biol. Baltimore*, 10, 181, 1938.

Bertrandy, C., Les apports pollutants non telluriques, *Oceanis*, 14(6), 767, 1988.

Best, M.A. and Thorpe, J.P., Particle size, clearance rate and feeding efficiency in marine Bryozoa, in *Biology and Palaeobiology of Bryozoans*, Hayward, P.J., Ryland, J.S., and Taylor, P.D., Eds., Olsen and Olsen, Fredensborg, 1994, 9.

Bhaud, M., Conditions d'établissement des larves de Eupolymnia nebulosa: acquis expérimentaux et observations en milieu naturel, Utilité d'une confrontation, *Oceanis*, 16(3), 181, 1990.

Bhosle, N.B., Sankaran, P.D., and Wagh, A.B., Carbohydrate sources of microfouling material developed on aluminium and stainless steel panels, *Biofouling*, 2, 151, 1990.

Bienfang, P.K. and Harrison, P.J., Sinking-rate response of natural assemblages of temperate and subtropical phytoplankton to nutrient depletion, *Mar. Biol.*, 83(3), 293, 1984.

Bingham, B.L. and Young, C.M., Larval behavior of the ascidian *Ecteinascidia turbinata* Herdman; an *in situ* experimental study of the effects of swimming on dispersal, *J. Exp. Mar. Biol. Ecol.*, 145(2), 189, 1991.

Blackburn, N. and Fenchel, T., Influence of bacteria, diffusion and shear on micro-scale nutrient patches, and implications for bacterial chemotaxis, *Mar. Ecol. Prog. Ser.*, 189(26), 1, 1999.

Blair, D.F., How bacteria sense and swim, *Annu. Rev. Microbiol.*, 49, 489, 1995.

Bonar, D.B., Morphogenesis at metamorphosis in opisthobranch molluscs, in *Settlement and Metamorphosis of Marine Invertebrate Larvae*, Chia, F.-S. and Rice, M.E., Eds., Elsevier/North Holland, New York, 1978, 177.

Bonar, D.B. et al., Control of oyster settlement and metamorphosis by endogenous and exogenous chemical cues, *Bull. Mar. Sci.*, 46(2), 484, 1990.

Borojević, R., Étude du developement et la differenciation cellulaire d'éponges calcaires calcinéennes (genres *Clathrina* et *Ascandra*), *Ann. Embryol. Morphol.*, 2(1), 15, 1969.

Bott, T.R., General fouling problems, in *Fouling Science and Technology*, Melo, L.F., Bott, T.R., and Bernardo, C.A., Eds., Kluwer, Dordrecht, 1988, 3.

Bousfield, E., Ecological control of the occurrence of barnacles in the Miramichi estuary, *Bull. Nat. Mus.* (Canada), 137, 1, 1955.

Bowden, K.F., *Physical Oceanography of Coastal Waters*, Ellis Horwood, Chichester, 1983, 302.

Bowes, H., Ship stranded at sea — were bacteria responsible? *Tairplay Int. Shipp. Weekly*, 302(5414), 37, 1987.

Boyle, P.J. and Mitchell, R., The microbial ecology of crustacean wood borers, in *Marine Biodeterioration: An Interdisciplinary Study*, Costlow, J.D. and Tipper, R.C., Eds., Naval Institute Press, Annapolis, MD, 1984, 17.

Braiko, V.D., *Obrastanie v Chernom more* (Fouling in the Black Sea), Nauk. Dumka, Kiev, 1985, 124.

Braiko, V.D., Theoretical aspects of biofouling, in *Izuchenie protsessov morskogo bioobrastaniya i razrabotka metodov bor'by s nim* (Studies of Marine Biofouling and Development of Methods of Its Control), Skarlato, O.A., Ed., Zoological Institute, Leningrad, 1987, 8.

Braiko, V.D. and Kucherova, Z.S., On the effect of the size of an experimental surface on the development of the fouling cenosis, in *3 s"ezd Vsesoyuznogo gidrobiologicheskogo obshchestva, mai 1976* (3rd Congr. All-Union Hydrobiological Society, May 1976), Vol. 2, Andrushaitis, G.P. et al., Eds., Zinante, Riga, 1976, 111.

Brancato, M.S. and Woollacott, R.M., Effect of microbial films on settlement of bryozoan larvae (*Bugula simplex*, *B. stolonifera* and *B. turrita*), *Mar. Biol.*, 71(1), 51, 1982.

Brewer, R.H., The influence of the orientation, roughness, and wettability of solid surfaces on the behavior and attachment of planulae of *Cyanea* (Cnidaria: Scyphozoa), *Biol. Bull.*, 166(1), 11, 1984.

Brownlee, C. et al., Signal transduction and early development in plants, in *Mar. Biol. Assoc. U.K., Annual Report 1994*, 1994, 45.

Bryan, P.J. et al., Induction of larval settlement and metamorphosis by pharmacological and conspecific associated compounds in the serpulid polychaete *Hydroides elegans, Mar. Ecol. Prog. Ser.*, 146(1–3), 81, 1997.

Bryers, J.D. and Characklis, W.G., Processes governing primary biofilm formation, *Biotechnol. Bioeng.*, 24, 2451, 1982.

Bultman, J.D., Griffith, J.R., and Field, D.E., Fluoropolymer coating for the marine environment, in *Marine Biodeterioration: An Interdisciplinary Study*, Costlow, J.D. and Tipper, R.C., Eds., Naval Institute Press, Annapolis, MD, 1984, 237.

Burke, R.D., The induction of metamorphosis of marine invertebrate larvae: stimulus and response, *Can. J. Zool.*, 61(8), 1701, 1983.

Burke, R.D., Pheromonal control of metamorphosis in the pacific sand dollar, *Dendraster excentricus, Science*, 225(4660), 442, 1984.

Burke, R.D., Pheromones and the gregarious settlement of marine invertebrate larvae, *Bull. Mar. Sci.*, 39, 323, 1986.

Burkovsky, I.V., *Ekologiya svobodnozhivushchikh infuzorii* (Ecology of Free-Living Ciliates), Moscow University, Moscow, 1984, 208.

Burkovsky, I.V., *Strukturno-funktsional'naya organizatsiya i ustoichivost' morskikh donnykh soobshchestv* (Structural and Functional Organisation and Stability of Marine Bottom Communities), Moscow University, Moscow, 1992, 208.

Burkovsky, I.V. and Kashunin, A.K., Resistance of the marine psammophilous ciliate community to copper stress at different stages of colonization, *Zh. Obshch. Biol.*, 6, 736, 1995.

Burridge, T.R., Lavery, T., and Lam, P.K.S., Effects of tributyltin and formaldehyde on the germination and growth of *Phyllospora comosa* (Labillardiere) C. Agardh (Phaeophyta: Fucales), *Bull. Environ. Contam. Toxicol.*, 55(4), 525, 1995.

Burton, D., Richardson, L.B., and Taylor, R., Control of colonial hydroid macrofouling by free-field ultrasonic radiation, *Science*, 223(4643), 1410, 1984.

Butman, C.A., Larval settlement of soft-sediment invertebrates: some predictions based on an analysis of near-bottom velocity profiles, in *Marine Interfaces Ecohydrodynamics*, Nihoul, J.C.J., Ed., Elsevier, Amsterdam, 1986, 487.

Butman, C.A., Larval settlement of soft-sediment invertebrates: the spatial scales of pattern explained by active habitat selection and the emerging role of hydrodynamical processes, *Oceanogr. Mar. Biol. Annu. Rev.*, 25, 113, 1987.

Buyanovsky, A.I. and Kulikova, V.A., Planktonic distribution of *Mytilus edulis* larvae and their settlement on collectors in Vostok Bay, the Sea of Japan, *Biol. Morya*, 6, 52, 1984.

Byrne, M. and Barker, M.F., Embryogenesis and larval development of the asteroid *Patiriella regularis* viewed by light and scanning electron microscopy, *Biol. Bull.*, 180(3), 332, 1991.

Cairns, J., Jr., Ed., *Artificial Substrates*, Ann Arbor Science, Ann Arbor, MI, 1982a, 279.

Cairns, J., Jr., Freshwater protozoan communities, in *Microbial Interactions and Communities*, Cairns, J., Jr., Ed., Academic Press, London, 1982b, 249.

Cairns, J., Jr. and Henebry, M.S., Interactive and noninteractive protozoan colonization processes, in *Artificial Substrates*, Cairns, J., Jr., Ed., Ann Arbor Science, Ann Arbor, MI, 1982, 23.

Cairns, J., Jr., Dickson, K.L., and Yongue, W.H., Jr., The consequences of nonselective periodic removal of portions of fresh-water protozoan communities, *Trans. Am. Microsc. Soc.*, 90(1), 71, 1971.

Caldwell, D.E., Surface colonization parameters from cell density and distribution, in *Microbial Adhesion and Aggregation. Dahlem Konferenzen, 1984*, Marshall, K.C., Ed., Springer-Verlag, Berlin, 1984, 125.

Caldwell, D.R. and Chriss, T.M., The viscous sublayer at the sea floor, *Science*, 205, 1131, 1979.

Callow, M.E., A world-wide survey of slime formation on anti-fouling paints, in *Algal Biofouling*, Evans, L.V. and Hoagland, K.D., Eds., Elsevier, Amsterdam, 1986, 1.

Callow, M.E. et al., Use of self-assembled monolayers of different wettabilities to study surface selection and primary adhesion processes of green algae (Enteromorpha), *Appl. Environ. Microbiol.*, 66, 3249, 2000.

Cameron, R.A., Introduction to the invertebrate larval biology workshop: a brief background, *Bull. Mar. Sci.*, 39(2), 145, 1986.

Cameron, R.A. and Hinegardner, R.T., Initiation of metamorphosis in laboratory cultured sea urchins, *Biol. Bull.*, 146(2), 335, 1974.

Camper, A.K. and McFeters, G.A., Biofouling in drinking water systems, in *Industrial Biofouling: Detection, Prevention, and Control*, Walker, J., Surman, S., and Jass, J., Eds., Wiley, Chichester, 2000, 13.

Carlton, J.T. and Geller, J.B., Ecological roulette: the global transport of nonindigenous marine organisms, *Science*, 261, 78, 1993.

Carlton, J.T. and Hodder, J., Biogeography and dispersal of coastal marine organisms: experimental studies on a replica of a 16th-century sailing vessel, *Mar. Biol.*, 121(4), 721, 1995.

Carlton, J.T et al., Remarkable invasion of the San Francisco Bay (California, USA) by the Asian clam *Potamocorbula amurensis*. I. Introduction and dispersal, *Mar. Ecol. Prog. Ser.*, 66, 81, 1990.

Caron, D.A. and Sieburth, J.M., Disruption fouling sequence on fiber glass-reinforced submerged in the marine environment, *Appl. Environ. Microbiol.*, 41(1), 268, 1981.

Carriker, M.R., Interrelation of functional morphology, behavior and autoecology in early stages of the bivalve *Mercenaria mercenaria*, *J. Elisha Mitchell Sci. Soc.*, 77, 168, 1961.

Cassidy, V.M., Copper-hulled boats run better, longer, *Mod. Met.*, 44(5), 36, 1988.

Cetrulo, G.L. and Hay, M.E., Activated chemical defenses in tropical versus temperate seaweeds, *Mar. Ecol. Prog. Ser.*, 207, 243, 2000.

Chabot, R. and Bourget, E., Influence of substratum heterogeneity and settled barnacle density on the settlement of cypris larvae, *Mar. Biol.*, 97(1), 45, 1988.

Chalmer, P.N., Settlement patterns of species in a marine fouling community and some mechanisms of succession, *J. Exp. Mar. Biol. Ecol.*, 58, 73, 1982.

Chamberlain, A.H.L., Algal settlement and secretion of adhesive materials, in *Proc. 3rd Int. Biodegradation Symp.*, Sharpley, J.M. and Kaplan, A.M., Eds., Applied Science, London, 1976, 443.

Chan, A.L.C. and Walker, G., The settlement of *Pomatoceros lamarckii* larvae (Polychaeta: Sabellida: Serpulidae): a laboratory study, *Biofouling*, 12, 71, 1998.

Chaplygina, S.F., Fouling hydroids in the northwestern Sea of Japan, in *Ekologiya obrastaniya v severo-zapadnoi chasti Tikhogo okeana* (Ecology of Fouling in the Northwestern Pacific), Kudryashov, V.A., Ed., Akademie Nauk, Vladivostok, 1980, 56.

Characklis, W.G., Biofilm development: a process analysis, in *Microbial Adhesion and Aggregation. Dahlem Konferenzen, 1984*, Marshall, K.C., Ed., Springer-Verlag, Berlin, 1984, 137.

Characklis, W.G., Zelver, N., and Turakhia, M., Microbial films and energy losses, in *Marine Biodeterioration: An Interdisciplinary Study*, Costlow, J.D. and Tipper, R.C., Eds., Naval Institute Press, Annapolis, MD, 1984, 75.

Chaudière, J., Some chemical and biochemical constraints of oxidative stress in living cells, in *Free Radical Damage and Its Control*, Rice-Evans, C.A. and Burdon, R.H., Eds., Elsevier, Amsterdam, 1994, 25.

Chet, I. and Mitchell, R., Control of marine fouling by chemical repellents, in *Proc. 3rd Int. Biodeterioration Symp.*, Sharpley, V.M. and Kaplan, A.M., Eds., Applied Science, London, 1976, 515.

Chevelot, L., Cochard, J.-C. and Yvin, J.-C., Chemical induction of larval metamorphosis of *Pecten maximus* with a note on the nature of naturally occurring triggering substances, *Mar. Ecol. Prog. Ser.*, 74, 83, 1991.

Chia, F.-S., Perspectives: settlement and metamorphosis of marine invertebrate larvae, in *Settlement and Metamorphosis of Marine Invertebrate Larvae*, Chia, F.-S. and Rice, M.E., Eds., Elsevier/North Holland, New York, 1978, 283.

Chia, F.-S. and Bickell, L.R., Mechanisms of larval attachment and the induction of settlement and metamorphosis in coelenterates: a review, in *Settlement and Metamorphosis of Marine Invertebrate Larvae*, Chia, F.-S. and Rice, M.E., Eds., Elsevier/North Holland, New York, 1978, 1.

Chia, F.-S. and Rice, M.E., Eds., *Settlement and Metamorphosis of Marine Invertebrate Larvae*, Elsevier, New York, 1978, 290.

Chia, F.-S., Buckland-Nicks, J., and Young, C.M., Locomotion of marine invertebrate larvae: a review, *Can. J. Zool.*, 62, 1205, 1984.

Chikadze, S.Z. and Railkin, A.I., Glucose suppresses settlement, attachment, and metamorphosis in the hydroid *Obelia loveni*, *Vestn. Sankt-Peterburg. Univ. Ser. 3 Biol.*, 3(17), 29, 1992.

Chikadze, S.Z. and Railkin, A.I., Interactions between larvae *Obelia loveni* and macroalgae are mediated by bacterial-algal films, in *31st European Marine Biology Symp., Russia*, Alinov, A.F. et al., Eds., Zoological Institute, St. Petersburg, 1996, 98.

Chipperfield, P.N.J., Observation of the breeding and settlement of *Mytilus edulis* L. in British waters, *J. Mar. Biol. Assn. U.K.*, 32(2), 449, 1953.

Christensen, B., Vedel, A., and Kristensen, E., Carbon and nitrogen fluxes in sediment inhabited by suspension-feeding (*Nereis diversicolor*) and non-suspension-feeding (*N. virens*) polychaetes, *Mar. Ecol. Prog. Ser.*, 192, 203, 2000.

Christie, A.O., Evans, L.V., and Shaw, M., Studies on the ship-fouling alga *Enteromorpha*. II. The effect of certain enzymes on the adhesion of zoospores, *Ann. Bot.*, 34(135), 467, 1970.

Christy, J.H., Adaptive significance of semilunar cycles of larval release in fiddler crabs (genus *Uca*): test of a hypothesis, *Biol. Bull.*, 163, 251, 1982.

Chuguev, Yu.P., Adhesion and its role in marine biofouling, in *Voprosy morskoi korrozii i bioobrastaniya* (Problems of Marine Corrosion and Biofouling), Kaplin, Ju.N., Ed., Akademie Nauk, Vladivostok, 1985, 52.

Clare, A.S., Marine natural product antifoulants: status and potential, *Biofouling*, 9, 211, 1996.

Clare, A.S. and Matsumura, K., Nature and perception of barnacle settlement pheromones, *Biofouling*, 15(1–3), 57, 2000.

Clare, A.S., Freet, R.K., and McClary, M., On the antennular secretion of the cypris of *Balanus amphitrite amphitrite*, and its role as a settlement pheromone, *J. Mar. Biol. Assn. U.K.*, 74(1), 243, 1994.

Cloney, R.A., Ascidian metamorphosis: review and analysis, in *Settlement and Metamorphosis of Marine Invertebrate Larvae*, Chia, F.-S. and Rice, M.E., Eds., Elsevier/North Holland, New York, 1978, 255.

Cochard, J.-C. et al., Induction de la metamorphose de la coquille Saint Jaques *Pecten maximus* L. par de derives de la tyrosine extracts de l'algue *Delesseria sanguinea* Lamouroux ou synthetiques, *Haliotis*, 19, 129, 1989.

Cole, H.A. and Knight-Jones, E.W., The setting behaviour of larvae of the European flat oyster *Ostrea edulis* L. and its influence on methods of cultivation and spat collection, *Fish. Invest. (London) Ser.* 2, 17(3), 1, 1949.

Cologer, C.P. and Preiser, H.S., Fouling and paint behavior on naval surface ships after multiple underwater cleaning cycles, *Marine Biodeterioration: An Interdisciplinary Study*, Costlow, J.D. and Tipper, R.C., Eds., Naval Institute Press, Annapolis, MD, 1984, 213.

Connell, J.H. and Slatyer, R.O., Mechanisms of succession in natural communities and their role in community stability and organisation, *Am. Nat.*, 111, 1119, 1977.

Cook, M., Composition of mussel and barnacle deposits at the attachment interface, in *Adhesion in Biological Systems*, Manly, R.S., Ed., Academic Press, New York, 1970, 139.

Cook, W.B., Colonization of artificial bare areas by microorganisms, *Bot. Rev.,* 22, 613, 1956.

Cooksey, B. et al., The attachment of microfouling diatoms, in *Marine Biodeterioration: An Interdisciplinary Study*, Costlow, J.D. and Tipper, R.C., Eds., Naval Institute Press, Annapolis, MD, 1984, 167.

Cooksey, K.E., Requirement for calcium in adhesion of a fouling diatom to glass, *Appl. Environ. Microbiol.*, 41(6), 1378, 1981.

Cooksey, K.E. and Cooksey, B., Adhesion of fouling diatoms to surfaces: some biochemistry, in *Algal Biofouling*, Evans, L.V. and Hoagland, K.D., Eds., Elsevier, Amsterdam, 1986, 41.

Cooksey, K.E. and Cooksey, B., Use of specific drugs in the dissection of the adhesive process in diatoms, in *Marine Biodeterioration*, Thompson, M.F., Sarojini, R., and Nagabhushanam, R., Eds., Oxford and IBH Publishers, New Delhi, 1988, 337.

Corpe, W.A., Attachment of marine bacteria to solid surfaces, in *Adhesion in Biological Systems*, Manly, R.S., Ed., Academic Press, New York, 1970, 73.

Corpe, W.A., Microfouling, in *Proc. 3rd Int. Congr. on Marine Corrosion and Fouling*, Acker, R.F. et al., Eds., Northwestern University Press, Evanston, IL, 1972, 598.

Corpe, W.A., Matsuuchi, L., and Armbruster, B., Secretion of adhesive polymers and attachment of marine bacteria to surfaces, in *Proc. 3rd Int. Biodeterioration Symp.*, Sharpley, V.M. and Kaplan, A.U., Eds., Applied Science, London, 1976, 433.

Cosson, R., Amiard-Triquet, C., and Grandier-Vazeille, X., Etude du metabolisme du cuivre, du zinc et de l'étain chez l'huitre et la moule en relation avec les pollutions dues a l'activité d'un port de plaisance, *Oceanis*, 15(4), 411, 1989.

Costa, P., Pitture ai siliconi, Tecnol. e trasp. mare, *Int. Mag. Adv. Mar. Technol. Transportation and Logistics,* 31(8), 76, 2000.

Costerton, J.W. et al., Microbial biofilms, *Annu. Rev. Microbiol.*, 49, 711, 1995.

Costlow, J.D. and Tipper, R.C., Eds., *Marine Biodeterioration: An Interdisciplinary Study*, Naval Institute Press, Annapolis, MD, 1984, 384.

Cragg, S.M., Swimming behaviour of the larvae of *Pecten maximus* (L.) (Bivalvia), *J. Mar. Biol. Assn. U.K.*, 60(3), 551, 1980.

Crisp, D.J., The behaviour of barnacle cyprids in relation to water movement over a surface, *J. Exp. Biol.*, 32(3), 569, 1955.

Crisp, D.J., The spread of *Elminius modestus* Darwin in northwest Europe, *J. Mar. Biol. Assn. U.K.*, 37, 483, 1958.

Crisp, D.J., Territorial behaviour in barnacle settlement, *J. Exp. Biol.*, 38(2), 429, 1961.

Crisp, D.J., Chemical factors inducing settlement in *Crassostrea virginica, J. Anim. Ecol.*, 36(2), 329, 1967.

Crisp, D.J., The role of the biologist in antifouling research, in *Proc. 3rd Int. Congr. on Marine Corrosion and Fouling*, Acker, R.F. et al., Eds., Northwestern University Press, Evanston, IL, 1973, 88.

Crisp, D.J., Factors influencing the settlement of marine invertebrate larvae, in *Chemoreception in Marine Organisms*, Grant, P.T. and Mackie, A.M., Eds., Academic Press, New York, 1974, 177.

Crisp, D.J., Settlement responces in marine organisms, in *Adaptation to the Environment: Essays on the Physiology of Marine Animals*, Newell, R.C., Ed., Butterworths, London, 1976, 83.

Crisp, D.J., Overview of research on marine invertebrate larvae, 1940-1980, in *Marine Biodeterioration: An Interdisciplinary Study*, Costlow, J.D. and Tipper, R.C., Eds., Naval Institute Press, Annapolis, MD, 1984, 103.

Crisp, D.J. and Barnes, H., The orientation and distribution of barnacles at settlement with particular reference to surface contour, *J. Anim. Ecol.*, 23(1), 142, 1954.

Crisp, D.J. and Meadows, P.S., The chemical basis of gregariousness in cirripedes, *Proc. R. Soc. London Ser. B*, 156(965), 500, 1962.

Crisp, D.J. and Meadows, P.S., Adsorbed layers: the stimulus to settlement in barnacles, *Proc. R. Soc. London Ser. B*, 158, 364, 1963.

Crisp, D.J. and Ryland, J.S., Influence of filming and of surface texture on the settlement of marine organisms, *Nature*, 185(4706), 119, 1960.

Crisp, D.J. and Williams, G.B., Effect of extracts from fucoids in promoting settlement of epiphytic polyzoa, *Nature* (London), 188(4757), 1206, 1960.

Crisp, D.J. et al., Adhesion and substrate choice in mussels and barnacles, *J. Colloid. Interface Sci.*, 104(1), 40, 1985.

Culliney, J.L., Settling of larval shipworms, *Teredo navalis* L. and *Bankia gouldi* Bartsch. is stimulated by humic material (Gelbstoff), in *3rd Int. Congr. on Marine Corrosion and Fouling,* Acker, R.F. et al., Eds., Northwestern University Press, Evanston, IL, 1973, 622.

Dahlem, C., Moran, P.J., and Grant, T.R., Larval settlement of marine sessile invertebrates on surfaces of different colour and position, *Ocean Sci. Eng.*, 9(2), 225, 1984.

Dalley, R. and Crisp, D.J., *Conchoderma*: a fouling hazard to ships underway, *Mar. Biol.*, 2, 141, 1981.

Dautov, S.Sh. and Nezlin, L.P., Ontogenetic changes of behavior and nervous system morphology in planktotrophic sea star larvae, in *Prostye nervnye sistemy* (Simple Nervous Systems), Part 1, Sakharov, D.A., Ed., Kazan University, Kazan, 1985, 55.

Davis, A.R., Variation in recruitment of the subtidal colonial ascidian *Podoclavella cylindrica* (Quoy and Gaimard): the role of substratum choice and early survival, *J. Exp. Mar. Biol. Ecol.*, 106(1), 57, 1987.

Davis, A.R., Butler, A.J., and van Altena, I., Settlement behaviour of ascidian larvae: preliminary evidence for inhibition by sponge allelochemicals, *Mar. Ecol. Prog. Ser.*, 72, 117, 1991.

de Jonge, V.N. and van den Bergs, J., Experiments on the resuspension of estuarine sediments containing benthic diatoms, *Estuarine Coastal Shelf Sci.*, 24(6), 725, 1987.

de Nys, R., Coll, J.C., and Price, I.R., Chemically mediated interactions between the red alga *Plocamium hamatum* (Rhodophyta) and the octocoral *Sinularia cruciata* (Alcyonacea), *Mar. Biol.*, 108(2), 315, 1991.

de Ruyter, E.D. et al., Growth inhibition of *Lobophora variegata* (Lamouroux) Womersley by scleractinian corals, *J. Exp. Mar. Biol. Ecol.*, 115(2), 169, 1988.

de Silva, P.H.D.P., Experiments on the choice of substrate by *Spirorbis* larvae (Serpulidae), *J. Exp. Biol.,* 39, 483, 1962.

Dean, T.A. and Hurd, L.E., Development in an estuarine fouling community: the influence of early colonists on later arrivals, *Oecologia*, 46, 295, 1980.

DeCourscy, P.J., Biological timing, in *The Biology of Crustacea, F. Behavior and Ecology,* Vernberg, F.J. and Vernberg, W.B., Eds., Academic Press, New York, 1983, 107.

Delgado, M., Jonge, de V.N., and Peletier, H., Experiments on resuspension of natural micro-phytobenthos populations, *Mar. Biol.*, 108(2), 321, 1991.

Denisenko, N.V., On the study of growth in bryozoans, in *3 Vsesoyuzn. Konf. po morskoi biologii, Sevastopol, 1988* (Abstracts of Papers, 3rd All-Union Conf. Marine Biology, Sevastopol, 1988), Vladivostok, 1988, 26.

Derjaguin, B.V., Selected works of B.V. Derjaguin. Vol. 1. Surface forces in thin films and disperse systems, in *Progress in Surface Science*, Davison, S.G. and Liu, W.-K., Eds., Pergamon Press, New York, 1992, 465.

Derjaguin, B.V., Churaev, N.V., and Muller, V.M., *Poverkhnostnye sily* (Surface Forces), Nauka, Moscow, 1985, 400.

Deslous-Paoli, J.-M., Toxicité de elements metalliques dissous pour les larves d'organismes marins: données bibliographiques, *Rev. Trav. Inst. Peshes Mar.*, 45(1), 73, 1981–1982.

Dexter, S.C., Influence of substratum critical surface tension on bacterial adhesion — *in situ* studies, *J. Colloid. Interface Sci.*, 70, 346, 1978.

Dirnberger, J.M., Dispersal of larvae with a short planktonic phase in the polychaete *Spirorbis spirillum* (Linnaeus), *Bull. Mar. Sci.*, 52, 898, 1993.

Dobretsov, S.V., Habitat Selection by Larvae of Mass Fouling Invertebrates of the White Sea, Cand. Sci. (Biol.) dissertation, St. Petersburg State University, St. Petersburg, 1998.

Dobretsov, S.V., Effect of macroalgae and biofilm on settlement of blue mussel (*Mytilus edulis* L.) larvae, *Biofouling*, 14, 153, 1999a.

Dobretsov, S.V., Macroalgae and microbial film determine substrate selection in planulae of *Gonothyraea loveni* (Allman, 1859) (Cnidaria: Hydrozoa), *Zoosyst. Rossica*, Suppl. 1, 109, 1999b.

Dobretsov, S.V. and Miron, G., Larval and post-larval vertical distribution of the mussel *Mytilus edulis* in the White Sea, *Mar. Ecol. Prog. Ser.*, 218, 179, 2001.

Dobretsov, S.V. and Qian, P.-Y., Effect of bacteria associated with the green alga *Ulva reticulata* on marine micro- and macrofouling, *Biofouling*, 18, 217, 2002.

Dobretsov, S.V. and Railkin, A.I., Correlative relationship between marine microfouling and macrofouling, *Russ. J. Mar. Biol.*, 20, 87, 1994.

Dobretsov, S.V. and Railkin, A.I., Effects of substrate features on settling and attachment of larvae in the blue mussel *Mytilus edulis* (Mollusca, Filibranchia), *Zool. Zh.*, 75(4), 499, 1996.

Dobretsov, S.V. and Railkin, A.I., The strategy of the substrate choice by larvae of *Mytilus edulis*, *Tr. Biol. Nauchno-Issled. Inst. St. Petersburg State Univ.*, 46, 53, 2000

Dobretsov, S.V. and Wahl, M., Recruitment preferences of blue mussel spat (*Mytilus edulis* L.) for different substrates and microhabitats in the White Sea (Russia), *Hydrobiologia*, 445, 27, 2001.

Dodds, W.K., Hydrodynamic constraints on evolution of chemically mediated interactions between aquatic organisms in unidirectional flows, *J. Chem. Ecol.*, 16, 1417, 1990.

Dogiel, V.A., Polyanskii, Yu.I., and Kheisin, E.M., *Obshchaya protozoologiya* (General Protozoology), Polyanskii, Yu.I., Ed., Akademie Nauk SSSR, Moscow, Leningrad, 1962, 592.

Dolgopol'skaya, M.A., Biological studies as the basis for design, development, and improvement of fouling control methods, in *Biologicheskie osnovy bor'by s obrastaniem* (Biological Foundations of Fouling Control), Vodyanitskii, V.A. and Dolgopol'skaya, M.A., Eds., Nauk. Dumka, Kiev, 1973, 5.

Dolgopol'skaya, M.A. et al., Biological action mechanisms of principal toxins used in anti-fouling paints, in *Biologicheskie osnovy bor'by s obrastaniem* (Biological Foundations of Fouling Control), Vodyanitskii, V.A. and Dolgopol'skaya, M.A., Eds., Nauk. Dumka, Kiev, 1973, 194.

Dorsett, L.C. et al., Ferreascidian: a highly aromatic protein containing 3,4-dihydroxyphenylala-nine from the blood cells of a solidobranch ascidian, *Biochemistry*, 26(25), 8078, 1987.

Dovgal, I.V., Flowage effect on the colonization of glass slides by Suctoria (Ciliophora), *Gidrobiol. Zh.*, 26(2), 37, 1990.

Dovgal, I.V., Hydrodynamic factors of evolution of the spatial structure of encrusting communities, *Paleontol. J.*, 32, 559, 1998a.

Dovgal, I.V., Origin and evolution of adhesive organelles in Infusoria (Ciliophora), *Vestn. Zool.*, 32, 18, 1998b

Dovgal, I.V., Morphological and ontogenetic changes in Protista accompanying transition to the sessile mode of life, *Zh. Obsch. Biol.*, 61, 291, 2000.

Dovgal, I.V. and Kochin, V.A., Adaptations of sessile protists to the running water factors, *Vestn. Zool.* (Ukraine), 29, 19, 1995.

Dovgal, I.V. and Kochin, V.A., Fluid boundary layer as an adaptive zone for sessile protists, *Zh. Obshch. Biol.*, 58(2), 67, 1997.

Duddrige, J.E. and Kent, C.A., Treatment of a medium for the reduction of biofouling, Claim 0197213, EPV, B 08 B 17/0, F 28 F 19/00, priority 10.04.85, publ. 15.10.86.

Duerden, J.E., Aggregated colonies in madreporian corals, *Am. Nat.*, 36, 461, 1902.

Duffy, J.E. and Hay, M.E., Seaweed adaptations to herbivory, *Bioscience*, 40(5), 368, 1990.

Duplakoff, S.N., Untersuchungen am Bewuchs im See Glubokoje, *Arb. Hydrobiol. Station am See "Glubokoje,"* 6, 20, 1925.

Durante, K.M., Larval behavior, settlement preference, and induction of metamorphosis in the temperate solitary ascidian *Molgula citrina* (Adler and Hancock), *J. Exp. Mar. Biol. Ecol.*, 145(2), 175, 1991.

Durkina, V.B., Morphofunctional changes in the progeny development of sea urchin caused by the influence of copper and zinc on the adults, *Biol. Morya*, 21(6), 398, 1995.

Eckelbarger, K.J., Metamorphosis and settlement in the Sabellariidae, in *Settlement and Metamorphosis of Marine Invertebrate Larvae*, Chia, F.-S. and Rice, M.E., Eds., Elsevier/North Holland, New York, 1978, 145.

Eckelbarger, K.J. and Chia, F.-Sh., Scanning electron microscopic observations of the larval development of the reef-building polychaete Phragmatopoma lapidosa, *Can. J. Zool.*, 54, 2082, 1976.

Eckman, J.E., A model of passive settlement by planktonic larvae onto bottoms of differing roughness, *Limnol. Oceanogr.*, 35(4), 887, 1990.

Eckman, J.E., Peterson, C.H., and Cahalan, J.A., Effects of flow speed, turbulence, and orientation on growth of juvenile bay scallops *Agropecten irradians concentricus* (Say), *J. Exp. Mar. Biol. Ecol.*, 132(2), 123, 1989.

Edel'kin, S.M. et al., A device for cleaning off fouling from ship hulls, Invention Certificate 1655838, USSR, B 63 B 59/04, 59/10, priority 30.03.89, publ. 15.06.91.

Effler, S.W. et al., Impact of zebra mussel invasion on river water quality, *Water Environ. Res.*, 68, 205, 1996.

Ehrhardt, J.P. and Seguin, G., *Le Plankton: Composition, Ecologie, Pollution*, Paris, 1978, 210.

Elfimov, A.S., Settlement of cypris larvae of *Solidobalanus fallax* and *Balanus spongicola* (Crustacea, Cirripedia) under laboratory conditions, *31st Europ. Mar. Biol. Symp., Russia*, Alimov, A.F. et al., Eds., Zoological Institute, St. Petersburg, 1996, 14.

Elfimov, A.S., Zevina, G.B., and Shalaeva, E. A., *Biologiya usonogikh rakov* (Biology of Cirripedes), Moscow University, Moscow, 1995, 128.

Elyakov, G.B. and Stonik, V.A., *Terpenoidy morskikh organizmov* (Terpenoids of Marine Organisms), Nauka, Moscow, 1986, 272.

Engel, S. and Pawlik, J.R., Allelopathic activities of sponge extracts, *Mar. Ecol. Prog. Ser.*, 207, 273, 2000.

Enikolopyan, N.S., Bogdanova, K.A., and Askarov, K.A., Metal complexes of porphyn and azoporphyn compounds as catalysts of oxidation by molecular oxygen, *Usp. Khim.*, 52(1), 20, 1983.

Ereskowsky, A.V., On the population structure and distribution of sponges in the tidal zone of eastern Murman, *Zool. Zh.*, 73(4), 5, 1994.

Etoh, H. et al., Stiblene glucosides isolated from *Eucalyptus rubida*, as repellents against the blue mussel *Mytilus edulis*, *Agric. Biol. Chem.*, 54(9), 2443, 1990.

Evans, L.V., Marine algae and fouling: a review, with particular reference to ship-fouling, *Bot. Mar.*, 24(4), 167, 1981.

Evans, S.M., Behaviour in polychaetes, *Q. Rev. Biol.*, 46(4), 379, 1971.

Evans, S.M., Leksono, T., and McKinnell, P.D., Tributyltin pollution: a diminishing problem following legislation limiting the use of TBT-based anti-fouling paints, *Mar. Pollut. Bull.*, 30(1), 14, 1995.

Fant, C. et al., Adsorption behavior and enzymatically or chemically induced cross-linking of mussel adhesive protein, *Biofouling*, 16, 119, 2000.

Faure-Fremiet, E., La diversification structurale des ciliés, *Bull. Soc. Zool. Fr.*, 77, 274, 1952

Fenchel, T., *Ecology of Protozoa*, Springer-Verlag, Berlin, 1987, 197.

Figueiredo, M.A. de O., Norton, T.A., and Kain (Jones), J.M., Settlement and survival of epiphytes on two intertidal crustose coralline alga, *J. Exp. Mar. Biol. Ecol.*, 213, 247, 1997.

Filenko, O.F., *Vodnaya toksikologiya* (Aquatic Toxicology), Moscow University, Moscow, 1988, 156.

Fischer, E.C. et al., Technology for control of marine biofouling — a review, in *Marine Biodeterioration: An Interdisciplinary Study*, Costlow, J.D. and Tipper, R.C., Eds., Naval Institute Press, Annapolis, MD, 1984, 261.

Flammang, P., Gosselin, P., and Jangoux, M., The podia, organs of adhesion and sensory perception in larval and post-metamorphic stages of the echinoid *Paracentrotus lividis* (Echinodermata), *Biofouling*, 12(1–3), 161, 1998.

Fletcher, M., The effects of culture concentration and age, time, and temperature on bacterial attachment to polystyrene, *Can. J. Microbiol.*, 23(1), 1, 1977.

Fletcher, M., The attachment of bacteria to surfaces in aquatic environments, in *Adhesion of Microorganisms to Surfaces*, Ellwood, D.C. et al., Eds., Academic Press, London, 1979, 87.

Fletcher, M., *Bacterial Adhesion: Mechanisms and Physiological Significance*, Plenum Press, New York, 1985, 476.

Fletcher, M. and Floodgate, G.D., An electron-microscopic demonstration of an acidic polysaccharide to solid surfaces, *J. Gen. Microbiol.*, 74(2), 325, 1973.

Fletcher, M. and Loeb, G.I., Influence of substratum characteristics on the attachment of a marine pseudomonads to solid surfaces, *Appl. Environ. Microbiol.*, 37(1), 67, 1979.

Fletcher, R.L., Post-germination attachment mechanisms in marine fouling algae, in *Proc. 3rd Int. Biodegradation Symp., 1975*, Sharpley, J.M. and Kaplan, A.M., Eds., Applied Science, London, 1976, 443.

Fletcher, R.L., Jones, A.M., and Jones, E.B.G., The attachment of fouling macroalgae, in *Marine Biodeterioration: An Interdisciplinary Study*, Costlow, J.D. and Tipper, R.C., Eds., Naval Institute Press, Annapolis, MD, 1984, 172.

Forman, H.J. and Boveris, A., Superoxide radical and hydrogen peroxide in mitochondria, in *Free Radicals in Biology, Vol. 5*, Pryor, W.A., Ed., Academic Press, New York, 1982, 65.

Foster, A.C., Current antifouling technologies, in *Biofouling Problems and Solutions*, Kjelleberg, S. and Steinberg, P., Eds., University of New South Wales, Sydney, 1994, 44.

Foster, B.A., Desiccation as a factor in the intertidal zonation of barnacles, *Mar. Biol.*, 8, 17, 1971.

Fraenkel, G.S. and Gunn, D.L., *The Orientation of Animals. Kineses, Taxes and Compass Reactions*, Dover, New York, 1961, 376.

Fraser, J., *Nature Adrift. The Story of Marine Plankton*, Foulis, G.T. and Co., London, 1962, 178.

Fréchétte, M., Butman, C.A., and Geyer, W.R., The importance of boundary-layer flows in supplying phytoplankton to the benthic suspension feeder, *Mytilus edulis* L., *Limnol. Oceanogr.*, 34(1), 19, 1989.

Frost, A.M., Main aspects of antifouling protection of ships by means of paint coatings, in *Vsesoyuzn. nauchno-tekhn. konf. "Zashchita sudov i tekhnicheskikh sredstv ot obrastaniya"* (All-Union Research Conf. "Antifouling Protection of Vessels and Technical Objects"), Lyublinskii, E.Ya., Ed., Sudostroenie, Leningrad, 1990, 98.

Frost, A.M., Polosov, B.V., and Simanovich, M.B., Antifouling varnishes and paints, *Varnishes and Paints and Their Applications*, 1999, 7–8, 56–57.

Fusetani, N., Marine metabolites which inhibit development of echinoderm embryos, in *Bioinorganic Marine Chemistry, Vol. 1*, Scheuer, P.J., Ed., Springer-Verlag, Berlin, 1987, 61.

Fusetani, N., The Fusetani biofouling project, *Biofouling*, 12, 3, 1998.

Gaines, S., Brown, S., and Roughgarden, J., Spatial variation in larval concentrations as a cause of spatial variation in settlement for the barnacle, *Balanus glandula, Oecologia (Berlin)*, 67, 267, 1985.

Galil, B.S., A sea under siege — alien species in the Mediterranean, *Biol. Invasions*, 2(2), 177, 2000.

Galkina, V.N., Kulakowski, E.E., and Kunin, B.L., The influence of mussel aquaculture in the White Sea on the environment, *Okeanologiya*, 22(2), 321, 1982.

Gallager, S.M. et al., Ontogenetic changes in the vertical distribution of giant scallop larvae, *Placopecten magellanicus*, in 9-m deep mesocosms as a function of light, food and temperature stratification, *Mar. Biol.*, 124(4), 679, 1996.

Gansloser, H. and Nissenbaum, B., Verfahren zum Herstellen eines algen-bewuchsabweisenden Unterwasseranstrichs, Claim 3222090, Germany, C 09 D 5/16, B 05 D 7/16, priority 11.06.82, publ. 15.12.83.

Gauthier, M.J. and Aubert, M., Chemical telemediators in the marine environment, in *Mar. Org. Chem. Evol., Compos., Interact. and Chem. Org. Matter Seawater*, Duursma, E.K. and Dawson, R., Eds., Elsevier, Amsterdam, 1981, 225.

Geesey, G.G. et al., Influence of calcium and other cations on surface adhesion of bacteria and diatoms: a review, *Biofouling*, 15, 195, 2000.

Geraci, S. et al., Field and laboratory efficacy of chlorine dioxide as antifouling in cooling systems of power plants, *Oebalia*, 19, 383, 1993.

Gerchakov, S.M. and Udey, L.R., Microfouling and corrosion, in *Marine Biodeterioration: An Interdisciplinary Study*, Costlow, J.D. and Tipper, R.C., Eds., Naval Institute Press, Annapolis, MD, 1984, 82.

Gerhart, D.J., Rittschof, D., and Mayo, S.W., Chemical ecology and the search for marine antifoulants: studies of a predator-prey symbiosis, *J. Chem. Ecol.*, 14(10), 1905, 1988.

Gerlach, S.A., *Marine Pollution: Diagnosis and Therapy*, Springer-Verlag, Berlin, 1981, 278.

Gessner, F., Die Limnologie des Naturschutzgebietes Seeon, *Arch. Hydrobiol.*, 47(4), 553, 1953.

Giese, A.C., Annual reproductive cycles of marine invertebrates, *Annu. Rev. Physiol.*, 21, 547, 1959.

Giese, A.C. and Pearse, J.S., Introduction: general principles, in *Reproduction of Marine Invertebrates*, Giese, A.C. and Pearse, J.S., Eds., Academic Press, New York, 1974, 251.

Gili, J.-M. and Coma, R., Benthic suspension feeders: their paramount role in littoral marine food webs, *Trends Ecol. Evol.*, 13, 316, 1998.

Gilyarov, M.S. et al., *Biologicheskii entsiklopedicheskii slovar'* (The Biological Encyclopaedic Glossary), Sovetskaya Entsiklopediya, Moscow, 1986, 831.

Gleibs, S. and Mebs, D., Sequestration of a marine toxin, *Coral Reefs,* 17, 338, 1998.

Goodbody, I., Inhibition of the development of a marine sessile community, *Nature*, 190(4772), 282, 1961.

Gorbenko, Yu.A., Formation of bacterial film on submerged plates coated with antifouling paints, *Tr. Sevastop. Biol. Stan.,* 16, 447, 1963.

Gorbenko, Yu.A., *Ekologiya morskikh mikroorganizmov perifitona* (Ecology of Microoorganisms of Marine Periphyton), Nauk. Dumka, Kiev, 1977, 252.

Gorbenko, Yu.A., On the degree of biodeterioration of rosin, paraffin, and antifouling paints, in *Obrastanie i biokorroziya v vodnoi srede* (Biofouling and Biocorrosion in Water), Reznichenko, O.G. and Starostin, I.V., Eds., Nauka, Moscow, 1981, 257.

Gorbenko, Yu.A., *Ekologiya i prakticheskoe znachenie morskikh mikroorganizmov* (Ecology and Practical Importance of Marine Microoogranisms), Nauk. Dumka, Kiev, 1990, 160.

Gorbenko, Yu.A. and Kryshev, I.I., *Statisticheskii analiz dinamiki ekosistemy morskikh mikroorganizmov* (Statistical Analysis of the Dynamics of a Marine Microbial Ecosystem), Nauk. Dumka, Kiev, 1985, 144.

Gordon, R., A retaliatory role for algal projectiles, with implications for the mechanochemistry of diatom gliding motility, *J. Theor. Biol.*, 126(4), 419, 1987.

Gordon, R. and Drum, R.W., A capillarity mechanism for diatom gliding locomotion, *Proc. Natl. Acad. Sci.*, 67(1), 338, 1970.

Gotelli, N.J., Spatial and temporal patterns of reproduction, larval settlement, and recruitment of the compound ascidian *Aplidium stellatum, Mar. Biol.,* 94, 45, 1987.

Goto, R. et al., Fatty acids as antifoulants in a marine sponge, *Biofouling*, 6, 61, 1992.

Goto, R. et al., Furospongolide, an antifouling substance from the marine sponge *Phyllospongia papyracea* against the barnacle *Balanus amphitrite, Nippon Suisan Gakkaishi, Bull. Jpn. Soc. Sci. Fish.*, 59, 1953, 1993.

Goudet, J.-L., Aquaculture: une industrie encore en devenir, *Sci. Technol.*, 37, 98, 1991.

Graham, K. and Sebens, K., The distribution of marine invertebrate larvae near vertical surfaces in the rocky subtidal zone, *Ecology,* 77, 933, 1996.

Granéli, W., Department of marine ecology in collaboration with the Royal Swedish Academy of Sciences, in *Marine Research 1991 1997: Information on Marine Research and Education*, Marine Research Centre, Göteborg University, Sweden, 1994, 38.

Gray, J.S., The attractive factor of intertidal sands to *Protodrilus symbioticus, J. Mar. Biol. Assn. U.K.*, 46(3), 627, 1966.

Green, N.P.O., Stout, G.W., and Taylor, D.J., *Biological Science*, Vol. 3, Cambridge University Press, London, 1988, 360.

Grimmek, D.W. and Sander, L.W., Economic and technical feasibility of copper-nickel sheathing of ship hulls, *Mar. Technol.*, 22(2), 142, 1985.

Grishankov, A.V., On the consortium structure of benthic communities from Solovetsky Bay (Onega Bay, the White Sea), *Vestn. Sankt-Peterburg. Univ. Ser. Biol.*, 17, 14, 1995.

Gromov, B.V. and Pavlenko, G.V., *Ekologiya bakterii: Uchebnoe posobie* (Ecology of Bacteria: A Handbook), Leningrad University, Leningrad, 1989, 248.

Gromov, F.N., Gorshkov, S.G., and Chernavin, V.N., Eds., *Chelovek i okean. Atlas okeanov* (Man and the Ocean: An Atlas of Oceans), Voenno-Morskoi Flot, St. Petersburg, 1996, 319.

Gschwend, P.M., MacFarlane, J.K., and Newman, K.A., Volatile halogenated organic compounds released to seawater from temperate marine macroalgae, *Science*, 227(4690), 1033, 1985.

Guerin, J.-P., Role des facteurs biotiques dans le determinisme de la fixation et de la metamorphose des larves meroplanctoniques, *Oceanis*, 8(5), 389, 1982.

Gurevich, E.S. and Dolgopol'skaya, M.A., General principles of development of antifouling methods, *Biol. Morya* (Kiev), 35, 3, 1975.

Gurevich, E.S. et al., *Zashchita ot obrastaniya* (Fouling Prevention), Nauka, Moscow, 1989, 271.

Gurin, I.S. and Azhgikhin, I.S., *Biologicheski aktivnye veshchestva gidrobiontov — istochnik novykh lekarstv i preparatov* (Bioactive Substances of Hydrobionts as Source of New Medicines and Preparations), Nauka, Moscow, 1981, 136.

Haderlie, E.C., A brief overview of the effects of macrofouling, in *Marine Biodeterioration: An Interdisciplinary Study,* Costlow, J.D. and Tipper, R.C., Eds., Naval Institute Press, Annapolis, MD, 1984, 163.

Hadfield, M.G., Metamorphosis in marine molluscan larvae: an analysis of stimulus and response, in *Settlement and Metamorphosis of Marine Invertebrate Larvae*, Chia, F.-S. and Rice, M.E., Eds., Elsevier/North Holland, New York, 1978, 165.

Hadfield, M.G., Settlement and recruitment of marine invertebrates: a perspective and some proposals, *Bull. Mar. Sci.*, 39, 418, 1986.

Hadfield, M.G., The D.P. Wilson lecture. Research on settlement and metamorphosis of marine invertebrate larvae: past, present and future, *Biofouling*, 12(1–3), 9, 1998.

Hadfield, M.G. and Pennington, J.T., Nature of the metamorphic signal and its internal transduction in larvae of the nudibranch *Phestilla sibogae, Bull. Mar. Sci.*, 46(2), 455, 1990.

Hadfield, M.G. and Scheuer, D., Evidence for a soluble metamorphic inducer in *Phestilla*: ecological, chemical and biological data, *Bull. Mar. Sci.*, 37(2), 556, 1985.

Hall, A., Copper accumulation in copper-tolerant and non-tolerant populations of the marine fouling alga, *Ectocarpus siliculosus*, *Bot. Mar.*, 24(4), 223, 1981.

Hall, A. and Baker, A.J.M., Settlement and growth of copper-tolerant *Ectocarpus siliculosus* (Dillw.) Lyngbye on different copper-based antifouling surfaces under laboratory conditions. I. Corrosion trials in sea water and development of an algal culture system, *J. Mater. Sci.*, 20(3), 1111, 1985.

Hamilton, W.A., Biofilms: microbial interactions and metabolic activities, in *Ecology of Microbial Communities,* Fletcher, M., Gray, T.R.G., and Jones, J.G., Eds., Cambridge University Press, Cambridge, 1987, 361.

Hannan, C.A., Polychaete larval settlement: correspondence of patterns in suspended jar collectors and in the adjacent natural habitat in Monterey Bay, California, *Limnol. Oceanogr.*, 26(1), 159, 1981.

Hannan, C.A., Planktonic larvae may act like passive particles in turbulent near-bottom flows, *Limnol. Oceanogr.*, 29(5), 1108, 1984.

Harborn, J.B., *Introduction in Ecological Biochemistry,* 4th ed., Academic Press, London, 1993, 318.

Harborn, R.S. and Kent, C.A., Aspects of cell adhesion, in *Fouling Science and Technology,* Melo, L.F., Bott, T.R., and Bernardo, C.A., Eds., Kluwer, Dordrecht, 1988, 125.

Harder, T. and Qian, P.-Y., Waterborne compounds from the green seaweed *Ulva reticulata* as inhibitive cues for larval attachment and metamorphosis in the polychaete *Hydroides elegans*, *Biofouling*, 16, 205, 2000.

Harington, C.R., A note on the physiology of the ship-worm (*Teredo norvegica*), *Biochem. J.*, 15, 736, 1921.

Harlin, M.M. and Lindbergh, J.M., Selection of substrata by seaweeds: optimal surface relief, *Mar. Biol.*, 40(1), 33, 1977.

Harms, J. and Anger, K., Seasonal, annual and spatial variation in the development of hard bottom communities, *Helgol. Meeresunters.*, 36, 137, 1983.

Harris, L.G. and Irons, K.P., Substrate angle and predation as determinants in fouling community succession, in *Artificial Substrates*, Cairns, J., Jr., Ed., Ann Arbor Science, Ann Arbor, MI, 1982, 131.

Harvey, M., Bourget, E., and Ingram, R.G., Experimental evidence of passive accumulation of marine bivalve larvae on filamentous epibenthic structures, *Limnol. Oceanogr.*, 40(1), 94, 1995.

Hay, C.H., Dispersal of sporophytes of *Undaria pinnatifida* by coastal shipping in New Zealand, and implications for further dispersal of *Undaria* in France, *Br. Phycol. J.*, 25, 301, 1990.

Hedgecock, D., Is geneflow from pelagic larval dispersal important in the adaptation and evolution of marine invertebrates?, *Bull. Mar. Sci.*, 39, 550, 1986.

Hellio, C., Bourgougnon, N., and le Gal, Y., Phenoloxidase (E.C. 1.14.18.1) from the byssus gland of *Mytilus edulis*: purification, partial characterization and application for screening products with potential antifouling activities, *Biofouling*, 16, 235, 2000.

Hentschel, E., Biologische Untersuchungen über den tierischen und pflanzlichen Bewuchs im Hamburger Hafen, *Mitt. Zool. Mus. Hamburg*, 33, 1, 1916.

Hentschel, E., Über den Bewuchs auf den treibenden Tangen der Sargassosee, *Mitt. Zool. Mus. Hamburg*, 38, 1, 1921.

Hentschel, E., Der Bewuchs an Seeschiffen, *Int. Rev. Gesamten Hydrobiol. Hydrogr.*, 11, 238, 1923.

Heslinga, G.A., Larval development, settlement and metamorphosis of the tropical gastropod *Trochus niloticus, Malacologia*, 20(2), 349, 1981.

Hick, D.W. and Tunnel, J.W., Invasion of the south Texas coast by the edible brown mussel, *Perna perna* (Linnaeus, 1758), *Veliger*, 36, 92, 1993.

Highsmith, R.C., Induced settlement and metamorphosis of sand dollar (*Dendraster excentricus*) larvae in predator-free sites: adult sand dollar beds, *Ecology*, 63(2), 329, 1982.

Highsmith, R.C. and Emlet, R.B., Delayed metamorphosis: effect on growth and survival of juvenile sand dollars (Echinoidea: Clypeasteroida), *Bull. Mar. Sci.*, 39(2), 347, 1986.

Hillman, R.E. and Nace, P.F., Histochemistry of barnacle cyprid adhesive formation, in *Adhesion in Biological Systems*, Academic Press, New York, 1970, 113.

Hills, J.M. and Thomason, J.C., On the effect of tile size and surface texture on recruitment pattern and density of the barnacle, *Semibalanus balanoides, Biofouling*, 13, 31, 1998.

Hills, J.M., Thomason, J.C., Milligan, J.L., and Richardson, M., Do barnacle larvae respond to multiple settlement cues over a range of spatial scales? *Hydrobiologia*, 375/376, 101, 1998.

Hines, A.H., The comparative reproduction ecology of three species of intertidal barnacles, in *Reproductive Ecology of Marine Invertebrates*, Stancyk, S.E., Ed., University of Carolina Press, Columbia, SC, 1979, 213.

Hirata, T., Succession of sessile organisms on experimental plates immersed in Nabeta Bay, Jzu Peninsula, Japan, *Mar. Ecol. Prog. Ser.*, 38(1), 25, 1987.

Hoare, K. and Davenport, J., Size-related variation in the sensitivity of the mussel, *Mytilus edulis*, to copper, *J. Mar. Biol. Assn. U.K.*, 74(4), 971, 1994.

Hoare, K., Beaumont, A.R., and Davenport, J., Variation among populations in the resistance of *Mytilus edulis* embryos to copper: adaptation to pollution? *Mar. Ecol. Prog. Ser.*, 120(1–3), 155, 1995.

Holland, D.E. and Walker, G., The biochemical composition of the cypris larvae of the barnacle *Balanus balanoides* L., *J. Cons. Int. Explor. Mer.*, 36, 162, 1975.

Holm, E.R. et al., Temporal and spatial variation in the fouling of silicone coatings in Pearl Harbor, Hawaii, *Biofouling*, 15, 95, 2000.

Holmström, C. and Kjelleberg, S., The effect of external biological factors on settlement of marine invertebrates and new antifouling technology, *Biofouling,* 8, 147, 1994.

Holmström, C., Rittchof, D., and Kjelleberg, S., Inhibition of settlement by larvae of *Balanus amphitrite* and *Ciona intestinalis* by a surface-colonizing marine bacterium, *Appl. Environ. Microbiol.,* 58(7), 2111, 1992.

Hoppe, H.-G., Attachment of bacteria: advantage or disadvantage for survival in the aquatic environment, in *Microbial Adhesion and Aggregation. Dahlem Konferenzen, 1984,* Springer-Verlag, Berlin, 1984, 283.

Horiguchi, T. et al., Imposex in Japanese gastropods (Neogastropoda and Mesogastropoda): effects of tributyltin and triphenyltin from antifouling paints, *Mar. Pollut. Bull.,* 31(4–12), 402, 1995.

Houghton, D.R., Toxicity testing of candidate antifouling agents and accelerated antifouling paint testing, in *Marine Biodeterioration: An Interdisciplinary Study,* Costlow, J.D. and Tipper, R.C., Eds., Naval Institute Press, Annapolis, MD, 1984, 255.

Houghton, D.R., Pearman, I., and Tierney, D., The effect of water velocity on the settlement of swarmers of *Enteromorpha* sp., in *Proc. 3rd Int. Congr. on Marine Corrosion and Fouling,* Acker, R.F. et al., Eds., Northwestern University Press, Evanston, MD, 1972, 682.

Huang, R. and Boney, A.D., Individual and combined interactions between littoral diatoms and sporelings of red algae, *J. Exp. Mar. Biol. Ecol.,* 85(2), 101, 1985.

Hudon, C., Bourget, E., and Legendre, P., An integrated study of the factors influencing the choice of the settling site of *Balanus crenatus* cyprid larvae, *Can. J. Fish. Aquat.,* 40(8), 1186, 1983.

Hurd, C.L., Harrison, P.J., and Druehl, L.D., Effect of seawater velocity on inorganic nitrogen uptake by morphologically distinct forms of *Macrocystis integrifolia* from wave-sheltered and exposed sites, *Mar. Biol.,* 126(2), 205, 1996.

Hurlbut, C.J., The effects of larval abundance, settlement and juvenile mortality on the depth distribution of a colonial ascidian, *J. Exp. Mar. Biol. Ecol.,* 150(2), 183, 1991.

Hurlbut, C.J., The adaptive value of larval behavior of a colonial ascidian, *Mar. Biol.,* 115(2), 253, 1993.

Hüttinger, K.J., Surface bound biocides — a novel possibility to prevent biofouling, in *Fouling Science and Technology,* Melo, L.F., Bott, T.R., and Bernardo, C.A., Eds., Kluwer, Dordrecht, 1988, 233.

Ilan, M. and Loya, Y., Sexual reproduction and settlement of the coral reef sponge *Chalinula* sp. from Red Sea, *Mar. Biol.,* 105, 25, 1990.

Il'ichev, V.D., Bocharov, B.V., and Gorlenko, M.V., Ekologicheskie osnovy zashchity ot bio-povrezhdenii (*Ecological Basis of Biodeterioration Control*), Nauka, Moscow, 1985, 263.

Il'ichev, V.D. et al., *Biopovrezhdeniya: Uchebnoe posobie dlya biologicheskikh spetsial'nostei vuzov* (Biodeterioration: Handbook for University Students), Vysshaya Shkola, Moscow, 1987, 352.

Il'in, I.N., Marine wood borers of the USSR and methods of their control, in *Obrastanie i biopovrezhdenie: ekologicheskie problemy* (Fouling and Biodeterioration as Ecological Problems), Il'in, I.N., Ed., Nauka, Moscow, 1992a, 21.

Il'in, I.N., Pelagic fouling in the tropical and subtropical ocean waters, in *Obrastanie i biopovrezhdenie: ekologicheskie problemy* (Fouling and Biodeterioration as Ecological Problems), Il'in, I.N., Ed., Nauka, Moscow, 1992b, 77.

Ina, K. et al., Isothiocyanates as an attaching repellent against the blue mussel *Mytilus edulis,* *Agric. Biol. Chem.,* 53(12), 3323, 1989.

Iserentant, R., Quelques réflexions sur la notion de periphyton, *Cah. Biol. Mar.,* 28, 297, 1987.

Isham, L.B. and Tierney, Some aspects of the larval development and metamorphosis of Teredo (Lyrodus) pedicellatus De Quatrefeges, *Bull. Mar. Sci. Gulf Caribb.*, 2, 574, 1953.

Itikawa, I., Antifouling formulation for marine vessels, Claim 59-206472, Japan, S 09 D 5/14, C 09 D 3/38, priority 11.05.83, N 58-80924, publ. 22.11.84.

Ivankovič, D. et al., Multiple forms of metallothionein from the digestive gland of naturally occurring and cadmium-exposed mussels, *Mytilus galloprovincialis, Helgol. Mar. Res.*, 56(2), 95, 2002.

Ivanova-Kazas, O.M., *Sravnitel'naya embriologiya bespozvonochnykh zhivotnykh. Prosteishie i nizshie mnogokletochnye* (Comparative Embryology of Invertebrates. Protists and Lower Multicellular Animals), Nauka, Novosibirsk, 1975, 372.

Ivanova-Kazas, O.M., *Sravnitel'naya embriologiya bespozvonochnykh zhivotnykh. Trokhofornye, shchupal'tsevye, shchetinkochelyustnye, pogonofory* (Comparative Embryology of Invertebrates. Trochozoa, Tentaculata, Chaetognatha, and Pogonophora), Nauka, Moscow, 1977, 312.

Ivanova-Kazas, O.M., *Ocherki po filogenii nizshikh khordovykh* (Essays on the Phylogeny of Lower Chordates), St. Petersburg University, St. Petersburg, 1995, 160.

Ivin, V.V., Fouling in *Laminaria japonica* mariculture, in ECOSET'95, The 6th Int. Conf. on Aquatic Habitat Enhancement, Japan International Marine Science Technology Federation, Tokyo, 1995, 495.

Ivlev, V.S., Ein Versuch zur experimentellen Erforschung der Ökologie der Wasserbiocönosen, *Arch. Hydrobiol.*, 25, 177, 1933.

Izral'yants, E.D., Kudinova, V.V., and Basova, L.S., Comparative tests of biological activity of organotin and organoarsenic toxins used in paints, *Lakokrasochnye Materially,* 1, 11, 1976.

Jackson, J.B.C. and Buss, L., Allelopathy and spatial competition among coral reef invertebrates, *Proc. Natl. Acad. Sci.*, 72(12), 5160, 1975.

Jackson, J.B.C., Competition on marine hard substrata: the adaptive significance of solitary and colonial strategies, *Am. Nat.*, 111(980), 743, 1977a.

Jackson, J.B.C., Habitat area, colonization, and development of epibenthic community structure, in *Biology of Benthic Organisms. Proc. 11th European Marine Biology Symp., Galway, 1976*, Keegan, B.F. et al., Eds., Pergamon Press, London, 1977b, 349.

Jackson, P.A. and Gill, J., Method and a system for counteracting marine biologic fouling of a hull or a submerged construction, Claim 0480936, EPV, B 63 B 59/04, B 06 B 1/02, priority 14.05.90, publ. 22.04.92.

Jensen, R.A. and Morse, D.E., Intraspecific facilitation of larval recruitment: gregarious settlement of the polychaete *Phragmatopoma californica* (Fewkes), *J. Exp. Mar. Biol. Ecol.*, 83(2), 107, 1984.

Jensen, R.A. and Morse, D.E., The bioadhesive of *Phragmatopoma californica* tubes: a silklike cement containing L-DOPA, *J. Comp. Physiol. Ser. B,* 158(3), 317, 1988.

Johnson, C.R., Induction of metamorphosis of coral larvae by coralline algae: evolution and altruism, in *6th Int. Conf. Coelenterate Biology*, den Hartog, J.C., Ed., Noordwijkerhout, the Netherlands, 1995, 56.

Johnson, C.R. and Mann, K.H., The crustose coralline algae, *Phymatolithon foslie* inhibits the overgrowth of seaweeds without relying on herbivores, *J. Exp. Mar. Biol. Ecol.*, 96(2), 127, 1986.

Jørgensen, C.B., *Biology of Suspension Feeding,* Pergamon Press, Oxford, 1966, 357.

Jumars, P.A., Self, R.F.L., and Nowell, A.R.M., Mechanics of particle selection by tentaculate deposit-feeders, *J. Exp. Mar. Biol. Ecol.*, 64, 47, 1982.

Kain, J.M., A view of the genus *Laminaria, Oceanogr. Mar. Biol. Annu. Rev.*, 17, 101, 1979.

Kaluzhny, S.V. and Ivanov, A.N., Biogenic corrosion of concrete and methods of its prevention, *Praktika Protivokorrosionnoi Zaschity* (Practice of Anticorrosion Protection), 1,
22, 2002.

Kamshilov, M.M., To the biology of cirripede larvae on the Eastern Murman, *Trudy Murm.
Biol. St.*, 4, 56, 1958.

Kaplan, H.M., Anesthesia in invertebrates, *Fed. Proc.*, 28(4), 1557, 1969.

Karsinkin, G.S., Versuch einer praktischen Lösung der Biocoenosenfrage, *Arb. Hydrobiol.
Station am See "Glubokoje,"* 6, 36, 1925.

Kashin, I.A., Zvyagintsev, A.Ju., and Maslennikov, S.I., Fouling of hydrotechnical structures
in the western part of Peter the Great Bay, the Sea of Japan, *Biol. Morya*, 26, 86, 2000.

Kashin, S.M. and Kuznetsova, I.A., Dynamics and mechanisms of *Balanus improvisus* settlement on different substrates, *Okeanologiya*, 25(5), 846, 1985.

Kasyanov, V.L., Planktotrophic larvae of bivalves: morphology, physiology, and behavior,
Biol. Morya, 3, 3, 1984a.

Kasyanov, V.L., Starfish larvae, their morphology, physiology, and behavior, *Biol. Morya*, 1,
3, 1984b.

Kasyanov, V.L., *Reproduktivnaya strategiya morskih dvustvorchatyh molluskov I iglokozhih*
(Reproductive Strategy of Marine Bivalves and Echinoderms), Akad. Nauk, Vladivostok, 1989, 179.

Kato, T. et al., Active components of *Sargassum tortile* affecting the settlement of swimming
larvae of *Coryne uchidai*, *Experientia*, 31(4), 433, 1975.

Kazar'yan, V.V. et al., Vertical structure of coastal waters of the Chupa Inlet of the White
Sea, in *4-ya nauchnaya sessiya Morskoi Biologicheskoi stantsii SPbGU* (4th Scientific
Session of the Marine Biological Station of St. Petersburg State University), Stogov,
I.A., Ed., St. Petersburg, 2003, 44.

Keats, D.W., Knight, M.A., and Pueschel, C.M., Antifouling effects of epithelial shedding in
three crustose coralline algae (Rhodophyta, Corallinales) on a coral reef, *J. Exp. Mar.
Biol. Ecol.*, 213, 281, 1997.

Keifer, P.A. and Rinehart, K.L., Jr., Renillafoulins, antifouling diterpenes from the sea pansy
Renilla reniformis (Octocorallia), *J. Org. Chem.*, 51(23), 4450, 1986.

Kelly, J.R. et al., The effects of tributylin within a *Thalassia* seagrass ecosystem, *Estuaries*,
13(3), 301, 1990.

Kempf, G., Tributylzinn adé- und dann? *Schiff und Hafen*, 53, 73, 2001.

Kent, C.A., Biological fouling: basic science and models, in *Fouling Science and Technology*,
Melo, L.F., Bott, T.R., and Bernardo, C.A., Eds., Kluwer, Dordrecht, 1988, 207.

Keough, M.J. and Downes, B.J., Recruitment of marine invertebrates: the role of active larval
choices and early mortality, *Oecologia*, 54, 348, 1982.

Ketchum, B.H., Factors influencing the attachment and adherence of fouling organisms, in
Marine Fouling and Its Prevention, Redfield, A.C. and Ketchum, B.H., Eds., U.S.
Naval Institute, Annapolis, MD, 1952, 230.

Khailov, K.M., *Ekologicheskij metabolizm v more* (Ecological Metabolism in the Sea), Nauk.
Dumka, Kiev, 1971, 252.

Khailov, K.M. et al., *Funktsional'naya morfologiya morskikh mnogokletochnykh vodoroslei*
(Functional Morphology of Marine Multicellular Algae), Nauk. Dumka, Kiev, 1992,
281.

Khailov, K.M. et al., Biological parameters of phytofouling in relation with physical parameters of the experimental 'reef' constructions in an eutrophic marine area, *Vodn.
Resurs.*, 21(2), 166, 1994.

Khailov, K.M., Prazukin, A.V., and Smolev, D.M., Development and growth of algal settlements on experimental objects, *Bot. Zh.*, 80(9), 21, 1995.

Khailov, K.M. et al., Concentration and functional activity of living matter in aggregations at different levels of organization, *Usp. Sovrem. Biol.* (Successes of Modern Biology), 119, 3, 1999.

Khalaman, V.V., Study of the fouling succession in the White Sea using the information index of species diversity, in *Ekologicheskie issledovaniya belomorskikh organizmov* (Ecological Studies of the White Sea Organisms), Berger, V.Ya., Ed., Zoological Institute, Leningrad, 1989, 34.

Khalaman, V.V., Fouling communities at mussel culture farms in the White Sea, *Biol. Morya,* 27(4), 268, 2001a.

Khalaman, V.V., Succession of fouling communities on artificial substrates at mussel culture farms in the White Sea, *Biol. Morya,* 27(6), 399, 2001b.

Khristoforova, N.K., *Bioindikatsiya i monitoring zagryazneniya morskikh vod tyazhelymi metallami* (Bioindication and Monitoring of Seawater Pollution with Heavy Metals), Nauka, Leningrad, 1989, 192.

Kirchman, D. and Mitchell, R., A biochemical mechanism for marine biofouling, in *Oceans 81: Conf. Rec., Boston, Mass., Sept. 16–18, 1981, Vol. 1,* New York, 1981, 537.

Kirchman, D. et al., Lectins may mediate in the settlement and metamorphosis of *Janua (Dexiospira) brasiliensis* Grube (Polychaeta: Spirorbidae), *Mar. Biol. Lett.,* 3(3), 131, 1982.

Kisseleva, G.A., Factors stimulating larval metamorphosis of a lamellibranch, *Brachyodontes lineatus* (Gmelin), *Zool. Zh.,* 45(10), 1571, 1966.

Kisseleva, G.A., Effect of substrate on settlement and metamorphosis of larvae of benthic animals, in *Donnye biotsenozy i biologiya bentosnykh organizmov Chernogo morya* (Bottom Biocenoses and the Biology of Benthic Organisms of the Black Sea), Vodyanitskii, V.A., Ed., Nauk. Dumka, Kiev, 1967a, 71.

Kisseleva, G.A., Settlement of *Polydora ciliata* (Jonston) larvae on various substrates, in *Donnye biotsenozy i biologiya bentosnykh organizmov Chernogo morya* (Bottom Biocenoses and the Biology of Benthic Organisms of the Black Sea), Vodyanitskii, V.A., Ed., Nauk. Dumka, Kiev, 1967b, 85.

Kitamura, H. and Hirayama, K., Effect of primary films on the settlement of larvae of a bryozoan *Bugula neritina, Bull. Jpn. Soc. Sci. Fish.,* 53(8), 1377, 1987a.

Kitamura, H. and Hirayama, K., Effect of cultured diatom films on the settlement of larvae of a bryozoan *Bugula neritina, Bull. Jpn. Soc. Sci. Fish.,* 53(8), 1383, 1987b.

Kjelleberg, S., Adhesion to inanimate surfaces, in *Microbial Adhesion and Aggregation. Dahlem Konferenzen, 1984,* Marshall, K.C., Ed., Springer-Verlag, Berlin, 1984, 51.

Kleemann, K., Evolution of chemically-boring Mytilidae (Bivalvia), in *The Bivalvia — Proc. of Memorial Symp. in Honour of Sir Charles Maurice Yonge, Edinburgh, 1986,* Morton, B., Ed., Hong Kong University Press, Hong Kong, 1990, 111.

Kleeman, K., Biocorrosion by bivalves, *P.S.Z.N.I.: Mar. Ecol.,* 17, 145, 1996.

Kloareg, B. and Quatrano, R.S., Structure of the cell walls of marine algae and ecophysiological functions of the matrix polysaccharides, *Oceanol. Mar. Biol. Rev.,* 26, 259, 1988.

Knight-Jones, E.W., Gregariousness and some other aspects of the setting behaviour of *Spirorbis, J. Mar. Biol. Assn. U.K.,* 30(2), 201, 1951.

Knight-Jones, E.W., Decreased discrimination during settling after prolonged planktonic life in larvae of *Spirorbis borealis* (Serpulidae), *J. Mar. Biol. Assn. U.K.,* 32, 337, 1953a.

Knight-Jones, E.W., Laboratory experiments on gregariousness during settlement in *Balanus balanoides* and other barnacles, *J. Exp. Biol.,* 30, 584, 1953b.

Knight-Jones, E.W. and Crisp, D.J., Gregariousness in barnacles in relation to the fouling of ships and to anti-fouling research, *Nature,* 171(4364), 1109, 1953.

Knight-Jones, E.W. and Morgan, E., Responses of marine animals to changes in hydrostatic pressure, *Oceanogr. Mar. Biol. Annu. Rev.*, 4, 267, 1966.

Konno, T., Effects of inclination of a rock surface on the distribution of sublittoral sessile organisms, *J. Tokyo Univ. Fish.*, 72(2), 99, 1986.

Konstantinov, A.S., *Obshchaya gidrobiologiya* (General Hydrobiology), Vysshaya Shkola, Moscow, 1979, 480.

Konstantinova, M.I., Characteristics of movement of pelagic larvae of marine invertebrates, *Dokl. Akad. Nauk SSSR*, 170(3), 726, 1966.

Konstantinova, M.I., Locomotion of polychaete larvae, *Dokl. Akad. Nauk SSSR*, 188(4), 942, 1969.

Kon-ya, K.M. et al., Indole derivates as potent inhibitors of larval settlement by the barnacle, *Balanus amphitrite, Biosci. Biotech. Biochem.*, 58, 2178, 1994.

Kon-ya, K.M., Shimidzu, N., Otaki, N., Yokoyama, A., Adachi, K., and Miki, W., Inhibitory effect of bacterial ubiquinones on the settling of barnacle, *Balanus amphitrite, Experientia*, 51, 153, 1995.

Korn, O.M., Larvae of barnacles of the order Thoracica, *Biol. Morya*, 2, 3, 1990.

Korovin, Ju.M. and Ledenev, A.V., Pecularities of bacteria development on metallic surfaces in ocean, in *Vsesoyuzn. nauchno-tekhn. konf. "Zashchita sudov i tekhnicheskikh sredstv ot obrastaniya"* (All-Union Research Conf. "Antifouling Protection of Vessels and Technical Objects"), Lyublinskii, E.Ya., Ed., Sudostroenie, Leningrad, 1990, 27.

Korte, F. et al., *Lehrbuch der Ökologishen Chemie: Grundlagen und Konzepte für die ökologishe Beurteilung von Chemikalien*, 3rd ed., Thieme Verlag, Stuttgart, 1992, 373.

Koshtoyants, Kh.S., *Osnovy sravnitel'noi fiziologii* (Basic Comparative Physiology), Nauka, Moscow, 1957, 635.

Kraak, M.H.S., Stuijfzand, S.C., and Admiral, W., Short-term exotoxicity of a mixture of five metals to the zebra mussel *Dreissena polymorpha, Bull. Environ. Contam. Toxicol.*, 63(6), 805, 1999.

Krumbein, W.E., Influences of organisms on their environment, *P.S.Z.N.I.: Mar. Ecol.*, 17, 1, 1996.

Krumbein, W.E. and Lapo, A.V., Vernadsky's biosphere as a basis of geophysiology, in *Gaia in Action,* Bunyard, P., Ed., Floris Books, Edinburgh, 1996, 116.

Kubanek, J. et al., Triterpene glycosides defend the Caribbean reef sponge *Erylus formosus* from predatory fishes, *Mar. Ecol. Prog. Ser.,* 207, 69, 2000.

Kubanin, A.A., Geographic distribution of bryozoans participating in marine fouling, in *Ekologiya obrastaniya v Severo-Zapadnoi chasti Tikhogo okeana* (Ecology of Fouling in the Northwestern Pacific), Kudryashov, V.A., Ed., Akademia Nauk, Vladivostok, 1980, 109.

Kucherova, Z.S., Diatoms, in *Biologicheskie osnovy bor'by s obrastaniem* (Biological Basis of Fouling Control), Vodyanitskii, V.A. and Dolgopol'skaya, M.A., Eds., Nauk. Dumka, Kiev, 1973, 47.

Kudryashov, V.A., Ed., *Biologiya anfel'tsii* (Biology of Ahnfeltia), Akademie Nauk SSSR, Vladivostok, 1980, 124.

Kuhn, A., Streit inder Schiffsfarbenindustrie, *Galvanotechnik*, 90(8), 2153, 1999.

Kulakowski, E.E., Studies of the mussel mariculture in the White Sea, in *Gidrobiologicheskie i ikhtiologicheskie issledovaniya na Belom more* (Hydrobiological and Ecological Research in the White Sea), Zoological Institute, Leningrad, 1987, 64.

Kulakowski, E.E., The biological bases of the mussel mariculture in the White Sea, *Tr. Zool. Inst. Ross. Akad. Nauk*, 50, 168, 2000.

Kulakowski, E.E. and Kunin, B.L., *Teoreticheskie osnovy kul'tivirovaniya midii v Belom more* (Theoretical Basis of Mussel Culture in the White Sea), Nauka, Leningrad, 1983, 35.

Kusakin, O.G. and Lukin, V.I., *Podvodnyi mir Kuril* (Underwater World of the Kurils), Dal'nauka, Vladivostok, 1995, 180.

Kwiatkowska, D. and Wichary, H., Korozja mikrobilogiczna w systemach technicznych: Przegląd literatury, *Ochr. Koroz.,* 44, 148, 2001.

Laban, C., Sponzen in de strijd tegen giftige scheepsverf, *Mens. Wet.*, 20(1), 12, 1993.

Laenko, Yu.I. et al., Lectins of marine invertebrates, *Usp. Sovr. Biol.*, 112(5–6), 785, 1992.

Lagadenc, Y., Conti, Ph., Retiere, C., Cabioch, L., and Dauvin, J.-C., Processus hydrodynamiques et recrutement de *Pectinaria koreni,* annélide polychète a cycle benthopélagique, en baie de Seine orientale, *Oceanis*, 16, 245, 1990.

Laius, Yu.A. and Kulakowski, E.E., Periphytonic microbial community of mussel mariculture in the White Sea: main stages of its development, in *Gidrobiologicheskie osobennosti yugo-vostochnoi chasti Kandalakshskogo zaliva v svyazi s marikul'turoi midii na Belom more* (Hydrobiological Features of Southeastern Kandalaksha Bay as Related to Mussel Mariculture in the White Sea), Andriyashev, A.P. et al., Eds., Zoological Institute, Leningrad, 1988, 74.

Lam, K.K.Y., Early growth of a pioneer recruited coral *Oulastrea crispate* (Scleractinia, Faviidae) on PFA-concrete blocks in a marine park in Hong Kong, China, *Mar. Ecol. Prog. Ser.*, 205, 113, 2000.

Lambert, W.J., Levin, D., and Berman, J., Changes in the structure of a New England (USA) kelp bed: the effect of an introduced species, *Mar. Ecol. Prog. Ser.*, p. 303, 1992.

Langston, W.J., Burt, G.R., and Mingjiang, Z., Tin and organotin in water, sediments, and benthic organisms of Poole Harbour, *Mar. Pollut. Bull.*, 18(12), 634, 1987.

Lapo, A.V., *Traces of Bygone Biospheres,* Mir Publishers, Moscow, 1987, 352.

Lapointe, L. and Bourget, E., Influence of substratum heterogeneity scales and complexity on a temperate epibenthic marine community, *Mar. Ecol. Prog. Ser.,* 189, 159, 1999.

Larman, V.N. and Gabbott, P.A., Settlement of cyprid larvae of *Balanus balanoides* and *Elminius modestus* induced by extracts of adult barnacles and other marine animals, *J. Mar. Biol. Assn. U.K.*, 55(1), 183, 1975.

Larman, V.N., Gabbott, P.A., and East, J., Physico-chemical properties of the settlement factor proteins from the barnacle *Balanus balanoides*, *Comp. Biochem. Physiol. Ser. B*, 72(3), 329, 1982.

Lau, S.C.K. and Qian, P.Y., Phlorotannins and related compounds as larval settlement inhibitors of the tube-building polychaete *Hydroides elegans*, *Mar. Ecol. Prog. Ser.,* 159, 219, 1997.

Lau, S.C.K. and Qian, P.Y., Inhibitory effect of phenolic compounds and marine bacteria on larval settlement of the barnacle *Balanus amphitrite amphitrite* Darwin, *Biofouling*, 16, 47, 2000.

Lav, S.C.K. and Qian, P.-Y., Larval settlement in the serpulid polychaete *Hydroides elegans* in response to bacterial films: an investigation of the nature of putative larval settlement cue, *Mar. Biol.,* 138(2), 321, 2001.

Le Tourneux, F. and Bourget, E., Importance of physical and biological settlement cues used at different spatial scales by the larvae of *Semibalanus balanoides*, *Mar. Biol.*, 97(1), 57, 1988.

Lebedev, E.M., Losses from biofouling and biodeterioration caused by the absence or technological breakdown of protection, in *Biologicheskie povrezhdeniya promyshlennykh i stroitel'nykh materialov* (Biodeterioration of Construction Materials), Lebedev, E.M. et al., Eds., Akad. Nauk, Moscow, 1973, 224.

Lebedev, E.M., Marine stone borers and their control, in *Obrastanie i biopovrezhdenie: ekologicheskie problemy* (Fouling and Biodeterioration as Ecological Problems), Il'in, I.N., Ed., Nauka, Moscow, 1992, 57.

Lebedeva, G.D., Biodamage of materials and aquatic toxicology, in *Izuchenie protsessov morskogo bioobrastaniya i razrabotka metodov bor'by s nim* (Study of Marine Biofouling and Development of Methods of Its Control), Skarlato, O.A., Ed., Zoological Institute, Leningrad, 1987, 12.

Lefèvre, M., Interference sur le recrutement benthique entre hydrodynamisme des masses et comportement larvaire, *Oceanis*, 16(3), 135, 1990.

Leitz, T., Induction of metamorphosis of the marine hydrozoan *Hydractinia echinata* Fleming, 1828, *Biofouling*, 12, 173, 1998.

Leitz, T. and Wagner, T., The marine bacterium *Alteromonas espejina* induces metamorphosis of the hydroid *Hydractinia echinata*, *Mar. Biol.*, 115(2), 173, 1993.

Lenihan, H.S., Oliver, J.S., and Stephenson, M.A., Changes in hard bottom communities related to boat mooring and tributylin in San Diego Bay: a natural experiment, *Mar. Ecol. Prog. Ser.*, 60(1–2), 147, 1990.

Lewis, C.A., A review of substratum selection in free-living and symbiotic cirripeds, in *Settlement and Metamorphosis of Marine Invertebrate Larvae*, Chia, F.-S. and Rice, M.E., Eds., Elsevier/North Holland, New York, 1978, 207.

Lewis, J.A., Impact of biofouling on the aquatic industry, in *Biofouling: Problems and Solutions*, Kjelleberg, S. and Steinberg, P., Eds., University of New South Wales, Sydney, 1994, 32.

Lignau, N.G., Fouling process in the sea, *Russ. Gidrobiol. Zh.*, 3(11–12), 280, 1924; 4(1–2), 1, 1925.

Lima-de-Faria, A., *Evolution without Selection: Form and Function by Autoevolution*, Elsevier, Amsterdam, 1988, 372.

Lin, G., Tian, M., and Wu, C., Ultrasonic removal of epibionts and silt in aquaculture of *Gracillaria*, *J. Fish. Chin.*, 12(1), 67, 1988.

Lindner, E., The attachment of macrofouling invertebrates, in *Marine Biodeterioration: An Interdisciplinary Study*, Costlow, J.D. and Tipper, R.C., Eds., Naval Institute Press, Annapolis, MD, 1984, 183.

Linskens, H.F., Adhäsion von Fortpflanzungszellen benthonischer Algen, *Planta (Berlin)*, 68(2), 99, 1966.

Lips, A. and Jessup, N.E., Colloidal aspects of bacterial adhesion, in *Adhesion of Microorganisms to Surfaces*, Ellwood, D.C. et al., Eds., Academic Press, London, 1979, 5.

Little, B. and Ray, R., A perspective on corrosion inhibition by biofilms, *Corrosion*, 58, 424, 2002.

Little, B.J., Succession in microfouling, in *Marine Biodeterioration: An Interdisciplinary Study*, Costlow, J.D. and Tipper, R.C., Eds., Naval Institute Press, Annapolis, MD, 1984, 63.

Little, B.J. and Wagner, P.A., Succession in microfouling, in *Fouling Organisms in the Indian Ocean: Biology and Control Technology*, Nagabhushanam, R. and Thompson, M.F., Eds., Oxford and IBH Publishing, New Delhi, 1997, 105.

Lock, M.A. et al., River epilithon: toward a structural-functional model, *Oikos*, 42(1), 10, 1984.

Loo, L.-O. and Rosenberg, R., Mytilus edulis culture: growth and production in western Sweden, *Aquaculture*, 35(2), 137, 1983.

Loosdrecht, van M.C.M. et al., Hydrophobic and electostatic parameters in bacterial adhesion, *Aquat. Sci.*, 52(1), 103, 1990.

Lovelock, J., *Gaia. A New Look at Life on Earth*, Oxford University Press, New York, 1979, 157.

Lubyanova, V.I. et al., Changes in some physiological functions of zebra mussels under ultrasonic treatment, Kiev, 1988, deposited at VINITI 20.09.88, N 7055-V 88.

Lucas, A., La nutrition des larves de bivalves, *Oceanis*, 8(5), 363, 1982.

Lukasheva, T.A., Ledenev, A.V., and Korovin, Ju.M., Complex studies of fouling and corrosion of metals in the Black Sea, in *Obrastanie i biopovrezhdenie: ekologicheskie problemy* (Fouling and Biodeterioration as Ecological Problems), Il'in, I.N., Ed., Nauka, Moscow, 1992, 161.

Luk'yanova, O.N. and Evtushenko, Z.S., Metallothioneins of marine invertebrates, *Biol. Morya*, 4, 3, 1982.

Lutsik, M.D., Panasyuk, E.N., and Lutsik, A.D., *Lektiny* (Lectins), Vishcha Shkola, Lvov, 1981, 156.

Lynch, W.F., The behavior and metamorphosis of the larvae of *Bugula neritina* (Linnaeus): experimental modification of the length of the free-swimming period and the responses of the larvae to light and gravity, *Biol. Bull. Mar. Biol. Lab. Woods Hole*, 92, 115, 1947.

Lyublinskii, E.Ya., *Chto nuzhno znat' o korrozii* (What Must Be Known about Corrosion), Lenizdat, Leningrad, 1980, 192.

Lyublinskii, E.Ya. and Yakubenko, A.R., Strategy and tactics of protection against marine fouling, in *Vsesoyuzn. nauchno-tekhn. konf. "Zashchita sudov i tekhnicheskikh sredstv ot obrastaniya"* (All-Union Research Conf. "Antifouling Protection of Vessels and Technical Objects"), Lyublinskii, E.Ya., Ed., Nauka, Sudostroenie, Leningrad, 1990, 5.

MacArthur, R.H. and Wilson, E.O., *The Theory of Island Biogeography*, Princeton University Press, Princeton, NJ, 1967, 208.

Makarova, N.L., Antifouling protection using beta nuclides, in *Vsesoyuzn. nauchno-tekhn. konf. "Zashchita sudov i tekhnicheskikh sredstv ot obrastaniya"* (All-Union Research Conf. "Antifouling Protection of Vessels and Technical Objects"), Lyublinskii, E.Ya., Ed., Nauka, Sudostroenie, Leningrad, 1990, 112.

Maki, J.S. and Mitchell, R., Involvement of lectins in the settlement and metamorphosis of marine invertebrate larvae, *Bull. Mar. Sci.*, 37(2), 675, 1985.

Maki, J.S. et al., Inhibition of attachment of larval barnacles, *Balanus amphitrite*, by bacterial surface films, *Mar. Biol.*, 97(2), 199, 1988.

Maki, J.S. et al., Effect of marine bacteria and their exopolymers on the attachment of barnacle cypris larvae, *Bull. Mar. Sci.*, 46(2), 499, 1990.

Maki, J.S. et al., Substratum/bacterial interactions and larval attachment: films and exopolysaccharides of Halomonas marina (ATCC 25374) and their effect on barnacle cyprid larvae, *Balanus amphitrite* Darwin, *Biofouling*, 16, 159, 2000.

Malakhov, V.V. and Medvedeva, L.A., *Embrional'noe razvitie dvustvorchatykh mollyuskov v norme i pri vozdeistvii tyazhelykh metallov* (Embryonic Development of Bivalves under Normal Conditions and under Influence of Heavy Metals), Nauka, Moscow, 1991, 134.

Maldonado, M. and Young, C.M., Effects of physical factors on larval behavior, settlement and recruitment of four tropical demosponges, *Mar. Ecol. Prog. Ser.*, 138(1–3), 169, 1996.

Mann, R., Nutrition in the Teredinidae, in *Marine Biodeterioration: An Interdisciplinary Study*, Costlow, J.D. and Tipper, R.C., Eds., Naval Institute Press, Annapolis, MD, 1984, 24.

Marfenin, N.N., Effect of flow rate on the growth of colonial hydroids (Hydrozoa, Thecaphora), *Dokl. Akad. Nauk SSSR*, 278(6), 1507, 1984.

Marfenin, N.N., *Fenomen kolonial'nosti* (The Phenomenon of Coloniality), Moscow University, Moscow, 1993a, 239.

Marfenin, N.N., *Funktsional'naya morfologiya kolonial'nykh gidroidov* (Functional Morphology of Colonial Hydrozoans), Zoological Institute, St. Petersburg, 1993b, 153.

Marfenin, N.N. and Kosevich, I.A., Biology of *Obelia loveni* (Allm.): colony formation, behavior and life cycle of the hydranths, and reproduction, *Vestn. Mosk. Univ. Ser. 16 Biol.*, 3, 16, 1984.

Marsden, J.R., Responses of planctonic larvae of the serpulid polychaete *Spirobranchus polycerus* var. *augeneri* to an alga, adult tubes and conspecific larvae, *Mar. Ecol. Prog. Ser.*, 71(3), 245, 1991.

Marshall, A. and Bott, T.R., Effectiveness of biocides, in *Fouling Science and Technology*, Melo, L.F., Bott, T.R., and Bernardo, C.A., Eds., Kluwer, Dordrecht, 1988, 591.

Marshall, K.C., *Interfaces in Microbial Ecology*, Harvard University Press, Cambridge, 1976, 156.

Marshall, K.C., Microorganisms and interfaces, *BioScience*, 30(4), 246, 1980.

Marshall, K.C., Stout, R., and Mitchell, R., Selective sorption of bacteria from seawater, *Can. J. Microbiol.*, 17(11), 1413, 1971.

Martin, A.T. and Foster, B.A., Distribution of barnacle larvae in Mahurangi Harbour, North Auckland, New Zealand, *J. Mar. Freshwater Res.*, 20(1), 67, 1986.

Mason, R.P., Free-radical intermediates in the metabolosm of toxic chemicals, in *Free Radicals in Biology, Vol. 5*, Pryor, W.A., Ed., Academic Press, New York, 1982, 223.

Mason, R.P., Rolfhus, K.R., and Fitzgerald, W.F., Methylated and elemental mercury cycling in surface and deep ocean waters of the North Atlantic, *Water Air Soil Pollut.*, 80(1–4), 665, 1995.

Matsumura, K. et al., Lentil lectin inhibits adult extract-induced settlement of the barnacle, *Balanus amphitrite*, *J. Exp. Zool.*, 280, 213, 1998.

Maximovich, N.V. and Shilin, M.B., The larvae of bivalve mollusks in plankton of the Chupa Inlet (the White Sea), in *Marine Plankton. Taxonomy, Ecology, Distribution II*, Stenanjants, S.D., Ed., Zoological Institute, St. Petersburg, 1993, 131.

McFeters, G.A. et al., Biofilm development and its consequences. Group report, in *Microbial Adhesion and Aggregation. Dahlem Konferenzen, 1984*, Marshall, K.C., Ed., Springer-Verlag, Berlin, 1984, 109.

McGregor, R. and Marr, J., Advanced coatings for marine use, in *AUSMARINE'98: The Third Ausmarine Conference, Fremantle, Nov. 3-5, 1998,* Baird Publishing, South Yarra, 1998, 183.

McIntire, C.D., Structural characteristics of benthic algal communities in laboratory streams, *Ecology*, 49(3), 520, 1968.

McKinney, F.K. and McKinney, M.J., Larval behaviour and choice of settlement site: correlation with environmental distribution pattern in an erect bryozoan, *Facies*, 29, 119, 1993.

Meadows, P.S., The attachment of bacteria to solid surfaces, *Arch. Mikrobiol.*, 75(4), 374, 1971.

Meadows, P.S. and Campbell, J.I., Habitat selection by aquatic invertebrates, *Adv. Mar. Biol.*, 10, 271, 1972.

Meadows, P.S. and Williams, G.B., Settlement of *Spirorbis borealis* Daudin larvae on surfaces bearing films of micro-organisms, *Nature*, 198(4880), 610, 1963.

Mel'nichuk, E.P., On the protective mechanism of toxin-free antifouling compositions, in *Materialy Vsesoyuznogo simpoziuma po izuchennosti Chernogo i Sredizemnogo morei, ispol'zovaniyu i okhrane ikh resursov. Chast 4* (All-Union Symp. on the Studies of the Black and Mediterranean Seas, Use and Protection of Their Resources, Part 4), Greze, V.N., Ed., Nauk. Dumka, Kiev, 1973, 83.

Merzlyak, M.N. and Sobolev, A.S., The role of superoxy anion-radicals and singlet oxygen in cell membrane pathology, in *Molekulyarnye mekhanizmy patologii kletochnykh membran. Biofizika* (Molecular Mechanisms of Cell Membrane Pathology. Biophysics), Vol. 5, Vladimirov, Ju. A., Ed., VINITI, Moscow, 1975, 118.

Metaxas, A., Behavior in flow: perspectives on the distribution and dispersion of meroplankton larvae in the water column, *Can. J. Fish. Aquat. Sci.*, 58(1), 86, 2001.

Michael, T. and Smith, C.M., Lectin probe molecular films in biofouling: characterization of early films on non-living and living surfaces, *Mar. Ecol. Prog. Ser.*, 119, 229, 1995.

Mihm, J.W., William, C.B., and Loeb, G.I., Effects of adsorbed organic and primary fouling films on bryozoan settlement, *J. Exp. Mar. Biol. Ecol.*, 54(2), 167, 1981.

Mileikovsky, S.A., Distribution of pelagic larvae of bottom invertebrates of the Norwegian and Barents Seas, *Mar. Biol.*, 1, 161, 1968a.

Mileikovsky, S.A., Some common features in the drift of pelagic larvae and juvenile stages of bottom invertebrates with marine currents in temperate regions, *Sarsia*, 34, 209, 1968b.

Mileikovsky, S.A., Types of larval development in marine bottom invertebrates, their distribution and ecological significance: a re-evaluation, *Mar. Biol.*, 10, 193, 1971.

Mileikovsky, S.A., The 'pelagic larvaton' and its role in the biology of the world ocean with special reference to pelagic larvae of marine bottom invertebrates, *Mar. Biol.*, 16, 13, 1972.

Mileikovsky, S.A., Speed of active movement of pelagic larvae of marine bottom invertebrates and their ability to regulate their vertical position, *Mar. Biol.*, 23, 11, 1973.

Millar, R.H., The biology of ascidians, *Adv. Mar. Biol.*, 9, 1, 1971.

Minchin, D. et al., Marine TBT antifouling contamination in Ireland, following legislation in 1987, *Mar. Pollut. Bull.*, 30(10), 633, 1995.

Minichev, Yu.S. and Seravin, L.N., Methodology and strategy of protection against marine fouling, *Biol. Morya*, 3, 63, 1988.

Miron, G., Boudreau, B., and Bourget, E., Intertidal barnacle distribution: a case study using multiple working hypotheses, *Mar. Ecol. Prog. Ser.*, 189, 205, 1999.

Mitchell, R. and Kirchman, D., The microbial ecology of marine surfaces, in *Marine Biodeterioration: An Interdisciplinary Study*, Costlow, J.D. and Tipper, R.C., Eds., Naval Institute Press, Annapolis, MD, 1984, 49.

Mitchell, R. and Maki, J.S., Microbial surface films and their influence on larval settlement and metamorphosis in the marine environment, in *Marine Biodeterioration: Advanced Techniques Applicable to the Indian Ocean*, Thompson, M.-F., Sarojini, R., and Nagabhushanam, R., Eds., Oxford and IBH Publishing, New Delhi, 1988, 489.

Mizobuchi, S. et al., Antifouling substances from a Palauan octocoral *Sinularia* sp., *Fish. Sci.*, 60, 345, 1994.

Moreno, C.A., Macroalgae as a refuge from predation for recruits of the mussel *Choromytilus chorus* (Molina, 1782) in southern Chile, *J. Exp. Mar. Biol. Ecol.*, 191(2), 181, 1995.

Morgan, S.G., Impact of planktivorous fishes on dispersal, hatching, and morphology of estuarine crab larvae, *Ecology*, 71, 1639, 1990.

Morgan, S.G., The timing of larval release, in *The Ecology of Marine Invertebrate Larvae*, McEdward, L. and Kennish, M.J., Eds., CRC Press, Boca Raton, FL, 1995, 157–191.

Morse, A.N.C. and Morse, D.E., Recruitment and metamorphosis of *Haliotis* larvae induced by molecules uniquely available at the surfaces of crustose red algae, *J. Exp. Mar. Biol. Ecol.*, 75, 191, 1984.

Morse, D.E., Recent progress in larval settlement and metamorphosis: closing the gaps between molecular biology and ecology, *Bull. Mar. Sci.*, 46(2), 465, 1990.

Morse, D.E., Molecular mechanisms controlling metamorphosis and recruitment in abalone larvae, in *Abalone of the World*, Shepherd, S.A. et al., Eds., Blackwell, Oxford, 1992, 107.

Morse, D.E. and Morse, A.N.C., Enzymatic characterization of the morphogen recognized by *Agaricia humilis* (scleractinian coral) larvae, *Biol. Bull.*, 181, 104, 1991.

Morse, D.E. et al., γ-Aminobutyric acid, a neurotransmitter, induces planktonic abalone larvae to settle and begin metamorphosis, *Science*, 204(4267), 407, 1979.

Morse, D.E. et al., Control of larval metamorphosis and recruitment in sympatric agariciid corals, *J. Exp. Mar. Biol. Ecol.*, 116, 193, 1988.

Morse, D.E. et al., Morphogen-based chemical flypaper for *Agarica humilis* coral larvae, *Biol. Bull.*, 186(2), 172, 1994.

Moshchenko, A.V. and Zvyagintsev, A.Ju., Distributional characteristics of macrofouling organisms on ocean-going ships of the Far East Sea basin, *Ocean Polar Res.*, 23, 323, 2001.

Moyse, J., Some observations on the swimming and feeding of the nauplius larvae of *Lepas pectinata* (Cirripedia: Crustacea), *Zool. J. Linn. Soc.*, 80(2–3), 323, 1984.

Moyse, J. and Hui, E., Avoidance by *Balanus balanoides* cyprids of settlement on conspecific adults, *J. Mar. Biol. Assn. U.K.*, 61, 449, 1981.

Müller, W.A. and Spindler, K.-D., Induction of metamorphosis by bacteria and by a lithium-pulse in the larvae of *Hydractinia echinata* (Hydrozoa), *Wilhelm Roux' Arch.*, 169(4), 271, 1972.

Mullineaux, L.S. and Butman, C.A., Recruitment of benthic invertebrates in boundary-layer flows: a deep-water experiment on Cross Seamount, *Limnol. Oceanogr.*, 35, 409, 1990.

Mullineaux, L.S. and Garland, E.D., Larval recruitment in response to manipulated field flows, *Mar. Biol.*, 116(4), 667, 1993.

Munteanu, N. and Maly, E.J., The effect of current on the distribution of diatoms settling on submerged glass slides, *Hydrobiologia*, 78, 273, 1981.

Myers, L.S., Jr., Free-radical damages of nucleic acids and their components by the direct absorption of energy, in *Free Radicals in Biology, Vol. 4*, Pryor, W.A., Ed., Academic Press, New York, 1980, 95.

Nair, N.B., Biodeterioration of cellulose materials in estuarine and insular biotopes, *Proc. Indian Nat. Sci. Acad. Part B*, 60(3), 213, 1994.

Naumov, D.V., Propp, M.V., and Rybakov, S.N., *Mir korallov* (The World of Corals), Gidrom-eteoizdat, Leningrad, 1985, 360.

Neal, A.L. and Yule, A.B., The tenacity of *Elminius modestus* and *Balanus perforatus* cyprids to bacterial films grown under different shear regimes, *J. Mar. Biol. Assn. U.K.*, 74(1), 251, 1994.

Nehring, S., Effecte von Tributylzinn (TBT) aus Antifoulinganstrichen auf Schneckenpopu-lationen an der deutschen Nordseeküste, *Hydrol. Wasserbewirtsch.*, 43(2), 66, 1999.

Neu, T.R., Microbial "footprints" and the general ability of microorganisms to label interfaces, *Can. J. Microbiol.*, 38, 1005, 1992.

Neumann, D., Entrainment of a semilunar rhythm by simulated tidal cycles of mechanical disturbance, *J. Mar. Biol. Ecol.*, 35, 73, 1978.

Neumann, R., Schmahl, G., and Hofmann, D., Bud formation and control of polyp morpho-genesis in *Cassiopea andromeda* (Scyphozoa), in *Developmental and Cellular Biol-ogy of Coelenterates*, Tardent, P. and Tardent, R., Eds., Elsevier/North Holland, Amsterdam, 1980, 217.

Nicolas, L., Robert, R., and Chevelot, L., Comparative effects of inducers on metamorphosis of the Japanese oyster scallop *Pecten maximus*, *Biofouling*, 12, 189, 1998.

Nishihira, M., Observations on the selection of algal substrata by hydrozoan larvae, *Sertu-larella miurensis*, in nature, *Bull. Mar. Biol. Station Asamushi*, 13, 34, 1967.

Nishihira, M., Experiments on the algal selection by the larvae of *Coryne uchidai* Stechow (Hydrozoa), *Bull. Mar. Biol. Stn. Asamushi*, 13, 83, 1968.

Nishikimi, M., Rao, N.A., and Yagi, K., The occurrence of superoxide action in the reaction of reduced phenazine methosulfate and molecular oxygen, *Biochem. Biophys. Res. Commun.*, 46, 849, 1972.

Noël, R.M., Composition anti-salissure pour adjonction aux revêtements des corps immerges et revêtement la contenant, Claim 2562554, France, C 09 D 5/14 claimed 08.04.84, publ. 11.10.85.

Nott, J. and Foster, B., On the structure of the antennular attachment organ of the cypris larva of *Balanus balanoides* (L.), *Philos. Trans. R. Soc. London Ser. B*, 256(803), 115, 1969.

O'Connor, N.J. and Richardson, D.L., Attachment of barnacle (*Balanus amphitrite* Darwin) larvae: responses to bacterial films and extracellular materials, *J. Exp. Mar. Biol. Ecol.*, 226, 115, 1998.

O'Toole, G.A., Kaplan, H.B., and Kolter, R., Biofilm formation as microbial development, *Annu. Rev. Microbiol.*, 54, 49, 2000.

Odum, E.P., *Basic Ecology, Vol. 2*, Saunders College Publishing, Philadelphia, 1983, 613.

Oehlmann, J., Stroben, E., and Fioroni, P., The morphological expression of imposex in *Nucella lapillus* (Linnaeus) (Gastropoda: Muricidae), *J. Mollusc. Study*, 57, 375, 1991.

Ohta, K., Matsumoto, H., and Nawamaki, T., A screening procedure for repellents against a sea snail, *Agric. Biol. Chem.*, 42(8), 1491, 1978.

Okamura, B., Particle size, flow velocity, and suspension feeding by the erect bryozoans *Bugula neritina* and *B. stolonifera*, *Mar. Biol.*, 105(1), 33, 1990.

Okano, K. et al., Enzymatic isolation and culture of cement secreting cells from cypris larvae of the barnacle *Megabalanus rosa*, *Biofouling*, 12, 149, 1998.

Oliveira, L., Walker, D.C., and Bisalputra, T., Ultrastructural, cytochemical, and enzymatic studies on the adhesive "plaques" of the brown algae *Laminaria saccharina* (L.) Lamour. and *Nereocystis luetkeana* (Nert.) Post. et Rupr., *Protoplasma*, 104(1–2), 1, 1980.

Olivier, F., Tremblay, R., Bourget, E., and Rittschof, D., Barnacle settlement: field experiments on the influence of larval supply, tidal level, biofilm quality and age on *Balanus amphitrite* cyprids, *Mar. Ecol. Prog. Ser.*, 199, 185, 2000.

Ong, E.E. and Din, Z.B., Cadmium, copper, and zinc toxicity to the clam *Donax faba* C., and the blood cockle *Anadara granosa*, *Bull. Environ. Contam. Toxicol.*, 66(1), 86, 2001.

Orlov, B.N. and Gelashvili, D.B., *Zootoksinologiya (yadovitye zhivotnye i ikh yady): Uchebnoe posobie* (Zootoxinology: Venomous Animals and Their Toxins. A Manual), Vysshaya Shkola, Moscow, 1985, 280.

Orlov, D.V., The role of biological interactions in the settlement of hydrozoan planulae, *Zool. Zh.*, 75(6), 811, 1996a.

Orlov, D.V., The ecological conditions of settlement of colonial hydrozoan planulae, *Zh. Obsch. Biol.*, 57(2), 112, 1996b.

Orlov, D.V. and Marfenin, N.N., Behavior and settlement of the White Sea hydroid polyp *Clava multicornis* (Athecata, Hydrozoa), *Vestn. Mosk. Univ. Ser. 16 Biol.*, 4, 24, 1993.

Orlov, D.V., Marfenin, N.N., and Railkin, A.I., Settlement of planulae of the White Sea hydroids *Gonothyraea loveni* (Allman) and *Laomedea flexuosa* (Hincks) (Hydroidea, Thecaphora) on the mucous bacterial film from the surface of intertidal algae, *Vestn. Mosk. Univ. Ser. 16*, 3, 47, 1994.

Oshurkov, V.V., Dynamics and structure of some fouling and benthic communities in the White Sea, in *Ekologiya obrastaniya v Belom more* (Ecology of Fouling in the White Sea), Berger, V.Ya. and Seravin, L.N., Eds., Zoological Institute, Leningrad, 1985, 44.

Oshurkov, V.V., Development and structure of some fouling communities in Avacha Bay, *Biol. Morya*, 5, 20, 1986.

Oshurkov, V.V., Succession and climax in some fouling communities, *Biofouling*, 6, 1, 1992.

Oshurkov, V.V., Succession and Dynamics of Upper Littoral Epibenthic Communities, Doctoral (Biology) dissertation, Zoological Institute, St. Petersburg, 1993, 427.

Oshurkov, V.V. and Oksov, I.V., Settlement of fouling larvae in Kandalaksha Bay of the White Sea, *Biol. Morya*, 4, 25, 1983.

Oshurkov, V.V. and Seravin, L.N., Development of fouling biocenoses in Chupa Inlet (the White Sea), *Vestn. Leningr. Univ. Ser. 3 Biol.*, 3, 37, 1983.

Osinga, R. et al., Measurement of sponge growth by projected body area and underwater weight, *Proc. 5th Int. Sponge Symp., Brisbane, 30 June 1999, Mem. Queensland Mus.*, 44, 419, 1998.

Osman, R.W., The establishment and development of a marine epifaunal community, *Ecol. Monogr.*, 47, 37, 1977.

Osman, R.W., Artificial substrates as ecological islands, in *Artificial Substrates*, Cairns, J., Jr., Ed., Ann Arbor Science, Ann Arbor, MI, 1982, 71.

Osman, R.W. and Whitlatch, R.B., Local control of recruitment in an epifaunal community and the consequences to colonization processes, *Hydrobiologia*, 375/376, 113, 1998.

Otto, J.J., The settlement of *Halyclystus* planulae, in *Settlement and Metamorphosis of Marine Invertebrate Larvae*, Chia, F.-S. and Rice, M.E., Eds., Elsevier/North Holland, New York, 1978, 13.

Ozmidov, R.V., *Gorizontal'naya turbulentnost' i turbulentnyi obmen v okeane* (Horizontal Turbulence and Turbulent Exchange in the Ocean), Nauka, Moscow, 1968, 200.

Paine, R.T., Marine rocky shores and community ecology: an experimentalist's perspective, in *Excellence in Ecology*, Kinne, O., Ed., Ecology Institute, Oldendorf, 1994, 152.

Panov, B.N., Triblat, I.N., and Vizhevskii, V.I., Hydrometeorological prerequisites of distribution of suspended organic matter and *Mytilus* larvae at the Gulf of Kerch, *Ekol. Morya (Kiev)*, 29, 46, 1988.

Pansini, M. and Pronsato, R., Etude des Spongiaires de substrats immerges Durant quatre ans, *Vie et Milieu*, 31, 77, 1981.

Pardo, J. et al., Purification of adhesive proteins from mussels, *Prot. Express. Purific.*, 1, 147, 1990.

Parsons, G.J., Dadswell, M.J., and Roff, J.C., Influence of biofilm on settlement of sea scallop *Placopecten magellanicus* (Gmelin, 1791) in Passamaquoddy Bay, New Brunswick, Canada, *J. Shellfish Res.*, 12, 279, 1993.

Partaly, E.M., To the study of the vertical structure of a marine fouling biocenosis, *Biol. Morya*, 4, 79, 1980.

Partaly, E.M., *Obrastanie v Azovskom more* (Fouling in the Sea of Azov), Renata, Mariupol, 2003, 378.

Patin, S.A., *Vliyanie zagryazneniya na biologicheskie resursy i produktivnost' Mirovogo okeana* (Effect of Pollution on the Biological Resources and Productivity of the World Ocean), Pishchevaya Promyshlennost, Moscow, 1979, 304.

Patrick, R., Factors affecting the distribution of diatoms, *Bot. Rev.*, 14(8), 473, 1948.

Paul, V.J. and Fenical, W., Natural products chemistry and chemical defense in tropical marine algae of the phylum Chlorophyta, in *Marine Chemistry*, Vol. 1, Scheuer, P.J., Ed., Springer-Verlag, Berlin, 1987, 1.

Paul, V.J. and van Alstyne, K.L., Chemical defense and chemical variation in some tropical Pacific species of *Halimeda* (Halimedaceae; Chlorophyta), *Coral Reefs*, 6(3/4), 263, 1988.

Pawlik, J.R., Chemical induction of larval settlement and metamorphosis in reef-building tube worm *Phragmatopoma californica* (Polychaeta: Sabellariidae), *Mar. Biol.*, 91(1), 59, 1986.

Pawlik, J.R., Natural and artificial induction of metamorphosis of *Phragmatopoma lapidosa* (Polychaeta: Sabellariidae), with crirtical look at the effects of bioactive compounds on marine invertebrate larvae, *Bull. Mar. Sci.*, 46(2), 512, 1990.

Pawlik, J.R., Chemical ecology of the settlement of benthic marine invertebrates, *Oceanogr. Mar. Biol. Annu. Rev.*, 30, 273, 1992.

Pawlik, J.R. and Butman, C.A., Settlement of a marine tube worm as a function of current velocity: interaction effects of hydrodynamics and behavior, *Limnol. Oceanogr.*, 38(8), 1730, 1993.

Pearce, C.M. et al., Settlement of larvae of the giant scallop, *Placopecten magellanicus*, in 9-m deep mesocosms as a function of temperature stratification, depth, food and substratum, *Mar. Biol.*, 124(4), 693, 1996.

Pearse, V., *Living Invertebrates*, Blackswell Scientific, Palo Alto, CA, 1987, 848.

Pechenik, J.A., Delayed metamorphosis by larvae of benthic marine invertebrates: does it occur? Is there a price to pay? *Ophelia*, 32, 63, 1990.

Pertsov, N.A. and Vilenkin, B.Ya., Primary production of phytofouling, *Dokl. Akad. Nauk SSSR*, 236(2), 494, 1977.

Peskin, A.V., Labas, J.A., and Tichonov, A.N., Superoxide radical production by sponges *Sycon* sp., *FEBS Lett.*, 434, 201, 1998.

Peterson, C.G. and Stevenson, R.J., Resistance and resilience of lotic algal communities: importance of disturbance timing and current, *Ecology*, 73(4), 1445, 1992.

Peterson, Ch.G., Hoagland, K.D., and Stevenson, R.J., Timing of wave disturbance and the resistance and recovery of a freshwater epilithic microalgal community, *J. N. Am. Benthol. Soc.*, 9(1), 54, 1990.

Peterson, C.G. et al., Mechanisms of benthic algal recovery following spates: comparison of simulated and natural events, *Oecologia*, 98, 280, 1994.

Pisano, E. and Boyer, M., Development pattern of an infralittoral bryozoan community in the western Mediterranean Sea, *Mar. Ecol. Prog. Ser.*, 27(1–2), 195, 1985.

Polikarpov, G.G. and Egorov, V.N., *Morskaya dinamicheskaya radiokhemoekologiya* (Dynamic Radiochemoecology of the Sea), Energoatomizdat, Moscow, 1986, 176.

Polishchuk, R.A., Response of fouling macrophytes to heavy metal ions, in *Biologicheskie osnovy bor'by s obrastaniem* (Biological Basis of Biofouling Control), Vodyanitskii, V.A. and Dolgopol'skaya, M.A., Eds., Nauk. Dumka, Kiev, 1973, 155.

Prater, B. and Hoke, R.A., A method for biological and chemical evaluation of sediment toxicity, in *Contamination and Sediments. Vol. 1*, Ann Arbor Science, Ann Arbor, MI, 1980, 483.

Preiser, H.S., Ticker, A., and Bohlander, G.S., Coating selection for optimal ship performance, in *Marine Biodeterioration: An Interdisciplinary Study*, Costlow, J.D. and Tipper, R.C., Eds., Naval Institute Press, Annapolis, MD, 1984, 223.

Price, H.A., Byssus thread strength in the mussel *Mytilus edulis*, *J. Zool.* (London), 194(2), 245, 1981.

Prosser, C.L. and Brown, F.A., *Comparative Animal Physiology*, 2nd ed., W.B. Saunders, Philadelphia, 1961, 688.

Protasov, A.A., Periphyton: terminology and main definitions, *Gidrobiol. Zh.*, 18(1), 9, 1982.

Protasov, A.A., *Metody issledovaniya perifitona* (Methods of periphyton studies), deposited at VINITI 17.03.87, N 2164-V87, Kiev, 1987, 36.

Protasov, A.A., *Presnovodnyi perifiton* (Freshwater Periphyton), Nauk. Dumka, Kiev, 1994, 307.

Puglisi, M.P., Paul, V.J., and Slattery, M., Biographic comparisons of chemical and structural defenses of the Pacific gorgonians *Annella mollis* and *A. reticulata*, *Mar. Ecol. Prog. Ser.*, 207, 263, 2000.

Punčochař, P., Studies of bacterial periphyton on hard submerged surfaces, in *Gidrobiologicheskie protsessy v vodoemakh* (Hydrobiological Processes in Water Bodies), Raspopov, I.M., Ed., Nauka, Leningrad, 1983, 45.

Qian, P.-Y. et al., Macrofouling in unidirectional flow: miniature pipes as experimental models for studying the effects of hydrodynamics on invertebrate larval settlement, *Mar. Ecol. Prog. Ser.*, 191, 141, 1999.

Qian, P.-Y., Rittschof, D., and Sreedhar, B., Macrofouling in unidirectional flow: miniature pipes as experimental models for studying the interaction of flow and surface characteristics on the attachment of barnacle, bryozoan and polychaete larvae, *Mar. Ecol. Prog. Ser.*, 207, 109, 2000.

Railkin, A.I., Distribution of diatoms over flat surfaces in a tangential flow, *Bot. Zh.*, 76(11), 1522, 1991.

Railkin, A.I., Behavioral and physiological responses of hydroid polyps and bivalves to some antifouling compounds, *Hydrobiol. J.*, 31, 66, 1995a.

Railkin, A.I., Negative chemotaxis and settlement suppression in hydroid planulae by a bacterial repellent, in *Knidarii. Sovremennoe sostoyanie i perspektivy issledovanii* (Cnidarians. Present State and Prospects of Research), Part II, Stepanjants, S.D., Ed., Zoological Institute, St. Petersburg, 1995b, 121.

Railkin, A.I., Heterotrophic flagellates on artificial substrates in the White Sea, *Tsitologiya*, 37(11), 951, 1995c.

Railkin, A.I., Benthos, periphyton and classification of ecological groups, *Vestn. Sankt-Peterburg. Univ. Ser. 3*, 3, 10, 1998a.

Railkin, A.I., The pattern of recovery of disturbed microbial communities inhabiting hard substrates, *Hydrobiologia*, 385, 47, 1998b.

Railkin, A.I., Relationships between *Mytilus edulis* and microorganisms, in *Tr. Biol. Nauchno-Issled. Inst. St. Petersburg State University*, 46, 65, 2000.

Railkin, A.I. and Chikadze, S.Z., The lectin-carbohydrate mechanism of adhesion of larvae *Obelia loveni* to microfouling films, in *6th Int. Conf. on Coelenterate Biology*, the Netherlands, Leeuwenhurst-Noodwijkerhout, 1995, 80.

Railkin, A.I. and Chikadze, S.Z., Lectin-carbohydrates interactions between hydroid larvae and microfouling films, *Zoosyst. Rossica*, Suppl. 1, 119, 1999.

Railkin, A.I. and Dobretsov, S.V., Effect of bacterial repellents and narcotizing substances on marine macrofouling, *Russ. J. Mar. Biol.*, 20, 16, 1994.

Railkin, A.I. and Dysina, T.Yu., Selection of natural substrates by larvae of *Molgula citrina* (Pleurogona, Molguloidea), *Zool. Zh.*, 76, 341, 1997.

Railkin, A.I. and Fateev, A.E., Standardization of biological tests. I. Hydrovane, a device for exposing experimental plates at a fixed angle to the current, *Vestn. Leningr. Univ. Ser. 3 Biol.*, 3(17), 11, 1990.

Railkin, A.I. and Seravin, L.N., Reversible blocking (narcosis) of motility and contractility of metazoans by calcium antagonist ions, *Zool. Zh.*, 68(6), 19, 1989.

Railkin, A.I. and Zubakha, M.A., Primary settlement of *Mytilus edulis* larvae, *Tr. Biol. Nauchno-Issled. Inst. St. Petersburg State Univ.*, 46, 65, 2000.

Railkin, A.I. et al., Toxic effect of porphyrin cobalt complexes on *Paramecium*, suppressed by superoxide dismutase, *Dokl. Akad. Nauk SSSR*, 274(5), 1257, 1984.

Railkin, A.I., Makarov, V.N., and Shoshina, E.V., The influence of substrate on settlement and attachment of *Laminaria saccharina* zoospores, *Biol. Morya*, 1, 37, 1985.

Railkin, A.I., Smirnov, B.R., and Onishchenko, V.A., Antifouling protection using reactive oxygen species, in *Izuchenie protsessov morskogo bioobrastaniya i razrabotka metodov bor'by s nim* (Studies of Marine Biofouling and Development of Prevention Methods), Zoological Institute, Leningrad, 1987, 85.

266 Marine Biofouling: Colonization Processes and Defenses

Railkin, A.I., Minichev, Yu.S., and Seravin, L.N., Chemobiological protection from marine fouling, in *Vsesoyuzn. nauchno-tekhn. konf. Zashchita sudov i tekhnicheskikh sredstv ot obrastaniya* (All-Union Research Conf. "Antifouling Protection of Vessels and Technical Objects"), Sudostroenie, Leningrad, 1990, 66.

Railkin, A.I. et al., Effect of bacterial repellents and barbiturates on microfouling, *Russ. J. Mar. Biol.*, 19, 325, 1993a.

Railkin, A.I., Pavlenko, G.V., and Skugarova, M.G., Suppression of the adhesion of bacterial pioneers of biofouling in the White Sea, *Mikrobiologiya*, 62(5), 951, 1993b.

Railkin, A.I., Ereskovsky, A.V., and Gonobobleva, E.L., First experimental evidences of the stimulation of settlement and metamorphosis of sponges by macroalgae (by the example of the sponge *Halisarca dujardini* and the brown alga *Fucus vesiculosus*), *Tr. Biol. Nauchno-Issled. Inst. St. Petersburg State Univ.*, 51, in press.

Raimondi, P.T., Settlement cues and determination of the vertical limit of an intertidal barnacle, *Ecology*, 69, 400, 1988.

Ramsay, G.G., Tackett, J.H., and Morris, D.W., Effect of low-level continuous chlorination on *Corbicula fluminea*, *Environ. Toxicol. Chem.*, 7(10), 855, 1988.

Rashevsky, N., Some remarks on the mathematical theory of nutrient of fishes, *Bull. Math. Biophys.*, 21(2), 161, 1959.

Rasmussen, E., A method of protection of the submerged part of the ship hull from fouling, Patent 413664 SSSR, 4302/69 Norway, B 63 B 59/00, priority 30.10.69a, publ. 30.01.74.

Rasmussen, Yu., A method of protection of the submerged part of the ship hull from fouling, Patent 503497 SSSR, 1954/69 Norway, B 63 B 59/00, priority 13.05.69b, publ. 15.02. 76.

Raymont, J.E.G., *Plankton and Productivity in the Oceans, Vol. 1. Phytoplankton*, 2nd ed., Pergamon Press, Oxford, 1980, 489.

Raymont, J.E.G., *Plankton and Productivity in the Oceans, Vol. 2. Zooplankton*, 2nd ed., Pergamon Press, Oxford, 1983, 824.

Razumov, A.S., Biofouling in drinking and technical water supply systems and methods of its control, in *Biologicheskoe obrastanie v sisteme pit'evogo i tekhnicheskogo vodosnabzheniya i mery bor'by s nim* (Biofouling in Drinking and Technical Water Supply and Methods of Its Control), Nauka, Moscow, 1969, 5.

Redfield, A.C. and Deevy, E.S., Jr., Temporal sequences and biotic successions, in *Marine Fouling and Its Prevention*, Redfield, A.C. and Ketchum, B.H., Eds., Naval Institute Press, Annapolis, MD, 1952, 42.

Redfield, A.C. and Ketchum, B.H., Ship resistance, in *Marine Fouling and Its Prevention*, Redfield, A.C. and Ketchum, B.H., Eds., U.S. Naval Institute, Annapolis, MD, 1952, 21.

Reed, C.G., Larval morphology and settlement of bryozoan, *Bowerbankia gracilis* (Vesicularioidea, Ctenostomata): structure and eversion of the internal sac, in *Settlement and Metamorphosis of Marine Invertebrate Larvae*, Chia, F.-S. and Rice, M.E., Eds., Elsevier/North Holland, New York, 1978, 41.

Reed, D.C., Amsler, Ch.D., and Ebeling, A.W., Dispersal in kelps: factors affecting spore swimming and competency, *Ecology*, 73, 1577, 1992.

Reinfelder, J.R., Jablonka, R.E., and Cheney, M., Metabolic responses to subacute toxicity of trace metals in a marine microalga (*Thalassiosira weissflogii*) measured by calorespirometry, *Environ. Toxicol. Chem.*, 19(2), 127, 2000.

Reisch, M., Maritime dilemma, *Chem. Eng. News*, 79, 16, 2001.

Reiswig, H.M., Particle feeding in natural populations of three marine demosponges, *Biol. Bull.*, 141, 568, 1971.

Reymers, N.F., *Azbuka prirody (mikroentsiklopediya biosfery)* (The ABC of Nature: A Small Encyclopedia of the Biosphere), Znanie, Moscow, 1980, 208.

Reznichenko, O.G., Classification and spatial characteristic of fouling biotopes, *Biol. Morya*, 4, 3, 1978.

Reznichenko, O.G., Long-term fouling in the lower epipelagial of the eastern Kamchatka area (the Pacific), in *Ekologiya massovykh vidov okeanicheskogo obrastaniya* (Ecology of the Mass Species of Oceanic Fouling), Reznichenko, O.G. and Tsikhon-Lukanina, E.A., Eds., Institute Okeanologii Akademii Nauk, Moscow, 1981, 76.

Reznichenko, O.G. and Tsikhon-Lukanina, E.A., An autecological review of *Conchoderma virgatum* (Crustacea: Cirripedia, Lepadidae), one of the dominant forms in oceanic fouling, in *Obrastanie i biopovrezhdenie: ekologicheskie problemy* (Fouling and Biodeterioration as Ecological Problems), Il'in, I.N., Ed., Nauka, Moscow, 1992, 124.

Reznichenko, O.G., Soldatova, I.N., and Tsikhon-Lukanina, E.A., Fouling in the World Ocean, in *Itogi nauki i tekhniki. Zoologiya bespozvonochnykh. Tom 4* (Advances in Science and Technology. Invertebrate Zoology. Vol. 4), Poznanin, L.P. and Starostin, I.V., Eds., VINITI, Moscow, 1976, 120.

Rice, A.L., Observations of the effects of changes of hydrostatic pressure on the behaviour of some marine animals, *J. Mar Biol. Assn. U.K.*, 44, 163, 1964.

Rice, E.L., *Allelopathy*, 2nd ed., Academic Press, Orlando, FL, 1984, 422.

Rice-Evans, C.A. and Burdon, R.H., Eds., *Free Radical Damage and Its Control*, Elsevier, Amsterdam, 1994, 392.

Riebesell, U., Comparison of sinking and sedimentation rate measurements in a diatom winter/spring bloom, *Mar. Ecol. Prog. Ser.*, 54(1–2), 109, 1989.

Riggio, S. and di Pisa, G., The patterns of settlement of benthic harbour communities in relation to substratum geometry, *Rapp. Proc. V. Reun. Commis. Int. Explor. Sci. Mer. Mediterr. Monaco*, 27(2), 177, 1981.

Riisgård, H.U. and Goldson, A., Minimal scaling of the lophophore filter-pump in ectoprocts (Bryozoa) exludes physiological regulation of filtration rate to nutritional needs. Test of hypothesis, *Mar. Ecol. Prog. Ser.*, 156, 109, 1997.

Riisgård, H.U. and Larsen, P.S., Filter-feeding in marine macro-invertebrates: pump characteristics, modeling and energy cost, *Biol. Bull.*, 70, 67, 1995.

Riisgård, H.U. and Manríquez, P., Filter-feeding in fifteen marine ectoprocts (Bryozoa): particle capture and water pumping, *Mar. Ecol. Prog. Ser.*, 154, 223, 1997.

Rilov, G. et al., Unregulated use of TBT-based antifouling paints in Israel (eastern Mediterranean): high contamination and imposex levels in two species of marine gastropods, *Mar. Ecol. Prog. Ser.*, 192, 229, 2000.

Ringelberg, J., Changes in light intensity and diel vertical migration: a comparison of marine and freshwater environments, *J. Mar. Biol. Assn. U.K.*, 75, 15, 1995.

Rittschof, D., Body odors and neutral-basic peptide mimics: a review of responses by marine organisms, *Am. Zool.*, 33, 487, 1993.

Rittschof, D., Natural product antifoulants: one perspective on the challenges related to coatings development, *Biofouling*, 15, 119, 2000.

Rittschof, D. and Bonaventura, J., Macromolecular cues in marine systems, *J. Chem. Ecol.*, 12(5), 1013, 1986.

Rittschof, D. and Costlow, J.D., Surface determination of macroinvertebrate larval settlement, in *Proc. 21st EMBS. Gdansk, 1986*, Styczynska-Jureqwicz, E., Ed., Polish Academie Science, Gdansk, 1989, 155.

Rittschof, D., Branscomb, S.E., and Costlow, J.D., Settlement and behavior in relation to flow and surface in larval barnacles, *Balanus amphitrite* Darwin, *J. Exp. Mar. Biol. Ecol.*, 82, 131, 1984.

Rittschof, D. et al., Inhibition of barnacle settlement and behavior by natural products from whip corals, *Leptogorgia virgulata* (Lamarck, 1815), *J. Chem. Ecol.*, 11(5), 551, 1985.

Rittschof, D. et al., Cues and context: larval responses to physical and chemical cues, *Biofouling*, 12, 31, 1998.

Robinson, M.G., Hall, B.D., and Voltolina, D., Slime films on antifouling paints [as] short-term indicators of long-term effectiveness, *J. Coat. Technol.*, 57(725), 35, 1985.

Rodriguez, S.R., Ojeda, F.P., and Inestrosa, N.C., Settlement of benthic marine invertebrates, *Mar. Ecol. Prog. Ser.*, 97, 193, 1993.

Roegner, G.C., Transport of molluscan larvae through a shallow estuary, *J. Plankton Res.*, 22, 1779, 2000.

Roegner, G.C. and Mann, R., Early recruitment and growth of the American oyster *Crassostrea virginica* (Bivalvia: Ostreidae) with respect to tidal zonation and season, *Mar. Ecol. Prog. Ser.*, 117, 91, 1995.

Rudyakov, Yu.A., *Dinamika vertikal'nogo raspredeleniya pelagicheskikh zhivotnykh* (Dynamics of the Vertical Distribution of Pelagic Animals), Nauka, Moscow, 1986, 135.

Rudyakov, Yu.A. and Tseitlin, V.B., Passive sinking rate in marine pelagic organisms, *Okeanologiya*, 20(5), 931, 1980.

Rudyakova, N.A., On the settlement and distribution of barnacles on moving ships, *Tr. Inst. Okeanol. Akad. Nauk SSSR*, 85, 77, 1967.

Rudyakova, N.A., *Obrastanie v severo-zapadnoi chasti Tikhogo okeana* (Fouling in the Northwestern Pacific), Nauka, Moscow, 1981, 68.

Rumrill, S.S., Population size-structure, juvenile growth, and breeding periodicity of the sea star *Asterina miniata* in Barkley Sound, British Columbia, *Mar. Ecol. Prog. Ser.*, 56, 37, 1989.

Rumrill, S.S. and Cameron, R.A., Effects of gamma-aminobutyric acid on the settlement of larvae of the black chiton *Katharina tunicata*, *Mar. Biol.*, 12(3), 243, 1983.

Russ, G.R., Overgrowth in a marine epifaunal community: competitive hierarchies and competitive networks, *Oecologia*, 53, 12, 1982.

Russell, D., Brancato, M.S., and Bennett, J., Comparison of trends in tributyltin concentrations among three monitoring programs in the United States, *J. Mar. Sci. Technol.*, 1, 230, 1996.

Rybal'skii, N.G. et al., *Ekologicheskie aspekty ekspertizy izobretenii: Spravochnik eksperta i izobretatelya* (Ecological Aspects of Evaluating Inventions: Reference Book for Experts and Inventors), Part 1, VINITI, Moscow, 1989, 450.

Ryland, J.S., Experiments on the selection of algal substrates by polyzoan larvae, *J. Exp. Biol.*, 36(4), 613, 1959.

Ryland, J.S., Physiology and ecology of marine bryozoans, *Adv. Mar. Biol.*, 14, 285, 1976.

Rzepishevsky, I.K., Kuznetsova, I.A., and Zevina, G.B., Settlement and life span of *Balanus balanoides* cyprid larvae under laboratory conditions, *Tr. Inst. Okeanol. Akad. Nauk SSSR*, 85, 91, 1967.

Rzhavsky, A.V., Revision of Januinae (Polychaeta, Srirorbidae) in seas of the USSR, *Zool. Zh.*, 70(8), 37, 1991.

Sammarco, P.W. and Coll, J.C., Lack of predictability in terpenoid function: multiple roles and intergration with related adaptations in soft corals, *J. Chem. Ecol.*, 16(1), 273, 1990.

Sand, W., Mechanisms of microbial deterioration, *Invest. Tecn. Pap.*, 37, 507, 2000.

Saroyan, J.R., Lindner, E., and Dooley, C.A., Attachment mechanism of barnacles, in *Proc. 2nd Int. Congr. on Marine Corrosion and Fouling*, Technical Chamber of Greece, Athens, 1968, 495.

Savilov, A.J., Ecological characteristics of the bottom invertebrate communities in the Sea of Okhotsk, *Tr. Inst. Okeanol. Akad. Nauk SSSR*, 46, 3, 1961.

Scheer, B.T., The development of marine fouling communities, *Biol. Bull.*, 89, 103, 1945.

Scheltema, R.S., Dispersal of phytoplanktotrophic shipworm larvae (Bivalvia: Teredinidae) over long distances by ocean currents, *Mar. Biol.*, 11, 5, 1971.

Scheltema, R.S., Biological interactions determining larval settlement of marine invertebrates, *Thalassia Jugosl.*, 10(1–2), 263, 1974.

Scheltema, R.S., Long-distance dispersal by planktonic larvae of shoal-water benthic invertebrates among central Pacific islands, *Bull. Mar. Sci.,* 39, 241, 1986a.

Scheltema, R.S., On dispersal and planktonic larvae of benthic invertebrates: an eclectic overview and summary of problems, *Bull. Mar. Sci.,* 39, 290, 1986b.

Scheltema, R.S. and Carlton, J.T., Methods of dispersal among fouling organisms and possible consequences for range extension and geographical variation, in *Marine Biodeterioration: An Interdisciplinary Study*, Costlow, J.D. and Tipper, R.C., Eds., Naval Institute Press, Annapolis, MD, 1984, 127.

Schlichting, H., *Boundary-Layer Theory*, McGraw-Hill, New York, 1979, 817.

Schmahl, G., Bacterially induced stolon settlement in the scyphopolyp of *Aurelia aurita* (Cnidaria, Scyphozoa), *Helgol. Meeresunters.*, 39, 33, 1985a.

Schmahl, G., Induction of stolon settlement in the scyphopolyps of *Aurelia aurita* (Cnidaria, Scyphozoa, Semaeostomeae), *Helgol. Meeresunters.*, 39, 117, 1985b.

Schmidt, G.H., Random and aggregative settlement in some sessile marine invertebrates, *Mar. Ecol. Prog. Ser.*, 9(1), 97, 1982.

Schoener, A., Artificial substrates in marine environments, in *Artificial Substrates,* Cairns, J., Jr., Ed., Ann Arbor Science, Ann Arbor, MI, 1982, 1.

Scott, H.W. et al., Monitoring dissolved copper concentrations in Chesapeake Bay, U.S.A., *Environ. Monitor. Assessment*, 11(1), 33, 1988.

Sears, M.A., Gerhart, D.J., and Rittschof, D., Antifouling agents from marine sponge *Lissodendoryx isodictyalis* Carter, *J. Chem. Ecol.*, 16, 791, 1990.

Sebens, K.P., Community ecology of vertical rock walls in the Gulf of Maine, U.S.A.: small-scale processes and alternative community states, in *The Ecology of Rocky Coasts*, Moore, P.G. and Seed, R., Eds., Hodder and Stoughton, London, 1985a, 346.

Sebens, K.P., The ecology of the rocky subtidal zone, *Am. Sci.*, 6, 548, 1985b.

Seed, R., Ecology, *Marine Mussels: Their Ecology and Physiology*, Cambridge University Press, New York, 1976, 13.

Seligo, A., Über den Ursprung Fischnahrung, *Mitt des Westpr. Fisch. V.,* 17(4), 52, 1905.

Seravin, L.N., Mechanisms and coordination of cellular locomotion, *Adv. Comp. Physiol. Biochem.*, 4, 37, 1971.

Seravin, L.N., Minichev, Yu.S., and Railkin, A.I., Studies of fouling and biodeterioration of man-made objects in the sea: some results and prospects, in *Ekologiya obrastaniya v Belom more* (Ecology of Fouling in the White Sea), Berger, V.Ya. and Seravin, L.N., Eds., Zoological Institute, Leningrad, 1985, 5.

Shadrina, L.A., Effect of high concentrations of active chlorine on juvenile barnacles, *Ekol. Morya*, 31, 77, 1989.

Shadrina, L.A., Ultrasonic antifouling protection: present state and prospects from the ecological viewpoint, Moscow, 1995, deposited at VINITI 01.11.95, N 2905-V95, 32.

Shcherbakov, P.S., Zobachev, Yu.E., and Kopelevich, V.A., Ultrasonic treatment is an efficient antifouling method, *Morsk. Flot*, 4, 44, 1972.

Shcherbakova, I.B. et al., Antifouling protection of ship cooling systems by periodic chlorination, *Sudostroenie*, 7, 24, 1986.

Shellenberger, J.S. and Ross, J.R.P., Antibacterial activity of two species of bryozoans from northern Puget Sound, *Northwest Sci.,* 72, 23, 1998.

Sherman, R.L., Gilliam, D.S., and Spieler, R.E., Site-dependent differences in artificial reef function: implications for coral reef restoration, *Bull. Mar. Sci.*, 69(2), 1053, 2001.

Sherratt, J.A., Lewis, M.A., and Fowler, A.C., Ecological chaos in the wake of invasion, *Proc. Natl. Acad. Sci.*, 92, 2524, 1995.

Shilin, M.B., Pelagic larvae of bottom invertebrates in plankton of Chupa Bay (the White Sea), in *Marine Plankton. Taxonomy, Ecology, Distribution*, Petrushevska, M.G. and Stenanjants, S.D., Eds., Zoological Institute, St. Petersburg, 1989, 132.

Shilin, M.B. et al., Quantity and settlement of planktonic larvae of fouling organisms in Kandalaksha Bay of the White Sea, *Okeanologiya*, 27(4), 652, 1987.

Shimidzu, N., Katsuoka, M., Mizobuchi, S., Ina, K., and Miki, W., Isolation of betabisabolene as a repellent substance against blue mussel from an octocoral *Sinularia* sp., *Nippon Suisan Geskhaishi, Bull. Jpn. Soc. Sci. Fish.*, 59, 1951, 1993.

Shinkarenko, N.V. and Aleskovskii, V.B., Chemical properties of singlet oxygen and its significance in biological systems, *Usp. Khim.*, 51(5), 713, 1982.

Shunatova, N.N. and Ostrovsky, A.N., Individual autozooidal behaviour and feeding in marine bryozoans, *Sarsia*, 86, 113, 2001.

Shuvalov, V.S., Pattern of vertical distribution of larvae of bottom-living invertebrates, in *Zakonomernosti raspredeleniya i ekologii pribrezhnykh biotsenozov* (Regularities of Distribution and Ecology of Coastal Marine Biocenoses), Golikov, A.N., Nauka, Leningrad, 1978, 32.

Sidorov, K.S., To the knowledge of the fauna of Kandalaksha Bay, the White Sea, in *Ekologiya morskikh organizmov* (Ecology of Marine Organisms), Khromov, V.M., Ed., Moscow University, Moscow, 1971, 79.

Sieburth, J.M. and Conover, J.T., Sargassum tannin, an antibiotic which retards fouling, *Nature*, 208(5005), 52, 1965.

Silvester, N.R. and Sleigh, M.A., The forces influencing on microorganisms at surfaces in flowing water, *Freshwater Biol.*, 15(4), 433, 1985.

Singarajah, K.V., Escape reactions of zooplancton: the avoidance of a pursuing siphon tube, *J. Exp. Mar. Biol. Ecol.*, 3, 171, 1969.

Siu, X., Factors influencing on development and survival of larval holoturian, *Oceanol. Limnol. Sin.*, 20, 314, 1989.

Skurlatov, Yu.I., Duka, G.G., and Miziti, A., *Vvedenie v ekologicheskuyu khimiyu: Uchebnoe posobie* (Introduction to Ecological Chemistry: a Manual), Vysshaya Shkola, Moscow, 1994, 400.

Sládečková, A., Limnological investigation method for the periphyton ('Aufwuchs') community, *Bot. Rev.*, 28, 286, 1962.

Slattery, M., Chemical cues in marine invertebrate larval settlement, in *Fouling Organisms of the Indian Ocean: Biology and Control Technology*, Nagabhushanam, R. and Thompson, M.F., Eds., Oxford and IBH Publishing, New Delhi, 1997, 135.

Smayda, T.J., The suspension and sinking of phytoplankton in the sea, *Annu. Rev. Oceanogr. Mar. Biol.*, 8, 353, 1970.

Smayda, T.J. and Boleyn, B.J., Experimental observation on the flotation of marine diatoms. II. *Skeletonema costatum* and *Rhizosolenia setigera*, *Limnol. Oceanogr.*, 11(1), 18, 1966a.

Smayda, T.J. and Boleyn, B.J., Experimental observation on the flotation of marine diatoms. III. *Bacteriastrum hyalinum* and *Chaetoceros lauderi*, *Limnol. Oceanogr.*, 11(1), 35, 1966b.

Smedes, G.W., Seasonal changes and fouling community interactions, in *Marine Biodeterioration: An Interdisciplinary Study*, Costlow, J.D. and Tipper, R.C., Eds., Naval Institute Press, Annapolis, MD, 1984, 155.

Smith, A.P. and Kretschmer, T.R., Electrochemical control of fouling, in *Marine Biodeterioration: An Interdisciplinary Study*, Costlow, J.D. and Tipper, R.C., Eds., Naval Institute Press, Annapolis, MD, 1984, 250.

Smith, C.A., A review of bright metalic powders for inks and paints, *Pigment Resin Technol.*, 13(2), 8, 1984.

Sokolova, E.V. and Markov, P.P., *Metody bor'by s razvitiem biologicheskikh obrastanii vodozabornykh sooruzhenii i sistem tekhnicheskogo vodosnabzheniya* (Methods of Biofouling Prevention in Water Intakes and Technical Water Supply Systems), VNII Gosstroya SSSR, Moscow, 1985, 53.

Sorokin, Yu.I., *Coral Reef Ecology*, Springer-Verlag, Berlin, 1993, 465.

Sousa, W.P., Experimental investigations of disturbance and ecological succession in a rocky intertidal algal community, *Ecol. Monogr.*, 49, 227, 1979a.

Sousa, W.P., Disturbance in marine intertidal boulder fields: the cross-equilibrium maintenance of species diversity, *Ecology*, 60, 1225, 1979b.

Sousa, W.P., The responses of a community to disturbance: the importance of successional age and species life histories, *Oecologia*, 45, 72, 1980.

South, G.R. and Whittick, A., *Introduction into Phycology*, Blackwell Scientific, Oxford, 1987, 341.

Srivastava, R.B., Gaonkar, S.N., and Karande, A.A., Biofilm characteristics in coastal waters of Bombay, *Proc. Indian Acad. Sci. Anim. Sci.*, 99, 163, 1990.

Starostin, I.V. and Permitin, Yu.E., Species composition and quantitative dynamics of macrofouling in seawater piping of metal works on the Sea of Azov, *Tr. Inst. Okeanol. Akad. Nauk SSSR*, 70, 124, 1963.

Steinberg, P.D., Schneider, R., and Kjelleberg, S., Chemical defenses of seaweeds against microbial colonization, *Biodegradation*, 8, 211, 1997.

Steinberg, P.D., de Nys, R., and Kjelleberg, S., Chemical inhibition of epibiota by Australian seaweeds, *Biofouling*, 12, 227, 1998.

Stemacek, V.S., Role of sinking in diatom life-history cycles: ecological, evolutionary and geological significance, *Mar. Biol.*, 84(3), 239, 1985.

Stevenson, R.J., Effects of current and conditions simulating autogenically changing microhabitats on benthic diatom immigration, *Ecology*, 64(6), 1514, 1983.

Stevenson, R.J., How currents on different sides of substrates in streams affect mechanisms of benthic algal accumulation, *Int. Rev. Ges. Hydrobiol.*, 69(2), 241, 1984.

Stevenson, R.J., Importance of variation in algal immigration and growth rates estimated by modelling benthic algal colonization, in *Algal Biofouling*, Evans, L.V. and Hoagland, K.D., Eds., Elsevier, Amsterdam, 1986, 193.

Stevenson, R.J., Benthic algal community dynamics in a stream during and after a spate, *J. N. Am. Benthol. Soc.*, 9(3), 277, 1990.

Stevenson, R.J. and Peterson, C.G., Variation in benthic diatoms (Bacillariophycea) immigration with habitat characteristics and cell morphology, *J. Phycol.*, 25, 120, 1989.

Stoecker, D., Relationship between chemical defense and ecology in benthic ascidians, *Mar. Ecol. Prog. Ser.*, 3(3), 257, 1980.

Strathmann, R.R., Larval settlement in echinoderms, in *Settlement and Metamorphosis of Marine Invertebrate Larvae*, Chia, F.-S. and Rice, M.E., Eds., Elsevier/North Holland, New York, 1978, 235.

Strathmann, R.R., What controls the type of larval development? Summary statement for the evolution session, *Bull. Mar. Sci.*, 39, 616, 1986.

Stroganov, N.S., Comparative sensitivity of hydrobionts to toxins, in *Itogi nauki i tekhniki. Obshchaya ekologiya. Biotsenologiya. Gidrobiologiya* (Advances in Science and Technology. General Ecology, Biocenology, and Hydrobiology), Vol. 3, Stroganov, N.S., Ed., VINITI, Moscow, 1976, 151.

Sukhotin, A.A. and Maximovich, N.V., Variability of growth rate in *Mytilus edulis* from Chupa Inlet (the White Sea), *J. Exp. Mar. Biol. Ecol.*, 176, 15, 1994.

Sullivan, B., Faulkner, D.J., and Webb, L., Siphonodictidine, a metabolite of the burrowing sponge *Siphonodictyon* sp. that inhibits coral growth, *Science*, 221(4616), 1175, 1983.

Sutherland, I.W., Bacterial exopolysacharides — their nature and production, in *Surface Carbohydrates of the Prokaryotic Cell,* Sutherland, I.W., Ed., Academic Press, London, 1977, 27–96.

Sutherland, J.P., Multiple stable points in natural communities, *Am. Nat.*, 108, 859, 1974.

Sutherland, J.P., Effect of *Schizoporella* (Ectoprocta) removal on the fouling community at Beaufort, North Carolina, USA, in *Ecology of Marine Benthos*, Coull, B.C., Ed., University of South Carolina Press, Columbia, 1977, 155.

Sutherland, J.P., The structure and stability of marine macrofouling communities, in *Marine Biodeterioration: An Interdisciplinary Study*, Costlow, J.D. and Tipper, R.C., Eds., Naval Institute Press, Annapolis, MD, 1984, 202.

Sutherland, J.P. and Karlson, R.H., Development and stability of the fouling community at Beaufort, North Carolina, *Ecol. Monogr.*, 47, 425, 1977.

Svane, I. and Petersen, J.K., On the problems of epibiosis, fouling and artificial reefs, a review, *Mar. Ecol.,* 22(3), 169, 2001.

Svane, I. and Young, C.M., The ecology and behaviour of ascidian larvae, *Oceanogr. Mar. Biol. Annu. Rev.*, 27, 45, 1989.

Svane, I., Havenhand, J.N., and Jorgensen, A.J., Effects of tissue extract of adults on metamorphosis in *Ascidia mentula* O.F. Muller and *Ascidiella scabra* (O.F. Muller), *J. Exp. Mar. Biol. Ecol.*, 110(2), 171, 1987.

Svidersky, V.L., The evolution of the nervous system and some problems of locomotion control in invertebrates, in *Evolutionary Physiology. Part I,* Kreps, E.M., Ed., Nauka, Leningrad, 1979, 24.

Swain, G.W., Nelson, W.G., and Preedeekanit, S., The influence of biofouling adhesion and biotic disturbance on the development of fouling communities on non-toxic surfaces, *Biofouling*, 12, 257, 1998.

Szewzyk, U. et al., Relevance of the exopolysaccharide of marine *Pseudomonas* sp. strain S9 for the attachment of *Ciona intestinalis* larvae, *Mar. Ecol. Prog. Ser.*, 75(2–3), 259, 1991.

Tamburri, M.N. et al., Chemical induction of larval settlement behavior in flow, *Biol. Bull.*, 191(3), 367, 1996.

Taniguchi, K., Kurata, K., and Suzuki, M., Chlorotannins of the brown alga *Ecklonia stolonifera* protect it from feeding of the gastropod *Haliotis discus hannai, Bull. Jpn. Soc. Sci. Fish.*, 57(11), 2065, 1991.

Tanimizu, M., Paint for Ship Bottoms, Patent 1345, Japan, S 09 D 5/16, priority 12.02.64.

Tapley, D.W., Buettner, G.R., and Shick, J.M., Free radicals and chemiluminescence as products of the spontaneous oxidation of sulfide in seawater, and their biological implications, *Biol. Bull.*, 196, 52, 2003.

Tarasov, N.I., About marine fouling, *Zool. Zh.*, 40(4), 477, 1961a.

Tarasov, N.I., Fouling in Soviet waters of the Sea of Japan, *Tr. Inst. Okeanol. Akad. Nauk SSSR*, 49, 3, 1961b.

Targett, N.M., Natural antifouling compounds from marine organisms: a review, in *Fouling Organisms in the Indian Ocean: Biology and Control Technology*, Nagabhushanam, R. and Thompson, M.F., Eds., Oxford and IBH Publishing, New Delhi, 1997, 85.

Targett, N.M. et al., Antifouling agents against the benthic marine diatom, *Navicula salinicola*, Homarine from the gorgonians *Leptogorgia virgulata* and *L. setacea* and analogs, *J. Chem. Ecol.*, 9, 817, 1983.

Targett, N.M. et al., Tropical marine herbivore assimilation of phenolic-rich plants, *Oecologia*, 103(2), 170, 1995.

ten Hallers-Tjabbes, C.C., Kemp, J.F., and Boon, JP., Imposex in whelks (*Buccinum undatum*) from the open North Sea: relation to shipping traffic intensities, *Mar. Pollut. Bull.*, 28(5), 311, 1994.

Teo, S.L.-M. and Ryland, J.S., Toxicity and palatability of some British ascidians, *Mar. Biol.*, 120, 297, 1994.

Terent'ev, A.P. et al., Method for protection of metal surfaces from biofouling in the aquatic medium, Invention Certificate, 218066, SSSR. Zayavl. 18.10.66, opubl. 7.05.68. NKI 85V, 1/20; 48D, 11/08.

Terry, L.A. and Edyvean, R.G.J., Microalgae and corrosion, *Bot. Mar.*, 24(4), 177, 1981.

The Webster University Dictionary of the English Language. Encyclopedic Edition, Webster University, Lexicon Publishing, St. Louis, MO, 1987, 844.

Thiébaut, E., Lagadeuc, Y., Olivier, F., Dauvin, J.C., and Retière, C., Do hydrodynamic factors affect the recruitment of marine invertebrates in a macrotidal area? *Hydrobiologia*, 375/376, 165, 1998.

Thirb, H.H. and Benson-Evans, K., The effect of different current velocities on the red alga *Lemanea* in a laboratory stream, *Arch. Hydrobiol.*, 96(1), 65, 1982.

Thomassen, S. and Riisgård, H.U., Growth and energetics of the sponge *Halichondria panicea*, *Mar. Ecol. Prog. Ser.*, 128, 239, 1995.

Thomason, J.C., Marrs, S.J., and Davenport, J., Antibacterial and antisettlement activity of the dogfish (*Scyliorhinus canicula*) eggcase, *J. Mar. Biol. Assn. U.K.*, 76, 777, 1996.

Thomason, J.C. et al., Hydrodynamic consequences of barnacle colonization, *Hydrobiologia*, 375/376, 191, 1998.

Thompson, J.E., Exudation of biologically-active metabolites in the sponge *Aplysina fistularis*. I. Biological evidence, *Mar. Biol.*, 88(1), 23, 1985.

Thompson, J.E., Walker, R.P., and Faulkner, D.J., Screening and bioassays for biologically-active substances from forty marine sponge species from San Diego, California, USA, *Mar. Biol.*, 88(1), 11, 1985.

Thompson, R.C., Norton, T.A., and Hawkins, S.J., The influence of epilithic microbial films on the settlement of *Semibalanus balanoides* cyprids — a comparison between laboratory and field experiments, *Hydrobiologia,* 375/376, 203, 1998.

Thorson, G., Reproductive and larval ecology of marine bottom invertebrates, *Biol. Rev.*, 25, 1, 1950.

Thorson, G., Fight and competition on the sea bed, in *Undersea Challenge*, Eaton, B., Ed., British Sub-Aqua Club, London, 1963, 22 and 48.

Thorson, G., Light as an ecological factor in the dispersal and settlement of larvae of marine bottom invertebrates, *Ophelia*, 1, 167, 1964.

Tkhung, D.K., Morskoye obrastanie u beregov Vietnama (*Marine Fouling near the Vietnam Coast*), Avtoref. dis.... kand. biol. nauk, Akademie Nauk, Vladivostok, 1994, 25.

Todd, Ch.D., Larval supply and recruitment of benthic invertebrates: do larvae always disperse as much as we believe? *Hydrobiologia,* 375/376, 1, 1998.

Todd, J.S. et al., The antifouling activity of natural and synthetic phenolic acid sulfate esters, *Phytochemistry,* 34, 401, 1993.

Toonen, R.J. and Pawlik, J.R., Settlement of the tube worm *Hydroides diantus* (Polychaeta: Serpulidae): cues for gregarious settlement, *Mar. Biol.*, 126, 725, 1996.

Trager, G., Achituv, Y., and Genin, A., Effects of prey escape ability, flow speed, and predator feeding mode on zooplankton capture by barnacles, *Mar. Biol.*, 120, 251, 1994.

Trager, G.C., Hwang, J.-S., and Strickler, J.R., Barnacle suspension-feeding in variable flow, *Mar. Biol.*, 105(1), 117, 1990.

Tsukerman, A.M., Strategy of chemical protection from biodeterioration of materials and constructions, in *Nasekomye i gryzuni – razrushiteli materialov i tehnicheskih ustroistv* (Insects and rodents as destroyers of materials and technical devices), Naumov, N.P., Nauka, Moscow, 1983, 258.

Tsukerman, A.M. and Rukhadze, E.G., The role of a repellent effect in the functioning of antifouling chemical coatings, in *Izuchenie protsessov morskogo bioobrastaniya i razrabotka metodov bor'by s nim* (Studies of Marine Biofouling and Development of Methods of Its Control), Zoological Institute, Leningrad, 1987, 76.

Tsukerman, A.M. and Ruhadze, E.G., Modern lines of creating effective ecologically safe antifouling coatings on the base of silicon compounds, in *Biopovrezhdeniya, obrast-anie i zaschita ot nego: Klimaticheskie, biohemicheskie i ecotoksikologicheskie fac-tory* (Biodeterioration, Fouling and Defense against It: Climatic, Biochemical, and Ecological Factors), Bocharov, B.V., Ed., Nauka, Moscow, 1996, 96.

Tsurumi, K. and Fusetani, N., Effects of early fouling communities formed in the field on settlement and metamorphosis of cyprids of the barnacle, *Balanus amphitrite* Darwin, *Biofouling*, 12, 119, 1998.

Turner, R.D., An overview of research on marine borers: past progress and future direction, in *Marine Biodeterioration: An Interdisciplinary Study*, Costlow, J.D. and Tipper, R.C., Eds., Naval Institute Press, Annapolis, MD, 1984, 3.

Turpaeva, E.P., *Biologicheskaya model' soobshchestva obrastaniya* (Biological Model of a Fouling Community), Inst. Okeanol. Akad. Nauk SSSR, Moscow, 1987a, 127.

Turpaeva, E.P., On the possible ways of fouling control in technical water supply systems, in *Izuchenie protsessov morskogo bioobrastaniya i razrabotka metodov bor'by s nim* (Studies of Marine Fouling and Development of Control Methods), Skarlato, O.A., Ed., Zoological Institute, Leningrad, 1987b, 95.

Tushinsky, S.G. and Shinkar, G.G., Pollution and protection of natural waters, in *Itogi nauki i tekhniki. Okhrana prirody i vosproizvodstvo prirodnykh resursov* (Advances in Science and Technology. Nature Protection and Reproduction of Natural Resources), Vol. 12, Tsitsarin, G.V., Ed., VINITI, Moscow, 1982, 199.

Tyurin, A.N., Metamorphosis of the chiton *Ischnochiton hakodadensis* as a biological test for environment pollution, *Biol. Morya*, 20(1), 68, 1994.

Uhlenbruck, G., Bacterial lectins: mediators of adhesion, *Zentralbl. Bacteriol. Mikrobiol. Hyg.*, 263(4), 497, 1987.

Ulanovskii, I.B. and Gerasimenko, A.G., Influence of algae on corrosion of carbonaceous steel and effect of ultrasonic oscillations on intensity of algal photosynthesis, *Tr. Inst. Okeanol. Akad. Nauk SSSR*, 70, 246, 1963.

Underwood, G.J.C., Paterson, D.M., and Parkes, R.J., The measurement of microbial carbohy-drate exopolymers from intertidal sediments, *Limnol. Oceanogr.*, 40(7), 1243, 1995.

Urban, H.J., Modeling growth of different developmental stages in bivalves, *Mar. Ecol. Prog. Ser.*, 238, 109, 2002.

Uriz, M.J., Reproduccion en *Hymeniacidon sanguinea* (Grant, 1926): biologia de la larva y primeros estadios postlarvarios, *Inv. Pesq.*, 46(1), 29, 1982.

Usachev, I.N., New methods of fouling control in marine power stations, *Dokl. Vsesoyuzn. nauchno-tekhn. konf.* (All-Union Research Conf. "Antifouling Protection of Vessels and Technical Objects"), Sudostroenie, Leningrad, 1990, 51.

Usachev, I.N. and Strugova, Yu.N., Efficient methods of protection of marine power stations from biocorrosion, in *Aktual'nye problemy biologicheskikh povrezhdenii i zashchita materialov, izdelii i sooruzhenii* (Topical Problems of Biological Damage and Pro-tection of Materials, Objects, and Constructions), Il'ichev, V.D., Ed., Nauka, Moscow, 1989, 218.

Vacelét, J., Étude qualitative et quantitative des salissures biologiques de plaques expérimentales immergées en pleine eau. VI. Les éponges, *Téthys*, 10(2), 165, 1981.

Vagué, D., Duarte, C.M., and Marrasé, C., Phytoplankton colonization by bacteria: encounter probability as a limiting factor, *Mar. Ecol. Prog. Ser.*, 54(1–2), 137, 1989.

Valiela, I., *Marine Ecological Processes*, Springer-Verlag, New York, 1984.

Valiela, I., *Marine Ecological Processes*, 2nd ed., Springer, New York, 1995.

van Alstyne, K.L., Herbivory grazing increases polyphenolic defenses in the intertidal brown alga *Fucus distichus*, *Ecology*, 69, 655, 1988.

van Alstyne, K.L. et al., Activated defense systems in marine macroalgae: evidence for an ecological role for DMSP cleavage, *Mar. Ecol. Prog. Ser.*, 213, 53, 2001.

Vandermeulen, H. and De Wreede, R.E., The influence of orientation of an artificial substrate (transite) on settlement of marine organisms, *Ophelia*, 21(1), 41, 1982.

Vashchenko, M.A. et al., Toxic effect of mercury chlorides on gametes and embryos of the sea urchin *Strongylocentrotus intermedius*, *Biol. Morya*, 21(5), 333, 1995.

Veglia, A. and Vaissiere, R., Seasonal variations of heavy metal concentrations in mussels and sea-urchins sampled near a harbour area, *Rapp. Proc.-Verb. Reun. Commis. Int. Explor. Sci. Mer. Mediterr. Monaco*, 30(2), 113, 1986.

Vernadsky, V.I., *La Biosphere*, Felix Alcan, Paris, 1929.

Vernadsky, V.I., *The Biosphere*, Springer-Verlag, New York, 1998, 192.

Vilenkin, B.Ya. et al., Biomass growth in *Balanus balanoides* juveniles at different current velocities, *Dokl. Akad. Nauk SSSR*, 278(2), 502, 1984.

Visscher, J.P., Nature and extent of fouling of ships' bottom, *Bull. Bur. Fisher.*, 43(2), 193, 1928.

Visser, J., Adhesion and removal of particles. I, in *Fouling Science and Technology*, Melo, L.F., Bott, T.R., and Bernardo, C.A., Eds., Kluwer, Dordrecht, 1988a, 87.

Visser, J., Adhesion and removal of particles. II, in *Fouling Science and Technology*, Melo, L.F., Bott, T.R., and Bernardo, C.A., Eds., Kluwer, Dordrecht, 1988b, 105.

Vladimirov, Ya.A. et al., Free redicals in living systems, in *Itogi nauki i tekhniki. Biofisika. Tom 29* (Advances in Science and Technology. Biophysics. Vol. 29), Arhipenko, Ju.V., Ed., VINITI, Moscow, 1991, 252.

Vogel, S., *Life in Moving Fluids. The Physical Biology of Flow*, Willard Grant Press, Boston, MA, 1981, 341.

Wahl, M., Marine epibiosis. I. Fouling and antifouling: some basic aspects, *Mar. Ecol. Prog. Ser.*, 58(1–2), 175, 1989.

Wahl, M., Living attached: Aufwuchs, fouling, epibiosis, in *Fouling Organisms in the Indian Ocean: Biology and Control Technology*, Nagabhushanam, R. and Thompson, M.F., Eds., Oxford and IBH Publishing, New Delhi, 1997, 31.

Wahl, M. and Banaigs, B., Marine epibiosis. III. Possible antifouling defense adaptations in *Polysyncraton lacazei* (Giard) (Didemnidae, Ascidiaceae), *J. Exp. Mar. Biol. Ecol.*, 145(1), 49, 1991.

Wahl, M. and Hoppe, K., Interactions between substratum rugosity, colonization density and periwinkle grazing efficiency, *Mar. Ecol. Prog. Ser.*, 225, 239, 2002.

Wahl, M. and Lafargue, F., Marine epibiosis. II. Reduced fouling on *Polysyncraton lacazei* (Didemnidae, Tunicata) and proposal of an antifouling potential index, *Oecologia*, 82(1), 75, 1990.

Wahl, M. and Mark, O., The predominantly facultative nature of epibiosis: experimental and observational evidence, *Mar. Ecol. Prog. Ser.*, 187, 59, 1999.

Wahl, M., Jensen, P.R., and Fenical, W., Chemical control of bacterial epibiosis on ascidians, *Mar. Ecol. Prog. Ser.*, 110(1–3), 45, 1994.

Wahl, M., Kröger, K., and Lenz, M., Non-toxic protection against epibiosis, *Biofouling*, 12, 205, 1998.

Waite, A.M., Thompson, P.A., and Harrison, P.J., Does energy control the sinking rate of marine diatoms? *Limnol. Oceanogr.*, 37(3), 468, 1992.

Waite, J.H., Mussel beards: a coming of age, *Chem. Ind. (London)*, 17, 607, 1991.

Waite, J.H. and Tanzer, M.L., Polyphenolic substance of *Mytilus edulis*: novel adhesive containing L-Dopa and hydroxyproline, *Science*, 212(4498), 1038, 1981.

Waite, J.H., Housley, T.J., and Tanzer, M.L., Peptide repeats in a mussel glue protein. theme and variations, *Biochemistry*, 24(19), 5010, 1985.

Waite, J.H., Hansen, D.C., and Little, K.T., The glue protein of ribbed mussels (*Geukensia demissa*): a natural adhesive with some features of collagen, *J. Comp. Physiol. Ser. B*, 159(5), 517, 1989.

Walch, M. et al., Enhanced setting of the eastern oyster (*Crassostrea virginica*) on biofilms of specific marine bacteria, in *Abstr. 87th Annu. Meet. Am. Soc. Microbiol.*, Washington, D.C., 1987, 45.

Walker, G. and Yule, A.B., Temporary adhesion of the barnacle cyprid: the existence of an antennular adhesive secretion, *J. Mar. Biol. Assn. U.K.*, 64(3), 679, 1984.

Walker, J.T. and Percical, S.L., Control of biofouling in drinking water systems, in *Industrial Biofouling: Detection, Prevention, and Control*, Walker, J., Surman, S., and Jass, J., Eds., Wiley, Chichester, 2000, 55.

Walker, R.P., Thompson, J.E., and Faulkner, D.J., Exudation of biologically-active metabolites in the sponge *Aplysina fistularis*. I. Chemical evidence, *Mar. Biol.*, 88(1), 27, 1985.

Walters, L.J., Field settlement locations on subtidal marine hard substrata: is active larval exploration involved? *Limnol. Oceanogr.*, 37, 1101, 1992.

Walters, L.J. and Wethey, D.S., Surface topography influences competitive hierarchies on marine hard substrata: a field experiment, *Biol. Bull.*, 170, 441, 1986.

Walters, L.J. and Wethey, D.S., Settlement, refuges, and adult body form in colonial marine invertebrates: a field experiment, *Biol. Bull.*, 180, 112, 1991.

Walters, L.J., Hadfield, M.G., and Smith, C.M., Waterborne chemical compounds in tropical macroalgae: positive and negative cues for larval settlement, *Mar. Biol.*, 126, 383, 1996.

Wapstra, M. and Soest, R.W.M., Sexual reproduction, larval morphology and behaviour in demosponges from the southwest of the Netherlands, in *Taxonomy of Porifera*, Springer-Verlag, Berlin, 1987, 281.

Watanabe, M.M., Takeuchi, Y., and Takamura, N., Copper resistance in the benthic diatom, *Achnanthes minutissima*, *Res. Rep. Natl. Inst. Environ. Stud. Jpn.*, 114, 233, 1988

Watermann, B., Was kommt nach dem TBT-Verbot 2003? *Schiff Hafen*, 53, 43, 2001.

Weinberg, J.R. and Helser, T.E., Growth of the Atlantic surfclam, *Spisula solidissima*, from Georges Bank to the Delmarva Peninsula, USA, *Mar. Biol.*, 126, 663, 1996.

Weiner, R.M., Dagasan, L., and Labare, M.P., Specific bacterial exopolymers enhance the settlement of oyster larvae, in *Abstr. 86th Annu. Meet. Am. Soc. Microbiol.*, Washington, D.C., 1986, 186.

Weis, J.S. and Weis, P., Effects of chromated copper arsenate (CCA) pressure-treated wood in the aquatic environment, *AMBIO*, 24(5), 269, 1995.

Weitzel, R.L., Periphyton measurements and applications, in *Methods and Measurements of Periphyton Communities: a Review*, Weitzel, R.L., Ed., American Society for Testing and Materials, ASTM STP 690, Philadelphia, PA, 3, 1979.

Wethey, D.S., Spatial pattern in barnacle settlement: day to day changes during the settlement season, *J. Mar. Biol. Assn. U.K.*, 64, 687, 1984.

Wethey, D.S., Ranking of settlement cues by barnacle larvae: influence of surface contour, *Bull. Mar. Sci.*, 39, 393, 1986.

White, D.C., Chemical characterization of films, in *Microbial Adhesion and Aggregation. Dahlem Konferenzen, 1984*, Marshall, K.C., Ed., Springer-Verlag, Berlin, 1984, 159.

White, D.C. and Benson, P.H., Determination of the biomass, physiological status, community structure, and extracellular plaque of the microfouling film, in *Marine Biodeterioration: An Interdisciplinary Study*, Costlow, J.D. and Tipper, R.C., Eds., Naval Institute Press, Annapolis, MD, 1984, 68.

Whitford, L.A., The current effect and growth of freshwater algae, *Trans. Am. Microsc. Soc.*, 79(3), 302, 1960.

Whorff, J.S., Whorff, L.L., and Sweet, M.H., Spatial variation in an algal turf community with respect to substratum slope and wave height, *J. Mar. Biol. Assn. U.K.*, 15(2), 429, 1995.

Wieczorek, S.K., Effects of microbial surface films on settlement of barnacle, *Balanus amphitrite amphitrite*, larvae, *Mar. Biol. Assn. U.K. Annu. Rep., 1994*, p. 32, 1994.

Wieczorek, S.K. and Todd, C.D., Inhibition and facilitation of settlement of epifaunal marine invertebrate larvae by microbial biofilm cues, *Biofouling*, 12, 81, 1998.

Wieczorek, S.K., Clare, A.S., and Todd, C.D., Inhibitory and facilitatory effects of microbial films on settlement of *Balanus amphitrite amphitrite* larvae, *Mar. Ecol. Prog. Ser.*, 119, 221, 1995.

Wielsputz, C. and Saller, U., The metamorphosis of the parenchymula-larva of *Ephydatia fluviatilis* (Porifera, Spongillidae), *Zoomorphology*, 109(4), 173, 1990.

Wildish, D.J. and Kristmanson, D.D., Control of suspension feeding bivalve production by current speed, *Helgol. Meeresunters.*, 39, 237, 1985.

Wildish, D.J. and Saulnier, A.M., The effect of velocity and flow direction on the growth of juvenile and adult giant scallops, *J. Exp. Mar. Biol. Ecol.*, 155(1), 133, 1992.

Williams, G.B., The effect of extracts from *Fucus serratus* in promoting the settlement of *Spirorbis borealis*, *J. Mar. Biol. Assn. U.K.*, 44(2), 397, 1964.

Williams, G.B., Observations on the behavior of the planulae larvae of *Clava squamata*, *J. Mar. Biol. Assn. U.K.*, 45, 257, 1965.

Williams, G.B., Aggregations during settlement as a factor in the establishment of coelenterate colonies, *Ophelia*, 15(1), 57, 1976.

Wilsanand, V., Wagh, A.B., and Bapuji, M., Antibacterial activities of anthozoan corals on some marine microfoulers, *Microbios*, 99, 137, 1999.

Wilsanand, V., Wagh, A.B., and Bapuji, M., Antibacterial activities of octocorals on some marine microfoulers, *Microbios*, 104, 131, 2001.

Wilson, D.P., The settlement of *Ophelia bicornis* Savigny larvae, *J. Mar. Biol. Assn. U.K.*, 32(1–2), 209, 1953.

Wilson, D.P., The attractive factor in the settlement of *Ophelia bicornis* Savigny, *J. Mar. Biol. Assn. U.K.*, 33(2), 361, 1954.

Wilson, D.P., The role of microorganisms in the settlement of *Ophelia bicornis* Savigny, *J. Mar. Biol. Assn. U.K.*, 34(3), 531, 1955.

Wilson, D.P., The settlement behaviour of the larvae of *Sabellaria alveolata* (L.), *J. Mar. Biol. Assn. U.K.*, 48(2), 387, 1968.

Winston, G.W. and di Giulio, R.T., Prooxidant and antioxidant mechanisms in aquatic organisms, *Aquat. Toxicol.*, 19, 137, 1991.

Winston, J.E., Feeding of marine bryozoans, in *Biology of Bryozoans*, Woollacott, R.M. and Zimmer, R.L., Eds., Academic Press, London, 1977, 233.

Witte, U. et al., Particle capture and deposition by deep-sea sponges from the Norwegian-Greenland Sea, *Mar. Ecol. Prog. Ser.*, 154, 241, 1997.

Woollacott, R.M., Environment factors in bryozoan settlement, in *Marine Biodeterioration: An Interdisciplinary Study*, Costlow, J.D. and Tipper, R.C., Eds., Naval Institute Press, Annapolis, MD, 1984, 149.

Woollacott, R.M., Structure and swimming behavior of the larva of *Haliclona tubifera* (Porifera: Demospongiae), *J. Morphol.*, 218, 301, 1993.

Wright, J.T., Benkendorff, K., and Davis, A.R., Habitat associated differences in temperate sponge assemblages: the importance of chemical defence, *J. Exp. Mar. Biol. Ecol.*, 213, 199, 1997.

Wright, J.T., de Nys, R., and Steinberg, P.D., Geographic variation in halogenated furanones from the red alga *Delisea pulchra* and associated herbivores and epiphytes, *Mar. Ecol. Prog. Ser.*, 207, 227, 2000.

Yakubenko, A.R. and Shcherbakova, I.B., A study of fouling of ship conduits containing outboard water, *Sudostroenie*, 12, 20, 1981.

Yakubenko, A.R. et al., Active methods of ship hull protection from fouling, *Sudostroenie*, 6, 46, 1981.

Yakubenko, A.R. et al., Method of protection of constructions from fouling by marine organisms, Invention Certificate 1110719, SSSR. MKI B 63 B 59/00. Zayavl. 23.05.83, opubl. 30.08.84.

Yakubenko, A.R. et al., Relation between the intensity of fouling in ambient water heat exchangers and their construction and operation parameters, *Sudostroenie*, 1, 18, 1984.

Yakubenko, A.R., Physico-chemical protection from marine fouling, in *Vsesoyuzn. nauchno-tekhn. konf. "Zashchita sudov i tekhnicheskikh sredstv ot obrastaniya"* (All-Union Research Conf. "Antifouling Protection of Vessels and Technical Objects"), Lyublinskii, E.Ya., Ed., Sudostroenie, Leningrad, 1990, 107.

Yamashita, K. et al., The nematocyst printing behavior during the settlement of actinula larvae of *Tubularia mesembryanthemum* (Hydrozoa), *Zool. Sci. Suppl.*, 10(6), 164, 1993.

Yan, Y., Dong, Y., and Yan, W., An ecological study of bouy fouling in Chzhan'zsyan Harbor, *Redat Haiyang Trop. Oceanol.*, 13(2), 68, 1994.

Yoshikawa, H. and Ohta, H., Interaction of metals and metallothionein, in *Biological Roles of Metallothionein*, Foulkes, E.C., Ed., Elsevier/North Holland, New York, 1982, 11.

Young, G.A., Response to, and selection between, firm substrata by *Mytilus edulis*, *J. Mar. Biol. Assn. U.K.*, 63(3), 653, 1983.

Young, G.A. and Crisp, D.J., Marine animals and adhesion, in *Adhesion. Vol. 6*, Allen, K.W., Ed., Applied Science, London, 1982, 19.

Young, L.Y. and Mitchell, R., Negative chemotaxis of marine bacteria to toxic chemicals, *Appl. Microbiol.*, 25(6), 972, 1973.

Yuki, S. and Tsuboi, M., Tyukogu tore. Formulations for fouling-resistant coatings, Claim 391571, Japan. MKI S 09 D 5/14, C 09 D 133/06. Claimed 04.09.89, N 1-227549, published 17.04.91.

Yule, A.B. and Crisp, D.J., Adhesion of cypris larvae of the barnacle, *Balanus balanoides*, to clean and arthropodin treated surfaces, *J. Mar. Biol. Assn. U.K.*, 63(1), 261, 1983.

Yule, A.B. and Walker, G., The temporary adhesion of barnacle cyprids: effects of some differing surface characteristics, *J. Mar. Biol. Assn. U.K.*, 64(2), 429, 1984.

Yushkin, N.P., We all have come from the crystal: a new hypothesis for the origin of life on Earth, *Sem' Dnei – Ekspress* [Weekly], August 4–10, 11–17, 18–24, 1995.

Yushkin, N.P., Mineralogical missions in abiogenesis and astrobiology, *Vestn. Inst. Geol. Komi Nauchn. Tsentra Ross. Akad. Nauk*, 11, 12, 1998.

Zaika, V.E., *Balansovaya teoriya rosta zhivotnykh* (The Balance Theory of Animal Growth), Nauk. Dumka, Kiev, 1985, 192.

Zainiddinov, Kh.T., A Device for Protection of Underwater Ship Hull from Biofouling, Invention Patent 1030255, USSR, B 63 B 59/04, priority 18.09.81, publ. 23.07.83.

Zainullin, R.G., A model of mussel culture taking into account the turbulent mixing of water, *Vestn. Sankt-Peterb. Univ. Ser. 3*, 1(3), 63, 1992.

Zaitsev, Yu.P., *Morskaya neistonologiya* (Study of Marine Neuston), Nauk. Dumka, Kiev, 1970, 264.

Zaitsev, Yu.P., Neuston of seas and oceans, in *The Sea Surface and Global Change*, Liss, P.S. and Duce, R.A., Eds., Cambridge University Press, Cambridge, 1997, 371.

Zann, L.P., *Living Together in the Sea*, T.H.F. Publications, Neptune, NJ, 1980, 416.

Zenkevich, L.A., *Morya SSSR. Ikh flora i fauna* (Seas of the USSR. Their Flora and Fauna), Gos. uch.-ped. izdat. Min. Prosveshcheniya RSFSR, Moscow, 1956, 424.

Zenkevich, L.A., *Izbrannye trudy. Tom 2. Biologiya okeana* (Selected Works. Vol. 2. Biology of the Ocean), Vinogradov, M.E., Ed., Nauka, Moscow, 1977, 244.

Zernov, S.A., To exploration of the Black Sea life, *Zap. Akad. Nauk Ser. 8*, 32, 1, 1914.

Zernov, S.A., *Obshchaya gidrobiologiya* (General Hydrobiology), Akademie Nauk SSSR, Moscow–Leningrad, 1949, 587.

Zevina, G.B., Fouling in the White Sea, *Tr. Inst. Okeanol. Akad. Nauk SSSR*, 70, 52, 1963.

Zevina, G.B., *Obrastaniya v moryakh SSSR* (Fouling in Seas of the USSR), Moscow University, Moscow, 1972, 215.

Zevina, G.B., The biological basis of marine macrofouling, in *Vsesoyuzn. nauchno-tekhn. konf. "Zashchita sudov i tekhnicheskikh sredstv ot obrastaniya"* (All-Union Research Conf. "Antifouling Protection of Vessels and Technical Objects"), Lyublinskii, E.Ya., Ed., Sudostroenie, Leningrad, 1990, 20.

Zevina, G.B., *Biologiya morskogo obrastaniya* (Biology of Marine Biofouling), Moscow University, Moscow, 1994, 135.

Zevina, G.B. and Lebedev, E.M., Marine biofouling, in *Biopovrezhdeniya materialov i izdelii* (Biodeterioration of Materials), Stroganov, N.S., Ed., Moscow University, Moscow, 1971, 88.

Zevina, G.B. and Negashev, S.E., Maximal biomass of coastal fouling in the South China Sea, in *Gidrobionty Yuzhnogo V'etnama* (Hydrobionts of Southern Vietnam), Pavlov, D.S. and Sbikin, Ju.N., Eds., Nauka, Moscow, 1994, 157.

Zevina, G.B. and Rukhadze, E.G., Fouling and its control in seas of the USSR, in *Obrastanie i biopovrezhdenie: ekologicheskie problemy* (Fouling and Biodeterioration as Ecological Problems), Il'in, I.N., Ed., Nauka, Moscow, 1992, 4.

Zevina, G.B., Zvyagintsev, A.Ju., and Negashev, S.E., *Usonogie raki poberezh'ya V'etnama i ikh rol' v obrastanii* (Cirripedes of the Vietnam Coast and Their Role in Biofouling), Akademie Nauk SSSR, Vladivostok, 1992, 144.

Zhdan-Pushkina, S.M., *Osnovy rosta kul'tur mikroorganizmov. Uchebnoe posobie* (Growth of Microbial Cultures: A Manual), Leningrad University, Leningrad, 1983, 188.

Zhuk, L.V., Study of the chemical nature of the "settlement cue" of the scallop *Patinopecten yessoensis* Jay, *Tezisy dokladov 4-go Vsesoyuznogo soveshchaniya po nauchno-tekhn. problemam marikul'tury* (Abstracts of Papers, 4th All-Union Workshop on Scientific and Techical Problems of Mariculture), Vladivostok, 1983, 157.

Zhuravleva, N.G. and Ivanova, L.V., On the behavior of Barents Sea sponges in marine aquaria, in *Povedenie vodnykh bespozvonochnykh* (Behavior of Aquatic Invertebrates), Mordukhai-Boltovskoi, F.D., Ed., Nauka, Yaroslavl, 1975, 21.

Zimmer-Faust, R.K. and Tamburri, M.N., Chemical identity and ecological implications of a waterborne, larval settlement cue, *Limnol. Oceanogr.*, 39(5), 1075, 1994.

ZoBell, C.E., The effect of solid surfaces upon bacterial activity, *J. Bacteriol.*, 46, 39, 1943.

ZoBell, C.E., *Marine Microbiology*, Chronica Botanica, Waltham, MA, 1946, 240.

ZoBell, C.E. and Allen, E.C., The significance of marine bacteria in the fouling of submerged surfaces, *J. Bacteriol.*, 29, 239, 1935.

Zviagintzev, D.G., *Vzaimodeistvie mikroorganizmov s tverdymi poverkhnostyami* (Interaction of Microorganisms with Hard Surfaces), Moscow University, Moscow, 1973, 176.

Zviagintzev, D.G. et al., Determining the force of adhesion of microbial cells to hard surfaces, *Mikrobiologiya*, 40(6), 1024, 1971.

Zvyagintsev, A.Ju., Port craft fouling in Aniva Bay (Sakhalin Island), in *Organizmy obrastaniya dal'nevostochnykh morei* (Fouling Organisms of the Far East Seas), Kudryashov, V.A. et al., Ed., Akademie Nauk SSSR, Vladivostok, 1981, 16.

Zvyagintsev, A.Ju., Ecology of Marine Fouling in the Northwestern Pacific, Doctoral (Biology) dissertation, Institute of Marine Biology, Vladivostok, 1999, 540.

Zvyagintsev, A.Ju., Fouling of ocean-going shipping and its role in the spread of exotic species in the seas of the Far East, *Sessile Organisms*, 17, 31, 2000.

Zvyagintsev, A.Ju. and Ivin, V.V., Fouling communities of the Seychelles Islands, *Atoll Res. Bull.*, 370, 1, 1992.

Zvyagintsev, A.Ju. and Ivin, V.V., Study of biofouling of the submerged structural surfaces of offshore oil and gas production platforms, *Mar. Technol. Soc. J.*, 29, 59, 1995.

Zvyagintsev, A.Ju. and Mikhailov, S.R., On the SCUBA method of studying the fouling of sea vessels, in *Ekologiya obrastaniya v severo-zapadnoi chasti Tikhogo okeana* (Ecology of Fouling in the Northwestern Pacific), Kudryashov, V.A., Ed., Akademie Nauk SSSR, Vladivostok, 1980, 17.

Zvyagintsev, A.Ju., Kashin, I.A., and Fadeev, V.I., Marine biofouling in coastal waters of Vietnam, in *Vsesoyuzn. nauchno-tekhn. konf. "Zashchita sudov i tekhnicheskikh sredstv ot obrastaniya"* (All-Union Research Conf. "Antifouling Protection of Vessels and Technical Objects"), Lyublinskii, E.Ya., Ed., Sudostroenie, Leningrad, 1990, 37.

Chemicals Index

A

α-Acetoxypukalide, 207; see also Pukalide
D-Acetyl-D-glucosamine, 130, 132
Acrylic acid, 199, 219
Aerothionin, 199–200
Alkaloids, 216
Ambiol A, defense against epibionts, 200
Aminoacids, 120
γ-Aminobutyric acid (*synonym* GABA), 127–128
ε-Aminocapronic acid, 128
δ-Aminovaleric acid, 127–128
Antibiotics, 216
Arsenic, 183–185, 190
   chlorophenoxarsine, 184
   organoarsenic, 183

B

Barbiturates, 213–215
   5–*p*-diethylamino anilinomethylene barbituric
      acid, 213–215
   5,5–diethylbarbituric acid, 213–215
Benzoic acid, 209–215
   repellent effect, 210
β-Bisabolene, 213
Blue dextran, 132
Bromopentane, 198

C

δ-Cadinen cyan, 200
Cadmium, 187, 189
Carbohydrates, 114, 131, 133
Carbomino acid, 184
Carboxylic acids, 184
Carotenoids, 191
Catalases, 218
Catechol, 198
Chlorine, industrial protection against biofouling,
      188–191, 193
   active chlorine, 191
   chlorination, 188
   chlorine dioxide, 188
   chlororganic compounds, 188–189
   ecological hazard, 193
   toxicity for organisms, 188, 191
Chloromertensine, 198–199

Clonidine, 213
Cobalt porphyrins or cobalt porphyrin complexes,
      218
   cobalt meso-(tetra-N-methylpiridine)-porphin
      complex, 218
   disodium salt of cobalt hematic acid IX
      complex, 218
   tetrasodium salt of cobalt complex of tetra-
      (*p*-sulphophenyl)-porphin, 218
Concanavalin A, 130, 132
Copper, 183–187, 189–194
   copper compounds, industrial protection
      against biofouling, 183–187
      catapins (organocopper biocides), 187
      copper-nickel alloys, 187
      copper sulphate, 186–187
      cuprous thiocyanate, 183
      oxides of copper, 183, 217
      toxicity for organisms, 184–186, 192
         ban on use, 194
p-Coumaric acid (*synonym* Zosteric acid), 213
Creosote, 21
Cu-SOD (superoxide dismutase), 218

D

2–Deoxy-D-glucose, 130
5–*p*-Diethylamino anilinomethylene barbituric
      acid, 213–215; see Barbiturates
5,5–Diethylbarbituric acid, 213–215; see
      Barbiturates
3,4–dihydroxyphenylalanin (*synonym* L-DOPA),
      126; see also L-DOPA; DOPA-
      containing proteins
2,6–Ditrebutyl-4–methylphenol, 126
L-DOPA, 119–121, 126; see
      3,4–dihydroxyphenilalanin
DOPA-containing proteins, 119–122
   aminoacid composition, 120
   underwater constructions, 121

E

Enzymes, 130–131, 134, 190
   inhibition by mercury and copper, 190
   matrix of microfouling film, 134
   trypsin, 130–131

12–Epi-deoxoscalarin, 207
Epoxypukalide, 201
Epoxy-δ-tocotrienol, 125
Eudistomins, 202–203

F

Fatty acids, 126, 207, 213, 216
Fe-SOD (superoxide dismutase), 218
L-Fucose, 132
Furan, 216
Furospongolide, 207

G

GABA (*synonym* γ-Aminobutyric acid), 127–128
D-Galactose, 103, 130, 132
Gallic acid, 198
Glycoproteids, 131, 133
Glycoproteins, 106, 108, 127
D-Glucose, 103, 130, 132
Glutathione peroxidases, 218
Glycogen, 191

H

Halogenoorganic compounds, 198
Heteronemin, 200
Hexaoxydiphenic acid, 198
Homarine, 201
Homoaerothionin, 199–200
Hydrocyanic acid, 217
Hydroquinone, 103
    negative chemotaxis of bacteria, 103

I

Idiadione, 200
Indole, 103, 209
Iodoethane, 198
Isonitrils, 207
Isothiocyanates, 184
    isothiocyanate derivatives with SCN group, 217

J

Jacaranone, 128–129

L

Lactone, 215
Lead, 185, 189–191, 217
    methylation, 189
    toxicity for algae, 185

Lectins, 130–133
    concanavalin A, 131
    lectin-carbohydrate hypothesis, 131–133
    peanut lectin, 130
Lipids, 216–217
    peroxidation of lipids, 217

M

D-Mannose, 130, 132
Medetomidine, 213
Mercury, 183, 185–186, 189–191
    global cycle, 189
    mercury chloride, 191
    mercury oxide, 186
    methylmercury, ecological hazard, 189
    toxicity for organisms, 191
Metallothioneins, 190
Methylation, 189
    heavy metals, 189
        comparative toxicity, 190–191
        ecological hazard, 189
    methylarsines (di- and trimethylarsines), 189
    methylmercury, 189
    methyltin, 189
α-Methyl-D-glycoside, 130
α-Methyl-D-mannoside, 130
6–Methylthiohexyl isothiocyanate, 213
Mucopolysaccharides, 110, 114, 118

N

N,N,N′,N′-Tetramethylethylenediamine, 209–212
Nucleotides, 191

O

Oxides of copper, 217
Oxocomplex, 218
Oxygen, 217–221
    oxygen anion-radical (*synonym* superoxide, see below), 218
    ozone; see Ozone
    radicals, 217
    reactive oxygen species, 217–221
        mechanism of action, 219–220
        superoxide, 217–220
            superoxide dismutase (SOD), 218–219
        toxicity, 217–220
        ecologically safe protection against biofouling, 220–221
        singlet oxygen, 217, 220
Ozone, 188–189, 217
    ecologically safe protection against biofouling, 217
    ozonation, 182

P

Palitoxin, 201
Pallescensin A, 200
Peroxidation of lipids, 217
Phenylthiourea, 103, 209
Phloroglucinol (*synonym*
          1,3,5–Trihydroxybenzene), 198, 207
Polyphenolic proteins, 118–121
    aminoacid composition, 120
Polyphenoloxidase, 118
Polyphenols, 118–119, 198, 215
Polysaccharides, 90, 106, 108, 114, 131, 198, 217
Polyvinylpirrolidone, 219
Porphyrins, 218–219
    cobalt porphyrins, 218
Proteins, 114, 118, 122–123, 126–128, 198, 217
Pukalide, 201; see also α-Acetoxypukalide
Pyridines, 215

Q

Quinone, o-quinone, 119–123
    quinone tanning proteins, 119–121; see also
          Polyphenolic proteins
       attachment mechanism of invertebrates,
          121–122

R

Reactive oxygen species, 217; see Oxygen
Renillafoulins, 201
D-Ribose, 103, 130

S

Salicylic acid, 184
Saponins, 201
Secondary metabolites, 213
Silver, 185
Siphonodictine, 200
SOD (superoxide dismutase), 218–219
    Cu-SOD, 218
    Fe-SOD, 218
Sodium periodate, 130, 132
Steroids, 216
Sterols, 216
Sugars, 103, 130–133
Sulfuric acid, 202
Superoxide; see Oxygen
Superoxide dismutase; see SOD; Oxygen

T

Tannic acid, 198, 203, 207, 209

Tannins, 197
Terpenes, 201
    diterpene-sugar esters, 217
    monoterpenes, 217
    terpenic compounds, 197
Terpenoids, 198
Thiocyanates, 184
Thiourea, 103
Tiols, 190
Tin, industrial protection against biofouling,
          183–187, 189–192
    tin compounds, 183–186, 191
       inorganic tin, 186
          toxicity, 186
       organotin compounds (tinorganics),
          183–186, 192–194
          ecological hazard, ban on use, 189,
             192–194
          methyltin, 189
          toxicity, 185–186, 192
          trialkylstannates, fouling-resistant
             concretes, 187
          trialkyltin, 183
          triaryltin, 183
          tributyltin (TBT), 183–184
          tributyltin fluoride, 183–184
          bis(tributyltin) oxide (TBTO), 184, 186
          triphenyltin chloride, 184
δ-Tocotrienol, 125
2,5,6–Tribromo-1–methylgramine, 203
Trichloroethylene, 198
1,3,5–Trihydroxybenzene (*synonym*
          Phloroglucinol), 198
Triterpene glycosides, 200
Trypsin, 130, 132

U

Ubiquinone, 216
    suppression of cyprid settlement, 216

V

N-Vinylpirrolidone, 219

X

D-Xylose, 132

Z

Zinc, 184–187, 189–193, 217
    zinc oxide as algicide, 184
    toxicity, 184–186
Zosteric acid (*synonym* p-Coumaric acid), 213

Taxonomic Index

Key to Taxonomic Index

BACTERIA

ALGAE

<u>Diatoms</u>
<u>Red algae</u>
<u>Brown algae</u>
<u>Green algae</u>
<u>Sea weeds</u>

ANIMALS

<u>Protists</u>
Heterotrophic flagellates
Ciliates

<u>Sponges</u>
<u>Coelenterates</u>
Hydroids
Scyphozoans
Corals
Actinia (Sea anemones)

<u>Polychaetes</u>
<u>Crustaceans</u>
Barnacles (Cirripedes)

<u>Mollusks</u>
Chitons
Gastropoda
Mussels (Bivalves)

<u>Bryozoans</u>
<u>Entoprocts</u>
<u>Echinoderms</u>
Ophiurs
Sea urchins
Starfishes

<u>Ascidians</u>
<u>Fishes</u>

285

BACTERIA (Figure 6.3)

Achromobacter sp., 20, 29
Alteromonas sp., 216
A. espejina, 70, 87
Bacillus sp., 20, 29, 216
Bacterium sp., 29, 104
Caulobacter sp., 30, 108
Deleya (Pseudomonas) marina, 92, 203–204
Escherichia sp., 216
E. coli, 103
Flavobacterium sp., 20, 29
Halomonas marina, 109
Hyphomicrobium sp., 30, 108
Micrococcus sp., 20, 29, 104
Pseudomonas sp., 20, 29, 90, 104, 107–108
P. atlantica, 203
P. marina, 130
P. pyocyanea, 104
Roseobacter sp., 90
Sarcina sp., 29
Serratia marcescens, 104
Synechococcus sp., 87
Vibrio sp., 20, 29, 216
V. campbelli, 203
V. vulnificus, 203

ALGAE

Diatoms (Figure 6.3)

Achnanthes sp., 30
Amphora sp., 30
A. coffeaeformis, 110
Bacillaria sp., 30
Berkeleya sp., 30
Biddulphia sp., 30
Cocconeis sp., 29
Fragilaria sp., 30
Grammotophora sp., 30
Licmophora sp., 29–30
Melosira sp., 30
Navicula sp., 29, 92, 202
Nitschia sp., 29
N. closterium, 203
Rhabdonema sp., 30
Stauroneis constricta, 203
Synedra sp., 30

Red algae (Figure 1.2, 5.6)

Ahnfeltia, 4, 81
A. tobuchiensis, 81
Callithamnion sp., 197
C. corymbosum, 125
Corallina officinalis, 96, 197
Gelidium coulteri, 35
Gigartina canaliculata, 35

G. stellata, 70, 203
Hildenbrandia, 127
Hydrolithon boergesenii, 79, 126
Lithophyllum, 127
Lithothamnium, 127
Nitophyllum punctatum, 96
Phycodris sp., 70
Phyllophora sp., 70
Ph. brodiae, 70
Plocamium hamatum, 198
Polysiphonia deusta, 96
P. harveyi, 96
Rhodymenia palmata, 87

Brown algae (Figure 1.2, 5.6)

Ascophyllum nodosum, 50, 78, 87
Chondrus crispus, 70, 96
Cystoseira barbata, 77, 80
Ecklonia stolonifera, 199
Ectocarpus sp., 101, 186
E. siliculosus, 96
F. distichus, 197
Fucus evanescens, 97
F. inflatus, 87
F. serratus, 70, 79, 217
F. vesiculosus, 70, 78, 87
Laminaria sp., 4, 34, 81, 115, 139
L. angustata var. *longissima*, 139
L. hyperborea, 81, 139
L. japonica, 10
L. saccharina, 10, 70, 84–85, 90, 95
Macrocystis pyrifera, 200
Padina sp., 197
Phyllospora comosa, 192
Sargassum tortile, 77, 125
Undaria pinnatifida, 15

Green algae (Figure 1.2)

Chlamydomonas, 87
Cladophora rupestris, 69, 78
Dunaliella sp., 131, 209
D. galbana, 87
Enteromorpha sp., 4, 96, 101, 116, 138, 186
E. intestinalis, 115
E. linza, 197
Halimeda sp., 199
U. reticulata, 198
Ulothrix sp., 186
Ulva, 4, 34
U. fasciata, 96
U. lactuca, 96
U. lobata, 87
Urospora, 97

Seaweeds

Zostera marina, 213

ANIMALS

Protists
Heterotrophic flagellates

Bodo, 30
Codonosiga, 30
Metromonas, 30
Monosiga, 30
Pteridomonas, 30
Spumella (*Monas*), 30

Ciliates

Paramecium caudatum, 218–219

Sponges (Figure 1.1)

Aplisina fistularis, 199–200
Axinella sp., 200, 207
Cliona sp., 22
Dysidea amblia, 200
Eurispongia sp., 200
Halichondria panicea, 70, 88, 196
Haliclona sp., 49
H. cinerea, 200
H. tubifera, 49
Halisarca dujardini, 70, 88
Leoselia idia, 207
Mycale sp., 22
M. cecila, 35
Ophlitaspongia seriata, 83
Phyllospongia papyracea, 207, 213
Siphonodictyon sp., 200

Coelenterates
Hydroids (Figures 1.1, 6.4, 6.5)

Clava multicornis, 50, 60, 88
Clava squamata, 114
Coryne uchidai, 77, 125
Dynamena pumila, 49, 60, 115, 210
Gonothyraea loveni, 49, 60, 63, 78, 87–88, 115, 117, 132, 191, 210
Hydractinia echinata, 63, 70, 87–88, 91, 117
Laomedea flexuosa, 60, 87–88
Obelia longissima, 60
Rhodymenia palmata, 87
Sertularella miurensis, 77
Tubularia sp., 185
T. crocea, 161
T. larynx, 51, 70

Scyphozoans

Aurelia aurita, 88, 158–160, 165
Cassiopea andromeda, 88, 126
Cyanea sp., 88, 116

Corals (Figure 1.1)

Acropora brüggemanni, 51
Agaricia humilis, 79, 126
A. tenuifolia, 79, 126
Galaxea aspera, 51
Leptogorgia virgulata, 201
L. setacea, 201
Montastrea cavernosa, 200
Pachycerianthus multiplicatus, 51
Palythoa toxica, 201
Parerythropodium fulvum fulvum, 114
Renilla reniformis, 201
Sinularia sp., 207, 213
S. cruciata, 198
Xenia macrospiculata, 114

Actinia (Sea anemones)

Metridium sp., 140

Polychaetes (Figures 1.1, 3.1, 4.2, 6.5)

Circeis spirillum, 47, 60
Dexiospira brasiliensis; see *Neodexiospira brasiliensis*
Eupolymnia nebulosa, 93
Harmatoë imbricata, 47–49
Hydroides dianthus, 48, 70, 161
H. elegans, 88–89, 92, 127, 198, 207
H. norvegica, 51
Janua brasiliensis; see *Neodexiospira brasiliensis*
Mercierella enigmatica, 15, 53
Neodexiospira (*Janua*) *brasiliensis*, 87, 130–131
Nereis zonata, 70
Ophelia bicornis, 48, 60, 87–88
Pectinaria coreni, 84
Phragmatopoma sp., 70, 126
Ph. californica, 123, 127
Ph. lapidosa, 64
Pigospio elegans, 60
Platinereis dumerilii, 70
Polydora sp., 60
P. antennata, 60
P. ciliata, 48–49, 53, 64, 88, 93–94
Pomatoceros lamarckii, 88
Protodrilus symbioticus, 87
Salmacina tribranchiata, 199–200
Scoloplos armiger, 49, 51
Spirorbis sp., 48, 60
S. borealis, 87–88, 101, 114
S. corallinae, 88
S. rupestris, 51
S. spirorbis, 79, 202
S. tridentatus, 88

Crustaceans

Eupagurus, 70, 87

Leptochelia dubia, 82
Limnoria lignorum, 21

Barnacles (Cirripedes)

(Figures 1.1, 3.1, 6.5, 6.9)

Balanus sp., 3, 31–32, 35, 184, 197
B. amphitrite, 49, 60, 83–84, 89–90, 92, 132,
         185–186, 198, 201, 203–204, 207
B. balanus, 48, 83
B. cariosus, 92
B. crenatus, 34, 48, 60, 97, 122
B. eburneus, 15, 49, 60, 122
B. improvisus, 15, 52, 60, 65, 138–139, 185
B. perforatus, 60, 116, 185
B. reticulatus, 12
B. spongicola, 70
Chirona evermanni, 95
Chtamalus dalli, 161
Ch. stellatus, 60, 83
Conchoderma sp., 79, 137
Elminius modestus, 49, 53, 60, 83, 89, 116
Lepas sp., 3, 79
L. hili, 83
L. pectinata, 49
Megabalanus rosa, 122
Megabalanus tintinnabulum, 12, 60, 174
Pollicipes spinosus, 60
Semibalanus balanoides, 47–49, 60–62, 65, 69–70,
        72, 80, 82–84, 89–90, 97, 100,
        115–116, 122–123, 127
Solidibalanus fallax, 70
Verruca sp., 185

Mollusks (Figures 1.1, 3.1, 4.4, 6.5)
Chitons

Ischnochiton hakodadensis, 192

Gastropoda

Archidoris pseudoargus, 217
Bittium reticulatum, 80
Haliotis sp., 3, 80
Haliotis discus, 199
H. rufescens, 127–128
Littorina littorea, 47
Littorina sitkana, 197
Megathura crenulata, 213
Monodonta neritoides, 209
Patella pontica, 118
Phestilla sibogae, 66, 70, 77, 127
Rissoa splendida, 80
Testudinalia tessellata, 47, 118
Tritonia plebeia, 15
Trochus niloticus, 80

Mussels (Bivalves)

Aequipecten opercularis, 70
Agropecten irradians, 167
Bankia, 21
Brachyodontes lineatus, 77
Corbicula fluminea, 191
Crassostrea gigas, 89–90, 92, 192
C. virginica, 49, 77, 82, 89–90, 92, 128
Dreissena sp., 15, 55
D. bugensis, 55
D. polymorpha, 55, 181
Geukensia demissa, 118
Haliotis rufescens, 128, 199
Heteronomia squamula, 34
Hiatella arctica, 34
Martesia, 21–22
Mercenaria mercenaria, 49, 51
Mizuhopecten yessoensis, 192
Modiolus modiolus, 118
Mytilus sp., 3, 31–32, 49
M. californianus, 118–119
M. edulis, 13, 33–34, 47, 52, 61–62, 66, 69, 82,
        89, 91, 94, 97–100, 116, 118–121,
        138, 158, 160–161, 184, 186,
        192–193, 207, 210, 213–214
M. galloprovincialis, 89, 94, 118, 192, 197
M. trossulus, 118
Ostrea, 3, 51
O. edulis, 49, 80
O. equestris, 35
Patinopecten yessoensis, 78, 128
Pecten sp., 186
P. maximus, 49, 128–129
Pinna nobilis, 118
Placopecten magellanicus, 89
Saccostrea commercialis, 89
Teredo sp., 49, 55, 78
T. navalis, 3, 21
T. pedicellatus, 49
Xylophaga, 21

Bryozoans (Figures 1.1, 4.5)

Alcyonidium sp., 66–67, 80
A. hirsutum, 70–71, 94
A. polyoum, 51, 70–71, 94
Bowerbankia sp., 49
B. pustulosa, 60
Bugula sp., 49, 60, 131
Bugula neritina, 86, 89–90, 92, 114, 204
B. pacifica, 203
B. simplex, 89–90
B. stolonifera, 89–90
B. turrita, 89–90, 161
Callopora craticula, 60
Celleporella hyalina, 50, 60, 70–71, 94

Conopeum sp., 43
C. seurati, 15
Cribrillina annulata, 60
Electra sp., *43*
E. pilosa, 47, 66–67
Frustrellidra hispida, 70–71, 94
Membranipora sp., 43, 49, 185
M. membranacea, 15
Philodophora pacifica, 199–200, 207
Schizoporella unicornis, 35, 161
Tricellaria occidentalis, 203
Watersipora cucullata, 60
Zoobotryon pellucidum, 203

Entoprocts
Loxocalyx sp., 202

Echinoderms (Figures 1.1, 4.6)
Ophiurs
Ophiopholus aculeata, 49

Sea urchins
Abracia punctulata, 89
Apostichopus japonicus, 89
Dendraster excentricus, 68, 72, 78, 82, 129
Echinarachnius parma, 72
Lytechinus pictus, 89
Paracentrotus lividius, 68
Strongylocentrotus droëbachiensis, 47
S. intermedius, 191–192
S. purpuratus, 22

Starfishes
Asterias rubens, 3, 47, 49, 211
Pisaster giganteus, 213

Ascidians (Figures 1.1, 4.7)
Aplidium californicum, 202
A. stellatum, 61, 140
Archidistoma psammion, 202
Ascidia mentula, 49
Ascidia nigra, 60
Botryllus gigas, 49
B. niger, 68
B. schlosseri, 50, 60–61, 202
Bowerbankia gracilis, 123
Ciona, 32
C. intestinalis, 49–50, 60, 89–90, 186
Clavelina lepadiformis, 68, 202
Cystodotes lobatus, 202
Dendrodoa grossularia, 51
Didemnum sp., 202
D. candidum, 61
Diplosoma listerianum, 61
Ecteinascidia turbinata, 49
Eudistoma olivaceum, 202
E. gladulosum, 202
Molgula citrina, 51, 70, 78, 89
M. complanata, 61
Morchellium argus, 202
Polysyncraton lacazei, 196, 202
Piura stolonifera, 123
Styela partita, 51
S. plicata, 35
S. rustica, 33–34, 70
Styelopsis grossularia, 49
Trididemnum sp., 202

Fishes
Blennius pholis, 83

Subject Index

A

Abiogenesis concept, 175
Accumulation, 25, 134, 137, 143
    mathematical models, 143, 145–148
Adaptive radiation, 176
Adhesion, 104, 112–114, 129; see also
          Attachment
    adhesives, 114, 170
    antiadhesive protection, 212–215
    bacteria, 104–109
    coefficient, 146
    diatoms, 109–111
    equation, 149
    forces, 104–105, 113, 144, 146
        mollusks, 121
        barnacles, 122–123
    inhibition, 213–215
    larvae, 113
    low-adhesion materials, 180–181
    macroalgal spores, 114
    macroorganisms, 113
    mechanisms, 104–111
    primary and secondary energy minimum, 105,
        106, 108
    protists, 111–112
    reviews, 103, 111
    and shearing stress, 146, 150; see also
        Biofouling, main (key) condition
Adhesives, 114–116, 118–124; see also
        DOPA-containing proteins
    ascidians, 123
    barnacles, 122
    bryozoans, 123
    echinoderms, 124
    hydroids, 116
    macroalgae, 114
        reviews, 115
    mollusks, 118–120
    polychaetes, 116, 123
    sponges, 116
Adsorption,
    ions, 103, 175
    molecules, 25, 175
    nutrients, 103
Aggregations, 77–78, 80–85, 127–129
    biological advantages, 82
    classification, 81
    induction, 77–85
        barnacles, 82–84
            arthropodins, 82–83
        echinoderms, 78, 129
        mollusks, 127–129
        polychaetes, 126–127
    monospecific aggregations, 81–85
        barnacles (cirripedes), 80, 82–84, 127
        echinoderms, 78, 129
        macroalgae, 81, 84–85
        mollusks, 77–78, 81–82, 128
Algae; see Autotrophic flagellates; Macroalgae
Algal bloom, 45
Anthropogenic substrates (surfaces, objects,
        materials), 12, 15, 100, 175–176; see
        also Artificial substrates
    cables, 1, 21
    nets, 13
    ropes, 13
Anticolonization protection, 204–225
    industrial protection, 204–207
    natural protection, 204–206
    ecologically safe protection, 206–225
        principles, 206–207
        methods, 207–221
            repellent protection, 207–212
            antiadhesive protection, 212–215
            biocidal protection with reactive
                oxygen species, 215–221
        discussion of prospects, 221–225
        chemical nature of antifoulants,
            223–224
        problems of antifoulants testing,
            223–224
Antifouling coatings, 179–180, 182–186
    biocides, 183–194; for details see Biocides
    types of coatings, 180, 183
        coatings with insoluble matrix, 183
        coatings with soluble matrix, 183
        ablative coatings (ABC), 180
        self-polishing copolymer (SPC),
            179–180, 183
Aquaculture, 12, 174; see Mariculture
Artificial substrates (materials or surfaces), 7, 9,
        12–13, 18–22, 32, 176; see also
        Anthropogenic substrates
Autotrophic flagellates, 29, 37–38

Arthropodins, 82–84, 127
Ascidians, 3, 68
    attachment, 68–69, 124
    defense against epibionts, 202–203
    feeding, 134–135
    larvac, 43, 68
    metamorphosis, 69
    selectivity during settlement, 70
    sensory systems (organs), 68–69
    settlement, 78–79, 82, 88–90
    swimming rate, 49
Attached (sessile) forms, 176; see Foulers, life
       forms
Attachment, 103–130; see also Adhesion
    adaptations, 111, 114–115
    adhesion force, 144, 146
       barnacle, 122–123
       mollusks, 121
    adhesives, 114–116, 118–124
    antiadhesive protection, 212–215
    appendages of larvae, 114–115
    attachment apparatus, 65–68, 118–119,
       121–124
    biological mechanisms, 113
    current (flow) velocity, 111, 124
    DLVO theory, 105–106, 110
    duration, 115
    glandular apparatus, 116–119
    inductors, 125–129
    inhibition, 130, 132, 212–215
       hydroids, 132
       mollusks, 213–215
       polychaetes, 130
    lectin-carbohydrate mechanism,
       130–133
    macroorganisms, 112–133
       animals, 63–69, 117 124
       algae, 115, 124–125
    main (key) condition, 150–151, 163
    microorganisms, 104–112
       bacteria, 104–109
       diatoms, 109–111
       protists, 111–112
    terminology, 112–114
    universal mechanisms, 129–130
    wettability, 115–116, 196–197
Attractants (attracting factors), 103, 125–129, 209;
       see also Inductors of settlement
    ascidians, 78
    bacteria, 103
    bivalves, 77
    chemotaxis, 209
    echinoderms, 78
    hydroids, 78
    mollusks, 78

    nudibranches, 77
    polychaetes, 77, 87
Autotrophic flagellates, 29

B

Bacterial-algal films, 29; see also Microfouling
    excessive development, 41
    matrix, 134
       as biochemically active system, 134
Bacteria, 26, 29
    adhesion, 103–109
       DLVO theory, 105–106
       mechanisms, 104–106
       selective settlement and adhesion, 103, 107
    antifoulants, 203–204
    chemokinesis, 209
    chemotaxis, 103
       attractants, 103
       repellents, 103, 209
    feeding, 134–135
    fouling of man-made structures, 17, 20
    growth, 136
    larvae, 43, 68
    settlement, 103
    swimming rate, 46
Ballast water, 55
Barnacles or cirripedes, 3, 47, 49, 65, 82–84, 117,
       122–123
    attachment, 64–65, 121–123
       maximum current rate, 137–138
    feeding, 134–135
    fouling of man-made structures, 12, 15, 18, 20,
       62, 100
    growth, 138
    larvae, 43, 47, 122
    metamorphosis, 118, 122–123, 127
    sedentary forms, 6
    selectivity during settlement, 69, 82–83
    sensory systems (organs), 60, 64–65
    settlement, 80, 82–84, 90, 94, 97, 127
    swimming rate, 49
Benthos, classification, 8
Bernal's hypothesis, 175; see also Abiogenesis
       concept
Bertalanffy equation, 154
Biocidal protection, 215–221; see also
       Chemobiocidal protection
Biocides for commercial protection, 183–194; see
       also Ecological consequences of
       commercial protection
    arsenic, 183–185, 190
    chlorine, 188 191, 193
       active chlorine, 191
       chlorination, 188

ecological hazard, 193
toxicity for organisms, 188, 191
copper, 183–187
ban on use, 194
leaching rate, 186
toxicity for organisms, 184–186, 192
copper-nickel alloys, 187
lead, 185, 190–191
mercury, 183, 185, 189–191
organotin, 183–187, 191–192
tributyltin (TBT), 183–184, 186
ban on use, 23, 189, 194
leaching rate, 186
toxicity for organisms, 185–186, 192
ozone, 190
silver, 185
Biocorrosion, 18–20
bacteria, 20
inhibition, 20
mechanisms, 19–20
Biodeterioration, 16–22
anticorrosion coating, 20
biocorrosion, 18–20
borers, 20–22
cables, 21
concrete structures, 21
wooden piles (structures), 21
financial investments, 22
materials, 18–22
concrete, 18, 21
metals, 18–19
wood, 18, 20–22
mechanical damage, 20
mechanisms, 19
Biofilms, 29, 85; see Microfouling
matrix, 41, 134
Biofouling (*synonym* of Biological fouling); see
also Colonization; Fouling
colonization, 25–28
evolution, 175–176
fouling, 25
general regularities, 169–176
main (key) condition, 150–151, 163
mathematical models, 143–155,
as process, 25–39
species diversity, 10, 12, 173
succession, 30–35
macrofouling, 31–35
microfoling, 28–31
Biogens, 137, 152, 155, 173
Borers, 3–5, 20–22; see also Foulers, life forms;
Life forms
Boundary layer or Stagnated layer, 143–145, 167
laminar layer, 144–145, 167
structure, 136

turbulent layer, 144, 167
viscous (sub)layer, 144, 167
hydrodynamic forces acting on propagules,
144, 146, 149–150
Brachiopods, 5
adaptations to soft grounds, 7
seston feeders, 5
Bryozoans, 3, 47, 67
adaptations to soft grounds, 7
attachment, 66–67, 123
defense against epibionts, 203
feeding, 134–135
fouling of man-made structures, 15–16, 18, 62
growth, 139–140
larvae, 43, 47, 66–67
selectivity during settlement, 66–67, 69, 71
sensory systems (organs), 66–67
settlement, 79, 82, 86, 90
swimming rate, 49
Byssus, 117–119, 121; see also Mollusks,
attachment

C

Catecholoxidase, 121
Cements, 116; see Adhesives
Chemobiocidal protection, 182–189
antifouling paints and enamels, 182–184, 187
paints with insoluble matrix, 183, 187
soluble paints, 183
ablative coatings (ABC), 180
self-polishing coatings (SPC), 180, 183
biocides, 183–189; for details see Biocides
chlorine, 188
coppercompounds, 183–187
copper-nickel alloys, 187
tributyltin (TBT), 183–184, 186
bulk (volume-based) protection, 189
cathodic and anodic protection, 187
electro-chemical chlorination, 188
ozonation, 189
superficial (surface-based) protection, 189
antifouling paints and enamels, 182–184
fouling-resistant concrete, 187
plating methods, 187
copper-nickel cladding, 187
Chemokinesis, 209
Chemotaxis, 78, 209
attachment, 103, 112
attractants, 103
bacteria, 130, 170
definition, 209
repellents, 103
bacteria, 103

Choice of habitat (substrate), 63–69, 69–73, 103,
        125–133
    selectivity during settlement, 69–73
    sensory systems of larvae, 63–69
        delay of larvae, 72
        generalists and specialists, 69–70
        multistage choice, 69
Ciliates, 29, 36,
Cirripedes; see Barnacles
Coatings; see Antifouling coatings
Collagen, 118
Colonization, 25–28, 169–173; see also
        Colonization processes
    cycles, 171–173
    as directed process, 170–172
    necessary conditions, 150
    phases, 171
    quantitative theory, see Mathematical models
    sufficient condition, 150
Colonization models, 143–155; see Mathematical
        models
Colonization processes, 25–35, 171–173; see
        Attachment; Development; Growth;
        Settlement; Transport
    accumulation, 27
    attachment, 27–28; see Attachment
    chronology, 26
    classification, 25–28
        by Characklis, 25
        by Wahl, 25–26
    development, 27–28, 58; see also
        Metamorphosis
    growth, 27–28; see Growth
    immigration, 27
    models, see Mathematical models
    settlement, 27–28, see Settlement
    similarity in micro- and macroorganisms, 27
    surface (biochemical) conditioning, 25, 27
    transport or transport by current (flow), 27–28
Communities of soft grounds; see also Emphyton,
        6–7
    adaptations of organisms, 7
    biomasses, 1, 10
    life forms, 6
    species composition, 7
Communities on man-made structures, 5, 12–13,
        15–19, 62; see also Man-made
        structures
Competition, 140, 193
    competition for resources (space and food), 35,
        52, 112, 195
    elimination of aboriginal species, 55
    interspecific competition, 82, 201
    protists, 112
    and succession, 33

Concentration (accumulation) of organisms on
        hard substrates, 13
    artificial materials, industrial objects, 12
    hard grounds, 10,
    limits of concentration of organisms, 174
    living organisms, 10
    microplankton near mariculture, 13
    natural non-living substrates, 10
    organisms and communities
        bacteria, 11–12, 134
        barnacles, 12
        fish (near hard substrates), 13
        invertebrates, 10–13
        macroalgae, 9–10
        microorganisms, 12
        mussels, 12–13
        polychaetes, 12
    protobionts, 175
    rate of concentration, 173
    reasons of concentration, 14, 172–174
    reefs, 10
    types of organisms concentration, 9
        bottom concentration, 10
        coastal concentration, 9–10, 173–174
        littoral concentration, 9
        man-made structure concentration,
            12
        oceanic concentration, 173
        planktonic concentration, 9
        Sargassian concentration, 9
Concept of metamorphic competence, 58
Conditioning of hard substrate, 25, 27, 175
Coral reefs, 1,
    abundance, biomass, composition, 10
    area, 1, 10
    concentration of organisms, 9–10
    species diversity, 2,
Corals, 1, 3, 9–10
    attachment, 114, 126
    defense against epibionts, 201
    feeding, 135
    gorgonarians, 5
    larvae (of cnidarians), 43,
    metamorphosis, 126
    reefs, 1–2, 10
    selectivity during settlement, 64
    sensory systems (organs), 63–64
    settlement, 80
    swimming rate of larvae, 49
Corrosion, 18–20; see also Biocorrosion
    corrosion zones, 19
    mechanisms, 19–20
Critical current velocity, 137–138
Current, 45, 53–55, 144
    coastal, 48, 53, 144

current velocity, 45, 48, 137–138
oceanic, 48, 54, 144–145

D

Density gradient, 160; see Gradient distribution of foulers
Defense against epibionts, 195–204
  chemical defense, 197–204
    ascidians, 202–203
    bryozoans, 203
    corals, 201–202
    macroalgae, 197–199
    sponges, 199–200
    toxic microorganisms, 203–204
      epiphytic bacteria, 203–204
      epiphytic diatoms, 203
  physical (mechanical) defense, 196–197
  reviews, 195
  terminology, 195, 197
Detachment, 25, 41–42, 113–114
  microorganisms, 41–42
  mollusks, 113
Detritus, 41, 175–176
  abundance of microorganisms, 41
  as a solid base for microfoulers, 175
Development, 27, 58; see also Metamorphosis
  hyponeuston, 50
  metamorphosis, 57, 116–118, 122–129
  ways of development, 42–43
Diatoms, 26, 29–30; see Foulers, microorganisms
Dispersal forms, 7, 8, 47; *synonym* Propagules, 41; see Larva; Microorganisms; Spores of macroalgae
  exchange between communities, 7–8, 173, 176
  macroorganisms, 7
  microorganisms, 8
  swimming velocity, 48–49
    reviews, 48
Dispersion, 8, 15, 28, 43, 52–53, 173
  current, 53
  larvae, 52
  mechanism in lecithotrophic larvae, 60
  offshore drift, 53
  review, 53
  transoceanic drift, 54–55
  vertical distribution of larvae, 52
DLVO theory (*synonym* theory of liophobic colloids), 105–106, 110
L-DOPA, 90, 119–120, 123, 126
DOPA-containing proteins, 123; see Polyphenolic proteins
Drift (drifting), 15, 48, 52–55
  offshore drift, 52–53
  transoceanic drift, 54–55

E

Echinoderms, 3, 47, 49, 67–68, 114, 123–124
  adaptations to soft grounds, 7
  attachment, 67–68, 114, 123–124
  larvae, 43, 47
  metamorphosis, 129
  selectivity during settlement, 72
  sensory systems (organs), 67–68
  settlement, 78, 82
  swimming rate, 49
Ecological consequences of commercial protection, 185, 189–194; see also Biocides
  anomalies of embryonic development, 192
  imposex, 192
  decreasing species diversity, 192–193
  disturbance at ecosystem level, 193
  pollution of marine environment, 192–193
  toxic-resistant species, 186
  toxicity of main biocides, 190–191
    chlorine, 189, 191
    copper, 183–186, 190
    lead, 185, 190
    tin, 185, 190
    zink, 185, 190
  transfer of toxicants along food chains, 190
Ecological groups, 5–8
  benthos, 8
    emphyton, 6–7
    periphyton, 6
      biofouling, 25
      epibenthos, 5
  nekton, 6
  neuston, 6, 14
  pelagos, 8
    plankton, 6, 41–42
  meroplankton, 42
Ecologically safe biocidal protection, 215–221
  cobalt complexes of porphyrins, 218–220
  ecological demands, 215
  natural biocidal antifoulants, 215–217
  reactive oxygen species, 217–220
    singlet oxygen, 217, 220
    superoxide, 218–221
Emphyton, 6–7
Epibenthos, 5
Epibiosis, 10
  epibiotic hierarchy, 140
  epibiotic potential, 140
Epibiotic communities, 5

F

Feeding, 134–137
  abiotic factors, 135
  biofilter stations, 135
  concentration of nutrients, 135
  current, 135
  diffusion restrictions, 135–136
  feeding rate, 135–136
    effect of current (flow), 135–138
      barnacles, 138
      hydroids, 137
      mollusks, 137
  filtration, 135
    active filtration, 134
    passive filtration, 134
  food chains, 13, 189–190, 205
  food concentration, 135
  food depletion, 136
  food flows, 155
  mathematical model, 152–153, 155
  microorganisms, 134
  predation, 137
  protists, 134
  Rashevsky model, 152
  sedimentation, 135
  supply of food, 137
  suspension feeders, 134, 137
    barnacles, 135
    mussels, 136
Flotsam, 9, 15, 176, 192
  area, 15
  biomass, 15
Foulers or Fouling organisms, 2, 11–12, 20, 26,
        29–31; see also Ascidians; Barnacles;
        Bryozoans; Coralas; Hydroids;
        Mollusks; Polychaetes; Sponges
  adaptations, 4–5, 7–8
  colonization cycles, 171–173
  dispersal forms (propagules), 8
  life cycle, 8
  life forms, 2,
    boring forms, borers, 3–5, 20–22
    sessile (attached) forms, 2–3, 5, 231
      as edifying (dominant) forms, 5,
        176
    vagile (motile) forms, 4–5, 113, 158, 171,
        176, 195
  macroorganisms, 2–6, 10, 12, 47
  microorganisms, 2, 11–12, 20, 26, 29–31, 109
Fouling, 5–6, 15–19, 25; see also Biofouling;
        Man-made structures
  abundance, 12–13
  biomass, 10–13, 137
    average value, 166

  equations, 145–148, 151, 166, 227
    maximum value, 137
  biodeterioration, 18–22
  corrosion, 18–20
  critical current velocity, 137–138
  edge effect, 156
  kinds of fouling, 25
  man-made structures, 15–20
  number of species, 10, 12
  occupied area, 10–12, 15
  spatial distribution, 9
    gradient distribution of foulers,
      156–166
    ships, 62–63
      horizontal distribution, 62
      vertical zonality, 61
  technical obstacles, 14–22
  terminology, 5–6, 25
Fouling in oceans, 11
  Atlantic Ocean, 15, 29, 54–55, 97
    Arcachon Bay (France), 193
    Bermudas, 202
    British Isles, 53
    Chesapeake Bay (U.S.), 193
    coast of Delaware, 34
    Dakar port, 138
    Gulf of Guinea, 138
    Gulf of Mexico, 29
    Miami Beach, 29
    Old Tampa Bay, Florida, 39
    southern New England, 35
    St. Lawrence estuary (Canada)
    Wales (Great Britain), 70
  Indian Ocean, 1
  Pacific Ocean, 15, 29
    Avachinsk Bay (Bering Sea), 33
    coast of California, 34
    Curaçao Island, 217
    harbor of Newport, 31, 78
    San-Diego Bay, California, 193
    southern coast of California, 34
Fouling in seas, 9
  Azov Sea, 15
  Baltic Sea, 18, 28 (St. Petersburg), 70
  Barents Sea, 50, 187
  Bering Sea, 33 (Avachinsk Bay)
  Black Sea, 15, 29, 50, 52–53, 77, 80, 94, 100,
        138, 188, 197, 222
  Caribbean islands, 200
  Caribbean Sea, 199, 201
  Caspian Sea, 15
  Mediterranean Sea (Palermo Harbor), 93
  Sea of Japan, 10, 13
  Sea of Okhotsk, 4, 10

South Chinese Sea, 12, 28 (port of Ho Chi
    Minh)
White Sea, 10, 12–13, 29, 33, 37, 50, 69–70,
    100
Fouling communities, 5, 7; see Fouling; Hard-
    substrate communities
Fungi, 2, 134
    nutrition, 134

G

Genetic exchange, 53
Geotaxis, 61
Glands; see Glandular apparatus
Glandular apparatus, 116–119
Gorgonarians, 5
Gradient distribution of foulers, 151, 158–166
    edge effect, 156, 160–161
    empirical equations, 164–166
    estimatiom of fouling, 166–167
    fouled objects, 156
        cylindrical objects, 156
        experimental plates, 158–161
        pipes, 156, 161–162
        ships, 156, 161–163
        vessels, 161
    foulers, 163
        bryozoans, 161
        cirripedes, 161
        diatoms, 159
        heterotrophic flagellates, 159
        hydroids, 161
        mussels, 158
        scyphistomae, 158–160
    main condition of fouling, 163
    patterns of distribution, 159, 163
Great Lakes, 55
Gregariousness, 80; see Aggregations
Growth, 133–140
    Bertalanffy equation, 154
    competition, 140
    concentration of biomass, 134, 141
    growth rate, 135, 138–139, 153
        artificial substrates, 136–137
        colonial species, 139
        effect of current (flow), 136–138
        solitary species, 139
    growth types, 138–140
        colonial growth, 139, 140
        individual growth, 138
        J-shaped growth, 138
        lateral growth, 140
        overgrowth, 140
            epibiotic potential, 140
    population growth, 27, 139

age-related changes, 138–139
laminar flow, 167
microorganisms, 138
turbulent flow, 167
S-shaped (-like) growth, 138–140
    ciliates, 139
    barnacles, 139
    growth phases, 138–139
mathematical model, 153–155
multilayered structure of communities, 134,
    139–140
review, 139

H

Hard bodies, 1, 231, see Hard substrates
Hard-substrate communities, 1–2, 4–6
    communities, 5, 7
        abundance, biomass, 11–13, 81
    epibiotic communities, 5
        abundance, biomass, 10
    evolution, 175–177
    food chains, 13
    fouling communities, 5
        abundance, biomass, 11–12, 15, 18
    history of study, 6
    life forms, 2–5
        borers or boring forms, 3–5
        sessile forms, 2–4
        vagile forms, 3–5
    as single ecological group, 4
    spatial structure, 4, 62
        gradient distribution of foulers, 62–63, 151,
            156
        multilayered structure, 4–5, 37, 112, 134,
            139–140
        vertical zonality, 62
    species composition, 7, 10
        similarities between communities, 7–8
    terminology, 6–7
Hard substrates (surfaces), 1–2; see also Artificial
            substrates
    area, 12, 176
    biodeterioration of materials, 18–22
    coatings, 20
    concrete, 18–19
    concentration (accumulation) of organisms, see
            above
    concentration of nutrients, 30, 103, 134, 175,
    conditioning, 25, 27, 175
    detritus, 175
    fouling abundance and biomass, 12
        barnacles, 12
        blue mussel *Mytilus edulis*, 13
    general characteristics, 1,

hydrophilic surfaces, 86, 108, 116
hydrophobic (low-adhesion) surfaces, 86, 105,
        108, 116, 180–181
    as islands, 32
living organisms, 2, 10
mechanical damage by foulers, 18
metals, 18
nektonic organisms, 2,
non-living natural surfaces, 5
reefs, 1
types of hard substrates, 28
wettability, 116
wood, 18–21
Heat exchangers, 16–17
Heterotrophic flagellates, 29, 37–38
    abundance, 29, 41
Hydrodynamics concept, 143–144, 156
Hydroids, 3, 117, 125, 132–133, 137–138
    attachment, 63, 116–117, 125, 132–133,
        137–138
        lectin-carbohydrate mechanism, 132–133
    biomass, 17–18
    feeding, 135
    fouling of man-made structures, 17–18, 62
    growth, 137–138
    larvae, 43, 47, 66
    metamorphosis, 117, 125
    selectivity during settlement, 64
    sensory systems (organs) (of cnidarians),
        63–64
    settlement, 77–78, 80, 87
    swimming rate, 49
Hydrovane, 156–157 (Figure 7.2)

I

Induction and stimulation of settlement, 75–96; see
        also Inductors
    classification, 75–77
    conspecific chemical induction, 80–85
        aggregations, 80–85
            advantages of monospecific
                aggregations, 82
            factor (inductor) of settlement in
                barnacles, 83–84
            gregariousness, terminology, 80
            reviews, 80
    combined influence of surface factors, 96–100
        hierarchy of factors, 97–100
    contact chemical induction, 79–80
        reviews, 79
        types of induction, 79
    distant chemical induction, 77–78
    by microfouling, 86–92
        negative influence on larvae, 92

        positive influence on larvae, 87–92
        reviews, 87
    physical surface factors, 93–96
        contour, 93–95
        roughness, 95–96
        shape of substrate, 93–94
        size of substrate, 94–95
        reviews, 76
        rugophily (rugophilic behavior), 95–96
            algal spores, 95–96
            larvae, 95
Inductors of attachment and metamorphosis,
        125–129
    barnacles, 127
    corals, 126
    hydroids, 125
    mollusks, 127–129
    polychaetes, 126, 132
    scyphozoans, 126
Inductors of settlement, 78–82, 84–90, 125–129
    ascidians, 78–79, 82, 88–90
    barnacles, 80, 82–84, 90
    bryozoans, 79, 82, 86, 90
    corals, 80, 126
    hydroids, 77–78, 81, 87–88, 125
    macroalgae, 84–85
    mollusks, 127–129
    polychaetes, 126, 132
    scyphozoans, 126
Industrial objects, 2; see also Man-made structures
    general characteristics, 2
Interfaces, 13–14, 175
    water-air, 13–14
        evolution on, 175
    water-hard surface, 2, 13–14
        concentration of food substances,
            175
        evolution on, 175
    water-soft ground, 13–14
Invasion, 15
    ecosystem disturbance, 15
    organisms, 15

K

Kinesis, 51, 82
    cyprid larvae, 82
    definition, 209

L

Larvae, 43, 47, 49, 64–68; see also Propagules
    attachment, 103–133, see above Attachment
    buoyancy, 43–46
    competence, 58

development,
  hyponeuston, 50
dispersion, 52–53
distribution in space, 50–52, 163–164
  horizontal distribution, 62
    redistribution after settlement, 164
  review, 51
  vertical distribution, 48–52, 61
    redistribution before settlement, 62
drift, 15, 52–55
feeding, 43, 52
food, 52, 134, 136
kineses, 51, 82
locomotion, 46–48
  review, 46
meroplankton, 42
metamorphosis, 69, 122, 125, 126–129
rafting, 15
reception, 63–69
responses to environment factors (kinesis and
    taxes), 48–52, 208–212
  light, 48–52
  gravity, 51
  hydrostatic pressure, 51, 59
reviews, 51, 184
sensory systems (organs), 63–69
  receptors, 63–66, 68–69
    chemoreceptors, 63–68
    mechanoreceptors, 64–66, 68
settlement, 69–73, 77–102; see below
    Settlement
swimming
  mechanisms, 46–48
  rate, 49
taxes, 48–52, 209
  barotaxis, 51
  chemotaxis, 78; see also Chemotaxis
  geotaxis, 50
  phototaxis, 50–52, 60, 63
  review, 51
  rheotaxis, 164
time of releasing into plankton, 42
types of larvae, 43, 47
vertical migrations, 52
ways of development, 42–43
Lectin-carbohydrate mechanism, 130–133
Lectins, 131
  review, 131
  mechanism of fouler attachment, 131, 131
Life forms of foulers, 2–5; see above Foulers, life
    forms

M

Macroalgae, 4, 84–85, 124–125, 135–137, 139,
    196–199
  attachment, 115, 124–125
    maximum current rate, 137–138
  coralline (calcareous) algae, 1, 7, 33, 80,
      196
    adaptations to soft grounds, 7
    coral reefs, 1
  defense against epibionts, 196–199
  development, 84–85
  fouling of man-made structures, 62
  growth, 139
  mariculture, 81
  photosynthesis (feeding), 135–137
  productivity on coral reefs, 10
  propagules, 42, 44
  settlement, 81, 84–85, 90
    selectivity during settlement, 84
  swimming rate of zoospores, 48
Macromolecular layer(s), 26
Macroorganisms, see Foulers or Fouling organisms
Man-made structures, 12–13, 15–18, 229; see also
    Industrial objects; Technical objects
  biohydrotechnical structures, 12
  buoys and beacons, 18
    fouling abundance, 13
    fouling biomass, 15, 18
  cables, 1, 21
  cooling systems, protection, 188
  drilling platforms, protection, 187
  fishing nets (nets, netting), 2, 13, 33
    mariculture, 13, 33
  fuel lines, 17
  heat exchangers, 17, 187, 223
    bacteria, 17
    protection, 187, 223
  high-speed boats, 16
  hydroelectric stations, protection, 188
  moorings, 2
  oil and gas platforms, 18
  piles, 19–20
  pipe(lines), 17–18, 100, 137, 156, 161–163,
      187–188, 225
    fouling biomass, 18, 165
    gradient distribution of foulers, 161–163,
        165
    protection, 187–188, 225, 229
  plants, 18
  power building, protection, 187–188
  ships (vessels), 1–2, 12, 15–17, 61–63,
      161–162, 166, 179–181, 186–187
    area, 12, 15

distribution of foulers, 61–63
  gradient distribution, 156, 161–162, 166
  extra fuel consumption, 16
  fouling abundance, 12
  fouling biomass, 15–16
  friction resistance, 16
  protection, 179–181, 186–188
  shipworm, 3, 21
    damage of wood, 21
    attraction to wood, 78
  shipwreck, 17
  speed loss, 16
  stationary structures, 17
Mariculture (aquaculture), 10, 12–14, 81, 128, 136–138, 174, 181, 221, 224
  antifouling protection, 181, 207, 224
    biological protection, 224
    ecologically safe protection, 207
    ultrasonic protection, 181
  biomass, 13, 136
  colonization processes, 100, 128, 141
  concentration of organisms, 12–13
  growth, 138
  macroalgae, 80
  mollusks, 12–13, 33, 127–128
  optimal environmental conditions, 136–137
  productivity, 137
  prospects, 174
  trade balance, 12
Marine snow, 45
Mathematical models, 143–155, 163–167, 227–230
  abundance equations, 147–148, 150
    maximum abundance, 149
  accumulation, 143, 145–148
  adhesion, 149
  biomass equations, 147–148, 150
    maximum biomass, 149
  feeding, 152–153
  growth, 153–155
  predictions, 151, 153–155
  protection against biofouling, 227–230
Meroplankton, 8, 42
Metallothioneins, 190
Metamorphosis, 57, 116–118, 122–129
  barnacles, 122
  hydroids, 117
  inductors, 125–129
    barnacles, 127
    corals, 126
    hydroids, 125
    mollusks, 127–129
    polychaetes, 126, 132

scyphozoans, 126
sea urchins, 129
polychaetes, 116
Microbial films, 31; see Microfouling
Microconditioning of the surface, 175; see Conditioning of hard substrate
Microfouling (communities), 29–31, 36–39, 85–92, 112, 134
  abundance, 29, 31, 41–42
  climax, 42
  concentratrion on hard substrates, 11–12
  hydrophilic or hydrophobic properties, 86
  induction (stimulation) of settlement, attachment, metamorphosis
    stimulation, 87–92
  inhibition, 92
  multilayered structure, 37, 112
  rate of development, 30, 85
  self-assembly, 36–39
  species composition, 29–30
  succession, 29–31
    growth processes, 30
  transport of nutrients, 134
Microorganisms, 2, 6, 8, 11–12, 103–112, 134, 138; see also Microfouling
  bacteria, 20, 29, 103–109, 136
    adhesion, 103–109
      selective adhesion, 107
    growth, 136
    physiological groups, 20, 30
      copiotrophs, 30, 175
      heterotrophs, 20
      oligotrophs, 20
      thiobacteria, 20
    species composition, 29
  diatoms, 29–30, 109, 110–111
    attachment, 109–111
      adhesive polymers and structures, 110–111
      inhibitors, 110
      mechanisms, 110
    centric forms, 30
    pennate forms, 29–30
    species composition, 29–30
  dispersal forms, 7
  protists, 29
    attachment, 111–112
    feeding, 134
    growth, 138
Mollusks, 3, 49, 117–121, 127–129
  adaptations to soft grounds, 6
  attachment, 66, 117, 118–121
  banks (beds), 13, 81, 135
  feeding, 134–135
    biofilters, 13

fouling of man-made structures, 21, 62, 100
growth, 137, 139
larvae, 43, 47, 63, 117
mariculture, 12–13, 128
metamorphosis, 127–129
sedentary forms, 7
selectivity during settlement, 69
sensory systems (organs), 66
settlement, 77–82, 91, 94
swimming rate, 49
Multilayered structure of hard-substrate
communities, 4–5, 37, 112, 134,
139–140
Mussel banks (beds), 13, 81, 135
biomass, 12

N

Natural inductors of colonization processes,
125–129
Nekton, 6
Neuston, 6, 14, 50, 190

O

Oceans; see Fouling in oceans
Organisms concentration; see Concentration
(accumulation) of organisms
Oxygen radicals, 7

P

Pelagic larvaton, 42; see Meroplankton
Pelagos, 8
Perception, 60
Periphyton, 6
Phototaxis, 48, 50–52, 60–61
Physical protection of man-made structures,
179–182
low adhesion-materials, 180–181
low-frequency vibration, 181
radiation, 181–182
self-polishing coatings, 179–180
ultrasonic methods, 181
Plankton, 6, 41–42
Polychaetes, 3, 79–81, 117–121, 126, 130–131,
134–135
adaptations to soft grounds, 7
attachment, 66, 117–121, 126, 130–131
choice (selection) of habitat, 64
feeding, 134–135
fouling of man-made structures, 15, 62
larvae, 43, 47, 63, 117
metamorphosis, 126

sedentary forms, 6
sensory systems (organs), 64
settlement, 79–81, 84, 87, 93
swimming rate, 49
Polyphenoloxidase, 118, 122
Polyphenolic proteins, 119; see DOPA-containing
proteins
aminoacid composition, 120
ascidians, 123
barnacles, 122
gene of polyphenolic protein, 121
polychaetes, 123
mollusks, 118–121
underwater constructions, 121
Primary succession, 28–35
microfouling succession, 28–31
growth processes, 30
mechanisms, 28, 30–31
stages, 29–30, 37
macrofouling succession, 31–35
classical or Scheer's scheme, 31–32
climax, 32,
mechanisms, 32–35
stages, 32
models of
Connell and Slatyer, 30–31
MacArthur and Wilson, 32
Oshurkov, 33–34
Propagules, 7, 41, 47; see also Dispersal forms;
Larvae; Microorganisms
models, 144
release into plankton, 41–42
sensitivity to toxicants, 184
swimming velocity, 48–49
reviews, 48
Protection against biofouling, 179–189, 195–230;
see also Anticolonization protection;
Chemobiocidal protection; Defense
against epibionts; Physical protection
antiadhesive protection, 212–215
anticolonization protection, 206–207, 221–225
biocidal protection, 215–221
biological protection, 224
chemical protection, 102, 179, 182, 186,
chemobiocidal protection, 182–189
classification, 179,
defense against epibionts, 195–204
ecologically safe protection, 195–225
general model, 227–230
mechanical protection, 179; see below physical
protection
underwater cleaning, 35
physical protection, 179–182

Q

Quantitative theory of colonization, 143–167; see
          also Mathematical models
Quinone tanning, 121, 123

R

Rafting, 15
Reactive oxygen species, 217–221
     reviews, 217
Recovery successions, 35–36
Recruitment, 7, 42–43
     compensation of larval losses, 73
     control by predators, 35
     larvae with pelagic development, 42–43
     reviews, 173
     sources of recruitment, 42–43, 174
Reefs, 1; see also Coral reefs
     artificial reefs, 13, 233
Repellents, 207–211
     bacteria, 130
     effect on microfouling, 212
     macrofoulers, 209–211
     terminology, 207–209

S

Seas; see Fouling in seas
Self-assembly of communities, 36–39
     immigration rate, 37
     macrofouling, 39
     microfouling, 36–39
          rate of self-assembly, 38
          recovery of abundance, 38
     stages, 37
     succession, 37–39
          recovery succession, 37–38
Sensory systems of larvae, 63–69
     ascidians, 68–69
     barnacles, 64–65
     bryozoans, 66–67
     cnidarians, 63
     echinoderms, 67–68
     hydroids, 63–64
     mollusks, 66
     polychaetes, 64
     sponges, 63
Sessile (attached) forms; see Foulers, dominants
Seston feeders, 5; see Feeding, suspension feeders
Settlement, 27–28, 57–102, 125–129
     distribution after settlement, 59–62
          gradient distribution of foulers, 156–166
          on ships, 62–63
               horizontal distribution, 62

          hydrodynamics, 63
          vertical zonality, 61
     redistribution after settlement, 164
     spatial orientation of substrates, 60–61
induction and stimulation of settlement, 75–96;
          see above Inductors, 125–129;
     combined influence of surface factors,
          96–100
          hierarchy of factors, 97–100
     conspecific chemical induction, 81–85
     contact heterospecific chemical induction,
          79–81
     distant chemical induction, 77–79
          aggregations, 80–85
     microfouling, 85–92
          review, 91
     physical surface factors, 93–96
          rugophilic behavior (rugophily), 95
inhibition of larval settlement, 91–92
     microfouling, 91–92
     repellent protection, 207–212
macroalgae, 84–85, 90, 96
     selectivity, 85, 96
macroorganisms, 59–100
     ascidians, 78–79, 89–90
     barnacles, 79–80, 89–90, 94, 97, 127
     bryozoans, 86, 89, 94
     corals, 80, 126
     hydroids, 77–78, 87–88, 91–92, 125
     mollusks, 77–78, 80–81, 89, 91, 94, 97–99,
          129
     polychaetes, 77–78, 87–88, 90, 93–94,
     scyphozoans, 88, 126
     sea urchins, 78, 89, 129
     sponges, 69–70, 88
microorganisms, 44–46, 103
     buoyancy, 44–46
     sedimentation, self-assembly, 37
     sinking rate, 44–46
reasons of settlement, 58
reviews, 51, 58–59
selectivity during settlement, 69–74
     delay of development, 72
     larval survival, 69
sensory systems of larvae, 63–69
taxes of larvae, 48–52
     barotaxis, 51, 61
     geotaxis, 50, 61
     phototaxis, 50–52, 60–61
technical objects, 100–102
terminology, 57–58, 112
Settlers, 139, 164
Shearing force, 146, 149
Shearing stress, 111, 149
Ships, 12, 15; see Man-made structures

Soft grounds, 6
    area, 9–10
    evolution, 176
Sponges, 3, 43–44, 46–47, 49, 63, 79, 135, 138,
    adaptations to soft grounds, 7
    attachment, 116
    defense against epibionts,199–200
    feeding, 135
    growth, 138
    larvae, 43, 46–47, 63, 117
    sedentary forms, 6
    selectivity during settlement, 69–70
    sensory systems (organs), 63
    settlement, 79, 81
    swimming rate, 49
Stokes law, 44, 46
Substrate selection; see Choice of habitat
    (substrate)
Succession, 28–35; see also Primary succession
    allogenic succession, 30
    alternate ways, 33–34
    autogenic succession, 31
    colonization processes, 28
    cyclic successions, 174
    generalized scheme, 34
    macrofouling, 31–35
        reviews, 31
    mechanisms, 30–34, 175
        microfouling, 30–31
        macrofouling, 32–35
    microfouling, 29–31
    models of Connell and Slatyer, 30–31
    rate of succession, 31
        macrofouling, 32
        microfouling, 31
    recovery succession, 35–36
        microfouling, 36
        microperiphyton, 35
        psammic ciliates, 36
    reviews, 31
    Scheer' concept, 31–33
    and self-assembly, 38–39
    species composition
        macroorganisms, 31–35
        microorganisms, 29–30
    stable and unstable conditions, 32–35

    stages, 28, 31–33, 37
        macrofouling, 31–32
        microfouling, 29
    variability, 32
Sugars, 103, 130, 132
Sulphated polysaccharides, 125

T

Tanning, 126; see Quinone tanning
Taxis, 209
Technical objects, 14, 100, 176; see also Man-
    made structures
Theory of functional morphology of algae, 154,
    174,
Theory of island biogeography, 32
Thickening of life, 9; see Concentration
    (accumulation) of organisms
Transport of foulers by current, 25, 27–28, 48–49,
    52–55
    as colonization process, 25, 27–28
    current velocity, 48
    offshore and oceanic drift, 52–55
    swimming rate of propagules, 48–49
        reviews, 46
Turbulent flow, 60, 144, 167, 232
Turbulent mixing, 45, 78, 136, 232

V

Vagile (motile) forms, 4–5, 113, 158, 171, 176,
    195; see also Life forms of foulers

W

Water exchange, 2, 6, 61, 163
Wettability, 86, 93, 99, 115–116, 196–197

X

Xenobiotics, 220

Y

Yeast, 29, 121, 199, 216

Printed and bound by CPI Group (UK) Ltd, Croydon, CR0 4YY

23/10/2024

01778238-0006